CW01238933

Yeast Biotechnology 3.0

Yeast Biotechnology 3.0

Editor

Ronnie G. Willaert

MDPI • Basel • Beijing • Wuhan • Barcelona • Belgrade • Manchester • Tokyo • Cluj • Tianjin

Editor
Ronnie G. Willaert
Vrije Universiteit Brussel
Belgium

Editorial Office
MDPI
St. Alban-Anlage 66
4052 Basel, Switzerland

This is a reprint of articles from the Special Issue published online in the open access journal *Fermentation* (ISSN 2311-5637) (available at: https://www.mdpi.com/journal/fermentation/special_issues/yeast3).

For citation purposes, cite each article independently as indicated on the article page online and as indicated below:

LastName, A.A.; LastName, B.B.; LastName, C.C. Article Title. *Journal Name* **Year**, *Article Number*, Page Range.

ISBN 978-3-03943-186-1 (Hbk)
ISBN 978-3-03943-187-8 (PDF)

Cover image courtesy of Ronnie Willaert.

© 2020 by the authors. Articles in this book are Open Access and distributed under the Creative Commons Attribution (CC BY) license, which allows users to download, copy and build upon published articles, as long as the author and publisher are properly credited, which ensures maximum dissemination and a wider impact of our publications.

The book as a whole is distributed by MDPI under the terms and conditions of the Creative Commons license CC BY-NC-ND.

Contents

About the Editor vii

Ronnie G. Willaert
Yeast Biotechnology 3.0
Reprinted from: *Fermentation* **2020**, *6*, 75, doi:10.3390/fermentation6030075 1

Charlotte Yvanoff, Stefania Torino and Ronnie G. Willaert
Robotic Cell Printing for Constructing Living Yeast Cell Microarrays in Microfluidic Chips
Reprinted from: *Fermentation* **2020**, *6*, 26, doi:10.3390/fermentation6010026 5

Anne-Céline Kohler, Leonardo Venturelli, Abhilash Kannan, Dominique Sanglard, Giovanni Dietler, Ronnie Willaert and Sandor Kasas
Yeast Nanometric Scale Oscillations Highlights Fibronectin Induced Changes in *C. albicans*
Reprinted from: *Fermentation* **2020**, *6*, 28, doi:10.3390/fermentation6010028 21

Konstantin Bellut, Maximilian Michel, Martin Zarnkow, Mathias Hutzler, Fritz Jacob, Jonas J. Atzler, Andrea Hoehnel, Kieran M. Lynch and Elke K. Arendt
Screening and Application of *Cyberlindnera* Yeasts to Produce a Fruity, Non-Alcoholic Beer
Reprinted from: *Fermentation* **2019**, *5*, 103, doi:10.3390/fermentation5040103 31

Laura Canonico, Enrico Ciani, Edoardo Galli, Francesca Comitini and Maurizio Ciani
Evolution of Aromatic Profile of *Torulaspora delbrueckii* Mixed Fermentation at Microbrewery Plant
Reprinted from: *Fermentation* **2020**, *6*, 7, doi:10.3390/fermentation6010007 55

Yeseren Kayacan, Thijs Van Mieghem, Filip Delvaux, Freddy R. Delvaux and Ronnie Willaert
Adaptive Evolution of Industrial Brewer's Yeast Strains towards a Snowflake Phenotype
Reprinted from: *Fermentation* **2020**, *6*, 20, doi:10.3390/fermentation6010020 65

Katrin Matti, Beatrice Bernardi, Silvia Brezina, Heike Semmler, Christian von Wallbrunn, Doris Rauhut and Jürgen Wendland
Characterization of Old Wine Yeasts Kept for Decades under a Zero-Emission Maintenance Regime
Reprinted from: *Fermentation* **2020**, *6*, 9, doi:10.3390/fermentation6010009 75

Miroslava Kačániová, Simona Kunová, Jozef Sabo, Eva Ivanišová, Jana Žiarovská, Soňa Felsöciová and Margarita Terentjeva
Identification of Yeasts with Mass Spectrometry during Wine Production
Reprinted from: *Fermentation* **2020**, *6*, 5, doi:10.3390/fermentation6010005 87

Alice Vilela
Modulating Wine Pleasantness Throughout Wine-Yeast Co-Inoculation or Sequential Inoculation
Reprinted from: *Fermentation* **2020**, *6*, 22, doi:10.3390/fermentation6010022 101

Pedro Miguel Izquierdo-Cañas, María Ríos-Carrasco, Esteban García-Romero, Adela Mena-Morales, José María Heras-Manso and Gustavo Cordero-Bueso
Co-Existence of Inoculated Yeast and Lactic Acid Bacteria and Their Impact on the Aroma Profile and Sensory Traits of Tempranillo Red Wine
Reprinted from: *Fermentation* **2020**, *6*, 17, doi:10.3390/fermentation6010017 121

Jared Muysson, Laurianne Miller, Robert Allie and Debra L. Inglis
The Use of CRISPR-Cas9 Genome Editing to Determine the Importance of Glycerol Uptake in Wine Yeast During Icewine Fermentation
Reprinted from: *Fermentation* **2019**, *5*, 93, doi:10.3390/fermentation5040093 135

Carmen Berbegal, Iuliia Khomenko, Pasquale Russo, Giuseppe Spano, Mariagiovanna Fragasso, Franco Biasioli and Vittorio Capozzi
PTR-ToF-MS for the Online Monitoring of Alcoholic Fermentation in Wine: Assessment of VOCs Variability Associated with Different Combinations of *Saccharomyces*/Non-*Saccharomyces* as a Case-Study
Reprinted from: *Fermentation* **2020**, *6*, 55, doi:10.3390/fermentation6020055 151

Mesfin Haile and Won Hee Kang
Antioxidant Properties of Fermented Green Coffee Beans with *Wickerhamomyces anomalus* (Strain KNU18Y3)
Reprinted from: *Fermentation* **2020**, *6*, 18, doi:10.3390/fermentation6010018 169

Alexander da Silva Vale, Gilberto Vinícius de Melo Pereira, Dão Pedro de Carvalho Neto, Cristine Rodrigues, Maria Giovana B. Pagnoncelli and Carlos Ricardo Soccol
Effect of Co-Inoculation with *Pichia fermentans* and *Pediococcus acidilactici* on Metabolite Produced During Fermentation and Volatile Composition of Coffee Beans
Reprinted from: *Fermentation* **2019**, *5*, 67, doi:10.3390/fermentation5030067 185

Khaled A. Selim, Saadia M. Easa and Ahmed I. El-Diwany
The Xylose Metabolizing Yeast *Spathaspora passalidarum* is a Promising Genetic Treasure for Improving Bioethanol Production
Reprinted from: *Fermentation* **2020**, *6*, 33, doi:10.3390/fermentation6010033 203

Susan Krull, Malin Lünsmann, Ulf Prüße and Anja Kuenz
Ustilago Rabenhorstiana—An Alternative Natural Itaconic Acid Producer
Reprinted from: *Fermentation* **2020**, *6*, 4, doi:10.3390/fermentation6010004 215

About the Editor

Ronnie G. Willaert, Associate Professor, has extensive expertise in yeast research (*Saccharomyces cerevisiae, S. pastorianus, Candida albicans*, and *C. glabrata*) and single-molecule biophysics (high-resolution microscopy, i.e., confocal laser microscopy, AFM, force spectroscopy, and scanning probe lithography), yeast space biology research and hardware development, protein science (yeast adhesins), cell (yeast) immobilization biotechnology, fermentation technology, and brewing science and technology.

Editorial

Yeast Biotechnology 3.0

Ronnie G. Willaert [1,2]

1. Research Group Structural Biology Brussels, Alliance Research Group VUB-UGent NanoMicrobiology (NAMI), IJRG VUB-EPFL BioNanotechnology & NanoMedicine (NANO), Vrije Universiteit Brussel, 1050 Brussels, Belgium; Ronnie.Willaert@vub.be
2. Department of Bioscience Engineering, University Antwerp, 2000 Antwerp, Belgium

Received: 23 July 2020; Accepted: 27 July 2020; Published: 29 July 2020

Keywords: *Saccharomyces cerevisiae*; non-*Saccharomyces* yeasts; fermentation-derived products; fermented beverages; wine; beer; coffee bean fermentation; flavor; itaconic acid production; bioethanol production; bioreactors; yeast micro- and nanobiotechnology

1. Yeast Biotechnology 3.0

This Special Issue is a continuation of the first and second "Yeast Biotechnology" Special Issue series of the journal *Fermentation* (MDPI). This issue compiles the current state-of-the-art of research and technology in the area of "yeast biotechnology" and highlights prominent current research directions in the fields of yeast micro- and nanobiotechnology, brewer's yeasts and beer fermentation, wine yeasts and wine fermentation, coffee bean fermentation and new developments in biochemicals production by yeasts. We very much hope that you enjoy reading it and are looking forward to the next Special Issue "Yeast Biotechnology 4.0" to appear in 2020–2021 (https://www.mdpi.com/journal/fermentation/special_issues/yeast_4).

2. Yeast Micro- and Nanobiotechnology

Living cell microarrays in microfluidic chips allow the non-invasive multiplexed molecular analysis of single cells. Yvanoff et al. [1] developed a simple and affordable perfusion microfluidic chip containing a living yeast cell array composed of a population of green fluorescent protein (GFP)-tagged *Saccharomyces cerevisiae* clones. Mechanical patterning in microwells and robotic piezoelectric cell dispensing in the microwells were combined to construct the cell arrays. The developed microfluidic technology has the potential to be easily upscaled to a high-density cell array, allowing one to perform dynamic systems biology (proteomics and localisomics) experiments on growing cells.

Yeast resistance to antifungal drugs is a major public health issue. Fungal adhesion onto the host mucosal surface is still a partially unknown phenomenon that is modulated by several actors, among which fibronectin plays an important role. Targeting the yeast adhesion onto the mucosal surface could lead to a highly efficient treatment for *Candida* infections. A nanoscale approach to study the behavior of the pathogenic yeast *C. albicans* was develop by Kohler et al. [2]. Using atomic force microscopy (AFM)-based detection of the nanoscale motions of the yeast cells, it was demonstrated that strongly adhering strains reduce their nanomotion activity upon fibronectin exposure, whereas low adhering *C. albicans* remain unaffected. These results open novel avenues to explore cellular reactions upon exposure to stimulating agents and to monitor, in a rapid and simple manner, the adhesive properties of *C. albicans*.

3. Brewer's Yeasts and Beer Fermentation

Due to changing lifestyle trends and legislation, there is a growing demand for non-alcoholic beers (NABs). In recent years, production methods have been improved and the use of non-*Saccharomyces* yeasts has been investigated. Non-*Saccharomyces* yeasts are interesting, since fruity ester aromas can be introduced. Bellut et al. [3] evaluated several *Cyberlindnera* strains for NAB production. It was demonstrated that the selected *Cyberlindnera subsuciens* was suitable to produce a fruity NAB. The outcome strengthens the position of non-*Saccharomyces* yeasts as a serious and applicable alternative to established methods in NAB brewing.

For some years, there has also been a new trend in using non-conventional yeasts to change the aroma profile of traditional beers. Canonico et al. [4] proposed the use of *Torulaspora delbrueckii* to obtain a beer with a distinctive aromatic taste. *S. cerevisiae/T. delbrueckii* mixed fermentations resulted in beers with increased concentrations of some aromatic compounds such as ethyl hexanoate, α-terpineol, and β-phenyl ethanol and an emphasized note of fruity/citric and fruity/esters notes.

Brewer's yeast flocculation is a well-appreciated characteristic of industrial brewer's strains, since it allows the removal of the cells from the beer in a cost-efficient way. However, many industrial strains are non-flocculent and genetic interference to increase the flocculation characteristics is not appreciated by the consumers. Optimization of the brewer yeast towards a more flocculating phenotype can lead to a more efficient beer production and a higher final beer quality. An attractive approach to enhance the attributes of microorganisms is the adaptive laboratory evolution (ALE) approach. Kayacan et al. [5] applied ALE to non-flocculent industrial *S. cerevisiae* brewer's strains using small continuous bioreactors and obtained an aggregative "snowflake" phenotype. It was demonstrated that ALE increased the sedimentation behavior and that no major flavor changes in the produced beer was detected.

4. Wine Yeasts and Wine Fermentation

Before cryopreservation was an established method to store wine yeasts, strain collections were stored at room temperature on agar slants in glass reagent tubes covered with vaspar and sealed with cotton plugs. Matti et al. [6] characterized 60 strains from the old wine yeast collection from the Geisenheim Yeast Breading collection and confirmed the suitability of storing yeasts by this old method. White wine fermentations and post-fermentation aroma analyses were performed. It was shown that this old strain collection bears treasures for direct use either in wine fermentations or for incorporation in yeast breeding programs aimed at improving modern wine yeasts.

Yeasts naturally occur in vineyards on the grapes and consequently in wines. Kačániová et al. [7] identified yeasts on 30 grape varieties and 60 wine samples. MALDI-TOF mass spectrometry was used for the identification of yeasts and a total of 1668 isolates were identified. The most isolated species from the grapes was *Hanseniaspora uvarum*, and from wine, it was *S. cerevisiae*.

The selection of the yeast(s) is one of the most important "tools" for modulating flavor and color in wines. Therefore, Vilela [8] reviewed the role of *Saccharomyces* and non-*Saccharomyces* yeasts, as well as lactic acid bacteria, on the perceived flavor and color of wines and the choice that winemakers can make by choosing to perform co-inoculation or sequential inoculation. This choice will help them to achieve the best performance in enhancing these wine sensory qualities, avoiding spoilage and the production of defective flavor or color compounds.

During the fermentation of wine, the malolactic fermentation (MLF), which is performed by lactic acid bacteria, takes a prominent role, since it influences the wine flavor and its microbiological stability. Izquierdo-Cañas et al. [9] studied the effects of simultaneous inoculation of a selected *S. cerevisiae* yeast strain with two different commercial strains of wine bacteria *Oenococcus oeni* at the beginning of the alcoholic fermentation on the kinetics of the MLF, wine chemical composition, and organoleptic characteristics in comparison with spontaneous MLF in Tempranillo grape must. It was shown that co-inoculation reduced the overall fermentation time by up to 2 weeks, resulting in a reduced volatile acidity. The fermentation-derived wine volatiles profile was distinct between the co-inoculated wines and spontaneous MLF and was influenced by the selected wine bacteria. Co-inoculation resulted in wines with very little lactic acid and buttery flavors.

Icewine is a sweet dessert wine that is fermented from the juice of naturally frozen grapes. The high concentration of sugars in Icewine juice results in considerable osmotic stress in the fermenting *S. cerevisiae*. Yeast can combat this stress by increasing the internal concentration of glycerol by activating the high osmolarity of the glycerol response to synthesize glycerol and by actively transporting glycerol into the cell from the environment. Muyssen et al. [10] investigated the role of the glycerol/H$^+$ symporter Stl1p in Icewine fermentations. Therefore, a strain of the common Icewine yeast *S. cerevisiae* K1-V1116 that lacks *STL1* was constructed using a developed CRISPR-Cas9-based genome editing method. The results demonstrate that glycerol uptake by Stl1p has a significant role during osmotically challenging Icewine fermentations, despite potential glucose downregulation.

There is a high interest in monitoring the changes in biochemical compounds that are changed during the alcoholic wine fermentation, since the management of the alcoholic fermentation is crucial in shaping the wine quality. Berbegal et al. [11] demonstrated the use of proton-transfer reaction-mass spectrometry coupled to a time-of-flight mass analyzer (PTR-ToF-MS) to monitor on-line volatile organic compounds (VOCs). The effect of multiple combinations of two *Saccharomyces* strains and two non-*Saccharomyces* strains (*Metschnikowia pulcherrima* and *Torulaspora delbrueckii*) on the content of VOCs in wine was assessed.

5. Coffee Bean Fermentation

Yeast fermentation of coffee beans improves the functionality of the beans and the quality of the coffee. Haile and Kang [12] evaluated the effect of green coffee bean fermentation with *Wickerhamomyces anomalu*. They demonstrated an improved functionality of the coffee beans, which was reflected into an increased total phenol and total flavoid content and a reduced total tannin content, and an improvement of the 2,2-diphenyl-1-picrylhydrazyl radical scavenging assay and the ferric reducing antioxidant power. Significant differences were also found in the superoxide dismutase activity.

The quality of coffee can also be improved by fermenting the mucilage layer of the coffee mixture with lactic acid bacteria and yeasts. da Silva Vale et al. [13] studied the effect of co-inoculation of *Pichia fermentans* and *Pediococcus acidilactici* on metabolite production during fermentation and the volatile composition of the coffee beans. They demonstrated an improved fermentation efficiency and a positive influence on the chemical composition of the coffee beans

6. New Developments in Biochemicals Production

Over the last few years, intense research has been focused on the generation of alternative renewable biofuels by fermenting agriculture waste using yeasts. One of the trends is the production of bioethanol by the fermentation of cellulosic and hemicellulosic biomass. The fermentation and assimilation of xylose (the second most abundant hemicellulosic carbohydrate) is still a bottleneck in the efficient production of bioethanol, since the conventional yeast *S. cerevisiae* cannot consume xylose. However, non-conventional yeasts, such as *Spathaspora passalidarum* and *Pichia stipitis,* can utilize xylose. Selim et al. [14] reviewed recent advances in xylose metabolizing yeasts, with special emphasis on *S. passalidarum* for improving bioethanol production.

Itaconic acid is an interesting biochemical for the polymer industry, since it can be produced from renewable substrates (such as lignocellulosic based hydrolysates) by fermentation and can replace petrochemical-based chemicals. It can be produced by the filamentous fungus *Aspergillus terreus*. Recently, alternative itaconic acid-producing yeasts such as the basidiomycetous yeasts of the family *Ustilaginaceae*, have been studied. Krull et al. [15] evaluated *Ustilago rabenhorstiana* as an alternative natural itaconic acid producer. By the optimization of media components and process parameters, a final itaconic acid concentration of 50 g L^{-1} using fed-batch fermentation was obtained. Moreover, itaconic acid was produced from different sugar monomers based on renewable feedstocks and the robustness against weak acids as sugar degradation products was confirmed.

Acknowledgments: The Belgian Federal Science Policy Office (Belspo) and the European Space Agency (ESA) PRODEX program supported this work. The Research Council of the Vrije Universiteit Brussel (Belgium) and the University of Ghent (Belgium) are acknowledged to support the Alliance Research Group VUB-UGhent

NanoMicrobiology (NAMI), and the International Joint Research Group (IJRG) VUB-EPFL BioNanotechnology and NanoMedicine (NANO).

Conflicts of Interest: The author declares no conflict of interest.

References

1. Yvanoff, C.; Torino, S.; Willaert, R.G. Robotic Cell Printing for Constructing Living Yeast Cell Microarrays in Microfluidic Chips. *Fermentation* **2020**, *6*, 26. [CrossRef]
2. Kohler, A.-C.; Venturelli, L.; Kannan, A.; Sanglard, D.; Dietler, G.; Willaert, R.; Kasas, S. Yeast Nanometric Scale Oscillations Highlights Fibronectin Induced Changes in *C. albicans*. *Fermentation* **2020**, *6*, 28. [CrossRef]
3. Bellut, K.; Michel, M.; Zarnkow, M.; Hutzler, M.; Jacob, F.; Atzler, J.J.; Hoehnel, A.; Lynch, K.M.; Arendt, E.K. Screening and Application of Cyberlindnera Yeasts to Produce a Fruity, Non-Alcoholic Beer. *Fermentation* **2019**, *5*, 103. [CrossRef]
4. Canonico, L.; Ciani, E.; Galli, E.; Comitini, F.; Ciani, M. Evolution of Aromatic Profile of Torulaspora delbrueckii Mixed Fermentation at Microbrewery Plant. *Fermentation* **2020**, *6*, 7. [CrossRef]
5. Kayacan, Y.; Van Mieghem, T.; Delvaux, F.; Delvaux, F.R.; Willaert, R. Adaptive Evolution of Industrial Brewer's Yeast Strains towards a Snowflake Phenotype. *Fermentation* **2020**, *6*, 20. [CrossRef]
6. Matti, K.; Bernardi, B.; Brezina, S.; Semmler, H.; von Wallbrunn, C.; Rauhut, D.; Wendland, J. Characterization of Old Wine Yeasts Kept for Decades under a Zero-Emission Maintenance Regime. *Fermentation* **2020**, *6*, 9. [CrossRef]
7. Kačániová, M.; Kunová, S.; Sabo, J.; Ivanišová, E.; Žiarovská, J.; Felsöciová, S.; Terentjeva, M. Identification of Yeasts with Mass Spectrometry during Wine Production. *Fermentation* **2020**, *6*, 5. [CrossRef]
8. Vilela, A. Modulating Wine Pleasantness Throughout Wine-Yeast Co-Inoculation or Sequential Inoculation. *Fermentation* **2020**, *6*, 22. [CrossRef]
9. Izquierdo-Cañas, P.M.; Ríos-Carrasco, M.; García-Romero, E.; Mena-Morales, A.; Heras-Manso, J.M.; Cordero-Bueso, G. Co-Existence of Inoculated Yeast and Lactic Acid Bacteria and Their Impact on the Aroma Profile and Sensory Traits of Tempranillo Red Wine. *Fermentation* **2020**, *6*, 17. [CrossRef]
10. Muysson, J.; Miller, L.; Allie, R.; Inglis, D.L. The Use of CRISPR-Cas9 Genome Editing to Determine the Importance of Glycerol Uptake in Wine Yeast During Icewine Fermentation. *Fermentation* **2019**, *5*, 93. [CrossRef]
11. Berbegal, C.; Khomenko, I.; Russo, P.; Spano, G.; Fragasso, M.; Biasioli, F.; Capozzi, V. PTR-ToF-MS for the Online Monitoring of Alcoholic Fermentation in Wine: Assessment of VOCs Variability Associated with Different Combinations of Saccharomyces/Non-Saccharomyces as a Case-Study. *Fermentation* **2020**, *6*, 55. [CrossRef]
12. Haile, M.; Kang, W.H. Antioxidant Properties of Fermented Green Coffee Beans with Wickerhamomyces anomalus (Strain KNU18Y3). *Fermentation* **2020**, *6*, 18. [CrossRef]
13. da Silva Vale, A.; de Melo Pereira, G.V.; de Carvalho Neto, D.P.; Rodrigues, C.; Pagnoncelli, M.G.B.; Soccol, C.R. Effect of Co-Inoculation with Pichia fermentans and Pediococcus acidilactici on Metabolite Produced During Fermentation and Volatile Composition of Coffee Beans. *Fermentation* **2019**, *5*, 67. [CrossRef]
14. Selim, K.A.; Easa, S.M.; El-Diwany, A.I. The Xylose Metabolizing Yeast Spathaspora passalidarum is a Promising Genetic Treasure for Improving Bioethanol Production. *Fermentation* **2020**, *6*, 33. [CrossRef]
15. Krull, S.; Lünsmann, M.; Prüße, U.; Kuenz, A. Ustilago Rabenhorstiana—An Alternative Natural Itaconic Acid Producer. *Fermentation* **2020**, *6*, 4. [CrossRef]

© 2020 by the author. Licensee MDPI, Basel, Switzerland. This article is an open access article distributed under the terms and conditions of the Creative Commons Attribution (CC BY) license (http://creativecommons.org/licenses/by/4.0/).

fermentation

Article

Robotic Cell Printing for Constructing Living Yeast Cell Microarrays in Microfluidic Chips

Charlotte Yvanoff [1,2,*], Stefania Torino [1,3] and Ronnie G. Willaert [1,2,4]

1. Research Group Structural Biology Brussels, Vrije Universiteit Brussel, 1050 Brussels, Belgium; Stefania.Torino@vub.vib.be (S.T.); Ronnie.Willaert@uantwerpen.be (R.G.W.)
2. Alliance Research Group VUB-UGent NanoMicrobiology (NAMI), International Joint Research Group VUB-EPFL NanoBiotechnology & NanoMedicine (NANO), Vrije Universiteit Brussel, 1050 Brussels, Belgium
3. Center for Structural Biology, Vlaams Instituut voor Biotechnologie (VIB), 1050 Brussels, Belgium
4. Department Bioscience Engineering, University Antwerp, 2000 Antwerp, Belgium
* Correspondence: Charlotte.Grabielle.Yvanoff@vub.be; Tel.: +32-26291846

Received: 31 January 2020; Accepted: 11 February 2020; Published: 14 February 2020

Abstract: Living cell microarrays in microfluidic chips allow the non-invasive multiplexed molecular analysis of single cells. Here, we developed a simple and affordable perfusion microfluidic chip containing a living yeast cell array composed of a population of cell variants (green fluorescent protein (GFP)-tagged *Saccharomyces cerevisiae* clones). We combined mechanical patterning in 102 microwells and robotic piezoelectric cell dispensing in the microwells to construct the cell arrays. Robotic yeast cell dispensing of a yeast collection from a multiwell plate to the microfluidic chip microwells was optimized. The developed microfluidic chip and procedure were validated by observing the growth of GFP-tagged yeast clones that are linked to the cell cycle by time-lapse fluorescence microscopy over a few generations. The developed microfluidic technology has the potential to be easily upscaled to a high-density cell array allowing us to perform dynamic proteomics and localizomics experiments.

Keywords: cell printing; piezoelectric dispensing; *Saccharomyces cerevisiae*; GFP-tagged yeast clone collection; living cell microarrays; microfluidic chip; dynamic single-cell analysis

1. Introduction

Cell assays have been miniaturized by growing cells in multiwell plates with increasing well number from 96 to 384 and 1536 wells and decreasing well volume from 280 µL down to 3 µL. These experiments are typically integrated in a robotic analysis platform. Major drawbacks of robotic platforms are the expense of the instrumentation, the cost of experimental consumables, closed systems (batch growth), and still relative medium throughput compared to recently developed microfluidic chips. As cell collections grow, further miniaturization of cell assays is needed to increase parallelism of the analyses. Cell microarrays provide an attractive solution, as they could increase the throughput significantly [1–3]. A cellular microarray consists of a solid support wherein small volumes of different biomolecules and cells can be displayed in defined locations, allowing the multiplexed interrogation of living cells and the analysis of cellular responses [4,5]. Living cell microarrays have been combined with microfluidic bioreactors, which provide multiple advantages for multiplex dynamic analyses and high-throughput screening [4,6]. Cellular arrays are emerging as important tools for functional genomics, drug discovery, toxicology, and stem cell research [4,7,8]. A major advantage of cell microarrays over microtiter plates is the opportunity to measure parameters on hundreds of individual single cells and average them, instead of measuring the parameters of a whole cell population. An interesting application where living yeast cell arrays are used is "dynamic proteomics" in which living cell microarrays using a fluorescent protein (e.g. GFP)-tagged yeast clone collection and automated time-lapse microscopy is used for the rapid acquisition of *in vivo* quantitative data about the dynamic proteome and protein subcellular localization ("localizomics"). This enables the identification of members of protein complexes and coregulated proteins, and the unravelling of signaling pathways such as exposure to stress compounds, antimicrobials, or mating conditions [9–13].

Mechanical cell patterning, where mechanical barriers capture the cells at specified spots, has been frequently used to create cellular microarrays [14] in microfluidic chips. Yeast cells have been mechanically patterned in (single-cell) microwells [15,16], microchambers [17,18], and mechanical trap barriers [19–21]. As for the creation of classical DNA microarrays, a fluid-dispensing device can be used to spot or "print" living cells in an array format [4]. Dispensing techniques are categorized as contact and non-contact dispensing [22,23]. Robotic contact printing (e.g. printing DNA microarrays) was initially used to print cells on a semi-solid agar growth medium [24]. Today, mostly non-contact-based devices are used to produce living cellular arrays [25–28]. Here, the fluid is ejected as a flying droplet or jet toward the surface from a short distance. One concept of non-contact printing is based on syringe–solenoid-driven printers, where a reservoir and a high-speed microsolenoid valve are connected to a high-resolution syringe [29,30]. Typically, droplet volumes of 10 to 20 nanoliter are the lower dispensing limit. Another concept is piezoelectrical dispensing, where a technology similar to the one used in an ink-jet printer is used [31,32]. A piezo actuator is fixed around a glass capillary close to the end of the tip. The squeezing of the capillary forced by the piezo actuation induces droplet ejection out of the capillary. The fast response time of the piezoelectric crystal permits fast dispensing rates (kHz range), and the small deflection of the crystal generates droplets from tens of picoliters to a few nanoliters. Robotic cell printing can be used to easily create a living cell microarray composed of a clone collection and at a positional xy accuracy of a few micrometer.

Here, we developed a perfusion microfluidic chip containing a living GFP-clone collection that can be used to perform dynamic proteomics and localizomics experiments and demonstrate its performance. We combined mechanical patterning and robotic cell printing to produce living cell arrays. Soft lithography was used to create a 102-microwell array in the epoxy resin SU-8 (Structured by UV-8) on top of a glass coverslip substrate. A commercial microchannel top plate was used to close the microfluidic chip. Piezo dispensing was optimized for the development of living yeast cell arrays in microfluidic chips. We demonstrated that a clone collection can successfully be printed into the microwell array and yeast cells can be grown in the microwells in continuous mode. Finally, time-lapse fluorescence imaging was performed using 6 selected GFP-tagged clones demonstrating that the approach is suitable to perform dynamic analyses of protein expression and protein localization in living cells.

2. Materials and Methods

2.1. Yeast Strain and Media

Clones from the *S. cerevisiae* Yeast GFP Clone Collection (ThermoFisher Scientific, Waltham, MA, USA) were used [33]. We selected 34 clones linked to the cell cycle (Table S1) and revived them from the cryo-stock on YPD-agar plates (yeast extract 10 g/L, peptone 20 g/L, dextrose 20 g/L, agar 15 g/L). Single colonies were then transferred to liquid cultures in Synthetic Complete (SC) medium supplemented with 2% (*m/v*) glucose. The yeast cells were grown overnight in a shaking incubator at 170 rpm and 30 °C. Prior to cell printing, the cultures were diluted in Phosphate Buffered Saline (PBS) to an optical density at 600 nm (OD_{600}) of 0.5.

2.2. Fabrication of the Microwell Substrate and the Microfluidic Chip

The microfluidic chip consists of two parts: a bottom glass slide with SU-8 microwells and a top plate containing a microfluidic channel with in- and outlet ports (Figure 1). The microwell substrate was produced in-house whereas the top plate was obtained commercially (sticky-Slide I 0.4 Luer, Ibidi, Gräfelfing, Germany). Microwells were fabricated by SU-8 photolithography on glass coverslips. Glass coverslips (75 × 25 mm, thickness of 170 μm; Ibidi, Gräfelfing, Germany) were cleaned in acetone

(Carl Roth, Karlsruhe, Germany) and 2-propanol (Carl Roth, Karlsruhe, Germany) for 15 minutes each, then rinsed with ultrapure water and finally blow-dried. The glass slides were then exposed to air plasma at 100 W, 50 kHz, for 5 minutes (Plasma System Cute, Femto Science, Dongtangiheung-Ro, Korea). The negative photoresist SU-8 2050 (Kayaku Advanced Materials, Westborough, MA, USA) was spin-coated onto the glass slides at 3500 rpm for 30 seconds in order to reach an approximative thickness of 50 µm. Next, the SU-8 was soft-baked for six minutes at 95 °C on a hot plate. The SU-8 slides were then aligned with the photomask (film photomask, Selba, Versoix, Switzerland) in the mask aligner UV-KUB3 (Kloé, Saint-Mathieu-de-Tréviers, France) and illuminated with 365 nm ultraviolet (UV) light (intensity of 35 mW/cm^2) for 10 seconds. Following UV exposure, the SU-8 slides were post-baked for six minutes at 95 °C on a hot plate. Finally, the slides were treated with SU-8 developer for seven minutes and next washed with 2-propanol and blow-dried. The dimensions of the wells were then evaluated by optical microscopy (Nikon Eclipse Ti2, Nikon, Tokyo, Japan) with a 10x objective (for the diameter and pitch distance) and with a 3D profilometer (Profilm 3D, Filmetrics, San Diego, CA, USA) for the well's depth. The microfluidic chip was assembled by pressing the sticky top plate (sticky-Slide I 0.4 Luer, Ibidi, Gräfelfing, Germany) to the microwell substrate.

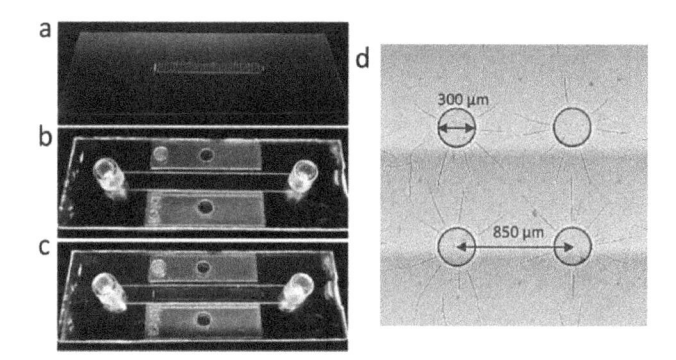

Figure 1. Construction of the microfluidic chip containing the cell microarray. (**a**) The bottom substrate: the SU-8 microwell array on the glass coverslip. (**b**) The top plate containing the channel, inlet and outlet (Ibidi, Gräfelfing, Germany). (**c**) The integrated microfluidic chip obtained by sticking the top plate to the bottom plate using double-sided sticky tape. (**d**) The microwells constructed by SU-8 UV-photolithography.

2.3. Cell Printing

Cell printing was performed using a non-contact iTWO-400 dispenser (M2 Automation, Berlin, Germany) (Figure S2a). The printer is established in an environmental enclosure for live cell printing that contains a HEPA filter and recirculating air is sterilized with a UV lamp. The environmental temperature can be controlled as well as the relative humidity. Moreover, the instrument deck contains a cooling system for source plates and target plates with dew point control, which prevents evaporation during spotting. Samples to be printed are manually dispensed in 384-multiwell plates (ShallowWell 384-multiwell plate, Thermo Fischer Scientific, Waltham, MA, USA), which are then mounted onto the "source plate" locations of the robot (Figure S2a). Likewise, the substrate to be printed is mounted onto the "target plate" locations of the robot (Figure S2a). The actual printing is performed with a piezo dispenser made of a borosilicate glass capillary surrounded by a piezo ceramic actuator (PDMD, M2 automation, Berlin, Germany), which is able to shoot pico- to nanoliter droplets at high frequency with high volume and position accuracies. Finally, the instrument deck is also equipped with a wash station, which enables to clean the tip of the piezo dispenser after sample aspiration and sample printing, in order to avoid cross-contamination. A typical printing run consists in aspirating the sample to be printed from the "source plate", washing the outside of the tip to get rid of contaminants that could impair the shooting, shooting the sample at its desired position on the "target plate" and finally dispensing the remaining sample at the washing station and washing the outside of the tip so that it is cleaned for the next sample. Each step of this procedure can be specified in the software controlling the robot. Additionally, the robot is equipped with a camera annexed to the piezo dispenser, which enables us to verify the successful printing of the samples.

Ultrapure water was printed with a pulse duration of 15 µs and an amplitude voltage of 75 V. The environmental temperature was around 28 °C, the relative humidity was controlled at 50% and the temperature of the target plate was at 15 °C. These parameters were maintained throughout each of the cell printing experiments. For optimization purposes, ultrapure water was printed on standard glass coverslips (24 × 24 × 0.17 mm) and in microwells of commercial nanotiter plates (Microfluidic ChipShop, Jena, Germany) made of the cyclo-olefin copolymer material Topas.

Prior to cell printing, the SU-8 microwells were coated with a solution of concanavalin A (Con A) (Con A from *Canavalia ensiformis*, Sigma Aldrich, Overijse, Belgium) at 2 mg/mL in H_2O, with 5 mM $CaCl_2$ and 5 mM $MnSO_4$. Sixty drops of Con A were printed per well with the parameters mentioned above and were left to incubate for 15 minutes before being allowed to dry.

Yeast suspensions were prepared at an OD_{600} of 0.5 in PBS and 40 droplets were printed into each well with the parameters mentioned above. Multiple yeast suspensions were successively printed in different wells of the microfluidic chip. Therefore, the piezo dispenser was thoroughly washed between each sample to avoid cross-contamination. The washing procedure consisted in first discarding the old sample by ejecting 30 µL of liquid at a flow rate of 30 µL/s using the syringe pump and next, flushing the outside of the piezo dispenser with ultrapure water for five seconds at the wash station. The yeast cells were left to sediment for 10 minutes. The microfluidic chip was closed and connected to a syringe filled with SC medium via silicone tubing. SC medium was gently perfused into the channel of the microfluidic chip with a syringe pump (KD Scientific, Holliston, MA, USA) at a flow rate of 25 µL/min. Finally, the complete set-up (microfluidic chip and syringe pump) was installed on a microscope for direct imaging or kept at 4 °C overnight to image the next day.

2.4. Microscopy

The microfluidic chip and syringe pump set-up were installed on a Nikon Eclipse Ti2 epifluorescence microscope (Nikon, Tokyo, Japan) for time-lapse imaging. The microfluidic chip was inserted into a temperature-controlled chamber (Ibidi, Gräfelfing, Germany), which was mounted onto an automated scanning stage (ProScan III, Prior Scientific Instruments, Cambridge, United Kingdom). The syringe pump was placed next to the microscope and the syringe was covered with a syringe heater (New Era Pump Systems, Farmingdale, NY, USA). Both the temperature controller and the syringe heater were set at 30 °C for the duration of the time-lapse experiments.

For the growth experiment with 34 clones, bright field images of yeasts were acquired every 20 minutes, for 18 hours with a 20x objective. At the final time point (18 h), the yeasts were also imaged with a 60x objective in bright field and fluorescence. The GFP fluorescence was observed by exciting the sample with a LED light source (pE-300^{white}, CoolLED, Andover, United Kingdom) and detecting it through a FITC filter. For the growth experiment with six clones, the yeasts were imaged in bright field and fluorescence and images were recorded every 30 minutes for three hours using a 60x objective.

2.5. Image Processing

Images acquired by the camera of the iTWO-400 dispenser were post-processed using Fiji [34]. More precisely, they were stitched together to form the pictures shown in Figure 2, Figure 3, and Figure 4 using the "Grid/Collection Stitching" plugin [35]. The bright field and GFP-fluorescent pictures were also processed with Fiji for background correction and manual stack alignment.

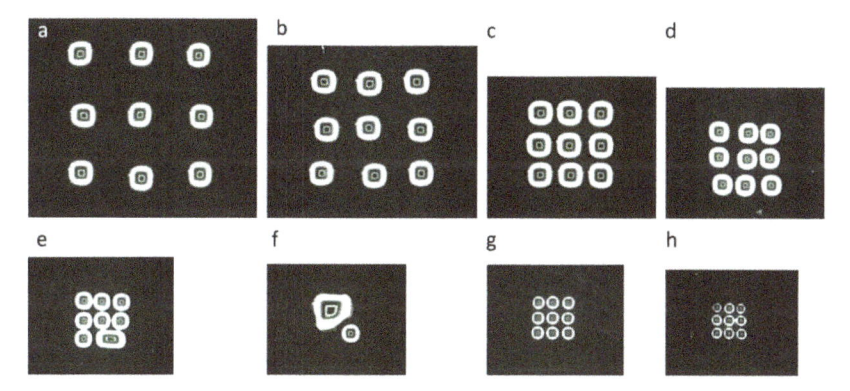

Figure 2. Optimization of the printing on a glass substrate. Water droplets were piezo dispensed (pulse duration of 15 µs and voltage of 75 V) as a 3 × 3 array on a glass coverslip: (**a**) 100 droplets at a frequency of 50 Hz with a pitch of 1000 µm; (**b**) 100 droplets at a frequency of 50 Hz with a pitch of 750 µm; (**c**) 100 droplets at a frequency of 50 Hz with a pitch of 500 µm; (**d**) 50 droplets at a frequency of 25 Hz with a pitch of 400 µm; (**e**) 25 droplets at a frequency of 25 Hz with a pitch of 300 µm; (**f**) 25 droplets at a frequency of 25 Hz with a pitch of 250 µm; (**g**) 10 droplets at a frequency of 5 Hz with a pitch of 250 µm; (**h**) five droplets at a frequency of 4 Hz with a pitch of 200 µm.

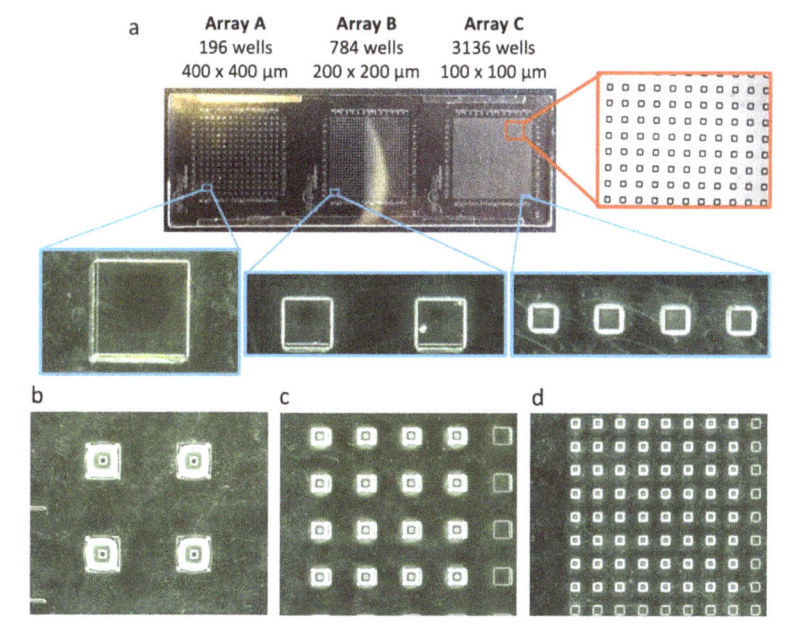

Figure 3. Optimization of piezo printing into microwells. (**a**) Commercial microtiter plate (Microfluidic ChipShop, Jena, Germany) containing three microwell arrays with square microwells of varying width; array A: width of 400 µm and pitch of 1125 µm, array B: width of 200 µm and pitch of 563 µm, array C: width of 100 µm and pitch of 281 µm. Water droplets were piezo dispensed (pulse duration of 15 µs, amplitude voltage of 75 V and frequency of 50 Hz) as (**b**) a 2 × 2 array in array A wells at 200 droplets/well, (**c**) a 4 × 4 array in array B wells at 25 droplets/well, (**d**) a 8 × 8 array in array C wells at five droplets/well.

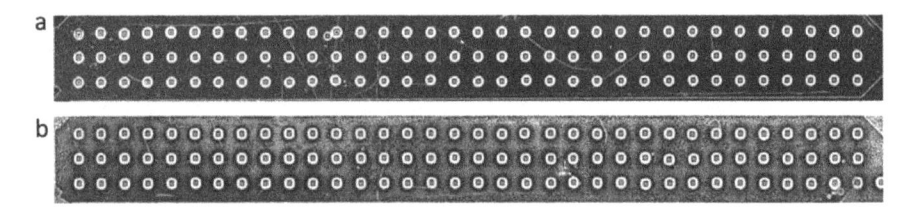

Figure 4. Con A and yeast cells dispensing into the microwells of the microfluidic ship (pulse duration of 15 µs, voltage of 75 V and frequency of 50 Hz): (**a**) 60 droplets of Con A were printed, (**b**) 40 droplets were printed from the yeast solution (OD$_{600}$ of 0.5).

3. Results

3.1. Construction of the Microfluidic Chip

The microfluidic chip (Figure 1c) was made of two parts: the microwell array on a glass slide (Figure 1a) and the top plate with a microfluidic channel (Figure 1b). The microwell array was produced in the epoxy-based photoresist SU-8 through a standard photolithography protocol. The mask design of the microwell array consisted of three rows of 34 wells with a well diameter of 300 µm and a pitch distance of 850 µm (Figure S1). Visual inspection of the microwells after photolithography showed that the SU-8 wells matched the mask's dimensions (Figure 1d). Furthermore, the measured well's depth was approximatively 50 µm (Figure S1b), which fits the expected thickness of the SU-8 layer, according to the protocol mentioned earlier. Altogether, these results prove the photolithography process to be successful and a quick method to produce the microwells in the microfluidic chip.

The overall length and thickness of the microwell array were 28.4 mm and 2.0 mm, respectively. These dimensions fit within the microfluidic channel of the top plate that was used to seal the microfluidic chip and connect the inlet and outlet tubes. The top plate was commercially available and consisted of a channel of 50 × 5 × 0.4 mm (l × w × h) surrounded by a double-sided tape, which enabled us to quickly and easily stick it onto the bottom substrate. Once the microfluidic chip was sealed (after filling with the cells), it was connected to silicone tubing and a syringe filled with SC growth medium. The medium was then perfused at a very low flow rate (25 µL/min) with a syringe pump in order to ensure nutrient renewal for cell growth.

3.2. Living-Cell Microarray Development

First, piezo dispensing parameters were optimized by printing a water droplet array on a glass substrate. Piezo dispensing is accomplished by applying rectangular voltage pulses to the piezo ceramic actuator (Figure S2c). For the duration of each pulse, the actuator tube contracts a few micrometers, thereby initiating a pressure wave that causes the ejection of a droplet. The amplitude and the duration of the pulse influence the volume and the velocity of the droplet. The higher the amplitude (applied voltage) and the longer the pulse, the bigger the droplet and the higher its velocity. The frequency of the pulses is directly correlated to the duration of the pulse and also influences the droplet's volume. Finally, the viscosity of the sample to dispense also has an influence on the printing parameters. Therefore, piezo dispensing needs to be optimized for each solution that has to be printed. In this work, yeast suspensions were prepared in PBS prior to printing.

Considering that PBS solution and ultrapure water have similar viscosities; we first optimized the piezo dispensing parameters with ultrapure water for simplicity reason. Typically, we started by determining the amplitude and duration of the voltage pulses that were giving a reliable ejection of water droplets. We obtained a stable shooting of ultrapure water with an amplitude of 75 V and a pulse duration of 15 µs. These parameters resulted in a droplet volume ranging between 70 to 80 pL (Figure S2b). Next, we created 3 × 3 arrays of water droplets with decreasing pitch distance on a glass substrate (Figure 2). We started by shooting 100 drops/spot at 50 Hz with a pitch distance of 1000 µm, 750 µm and 500 µm. We could not print arrays with pitch distances lower than 500 µm without merging of neighboring spots. Hence, we reduced the droplet number and printed 50 drops/spot at 25 Hz with a decreasing pitch distance from 400 to 300 and 250 µm. Surprisingly, printing half the droplet number did not allow halving the pitch distance as expected. Additional testing established that five drops/spot at 4 Hz with a pitch distance of 200 µm were the limit parameters at which a 3 × 3 array of water droplets could be stably printed on a glass slide. This optimization process aimed at determining the minimal spot sizes and pitch distances that can be stably printed in order to increase the throughput of the platform.

Secondly, we evaluated the filling of microwells with water droplets using piezo dispensing. Therefore, we initially used a commercially available microtiter plate (size of a microscope slide) that contains squared microwells with dimensions of 400 × 400 µm (array A), 200 × 200 µm (array B), 100 × 100 µm (array C), and height of 20 µm (Figure 3). The pulse duration (15 µs) and amplitude voltage (75 V) were as for printing on the glass substrate. With a frequency of 50 Hz, the number of droplets decreased from 200 drops per well for array A to 25 drops per well for array B and five drops per well for array C.

Once the piezo-dispensing parameters were established with ultrapure water, the printing protocol was optimized for living cells. First, the microwells were filled with 60 droplets of Con A at a pulse duration of 15 µs, amplitude voltage of 75 V and frequency of 50 Hz (Figure 4a). Con A is a lectin, which binds to the mannose glycans at the yeast cell wall [36]. Hence, Con A was used as a coating, which anchored the yeast cells to the bottom of the microwells so that they were not flushed away when SC medium was perfused during continuous cultivation. Next, the density of printed yeast cells per well was optimized. The final objective of this study consisted in developing a method that allowed monitoring the growth of single yeast cells as well as their expression of GFP-tagged proteins by time-lapse fluorescence microscopy. Therefore, the printing process had to deliver only a few cells per well. To do so, two parameters were adapted: the cell density of the yeast suspensions to print and the number of droplets/well. The latter was evaluated first and 40 drops/well provided optimal filling of the microwell without overflowing (Figure 4b). Since this parameter was kept constant for all the living cell experiments, the cell density of the yeast suspensions was then optimized. Yeasts suspensions at OD_{600} of 0.8, 0.5, and 0.2 were tested (Figure S2d). The suspension at an OD_{600} of 0.8 proved difficult to print since this high cell concentration increased the viscosity of the sample, thereby preventing the droplet ejection. On the other hand, cell suspensions at an OD_{600} of 0.2 were possible to shoot but did not deliver yeast cells in every well. The suspension at an OD_{600} of 0.5 was then selected, as it could be reproducibly printed and guaranteed the presence of a significant number of single cells per well. All the yeast printing experiments were performed with the following optimized parameters: pulse duration of 15 µs, amplitude voltage of 75 V, and a frequency of 50 Hz.

Finally, since the objective was to print multiple clones from the GFP-tagged yeast collection in different wells of the microfluidic chip, a last printing procedure had to be optimized, namely the washing step between samples in order to avoid cross-contamination. The washing step includes the disposal of the printed sample followed by the cleaning of the outside of the piezo dispenser capillary. Discarding the sample can be done either by applying pressure to the liquid path of the robot or by ejecting it with a syringe pump. Indeed, the iTWO-400 robot is equipped with both a pressure unit and a syringe pump, which enable to flow liquid through the piezo dispenser on a passive or active basis, respectively. With the pressure unit, the liquid continuously flows at a fixed fluid flow rate for as long as the pressure is applied. With the syringe pump, a specified volume of liquid flows at a specified flow rate. Cleaning the outside of the piezo dispenser is performed at the washing station of the robot (Figure S2a), and only the washing time can be adapted. To evaluate the efficiency of the washing step on avoiding cross-contamination, wells from a 384-multiwell plate (MTP) were filled with both a yeast suspension in PBS (at OD_{600} of 0.5) and a colored ink according to a designed pattern (Figure 5a). The samples were then printed on solid agar medium in a petri dish and left to grow at 30 °C for 24 hours before evaluation. In the first experiment, the printed samples were discarded by means of pressure (at 450 mbar) for 10 seconds, and the outside of the piezo dispenser was cleaned for five seconds. Although most of the design was correctly printed, one cross-contaminated spot was visible (Figure 5b). The experiment was then repeated; however, the samples were discarded in the wash station by means of the syringe pump. The cleaning of the piezo dispenser was kept at five seconds. Using this procedure, no cross-contamination was observed (Figure 5c).

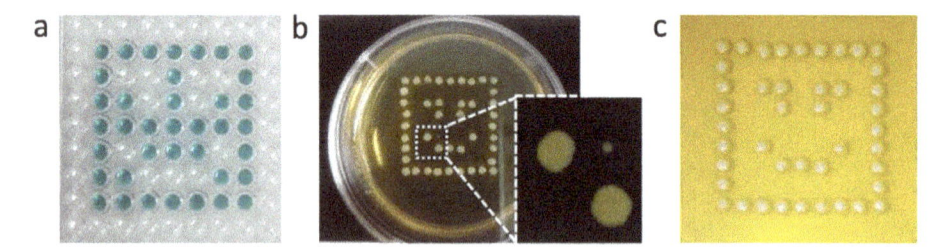

Figure 5. Evaluation of the washing protocol to avoid cross-contamination. (**a**) Printing pattern to evaluate the washing protocol of the piezo dispenser. Transparent wells contain a yeast solution in PBS, and the colored wells contain a blue dye (no cells). (**b**) Washing protocol based on pressure resulted in 1 cross-contamination colony. (**c**) Washing protocol based on the syringe pump resulted in spotting without cross-contamination.

3.3. Growth of Yeast Cells in the Microfluidic Chip

A selection of 34 GFP-tagged *S. cerevisiae* clones that are related to the cell cycle were printed as triplicates into the well array (Figure 4, S2e), and the chip was closed by sticking the microfluidic channel to the microwell substrate (Figure 1). Time-lapse microscopy was used to follow the growth during 18 h. An overview of all wells is presented in Figure S3 and some selected wells in Figure 6. At the end of the growth experiment, the GFP-tagged proteins were visualized.

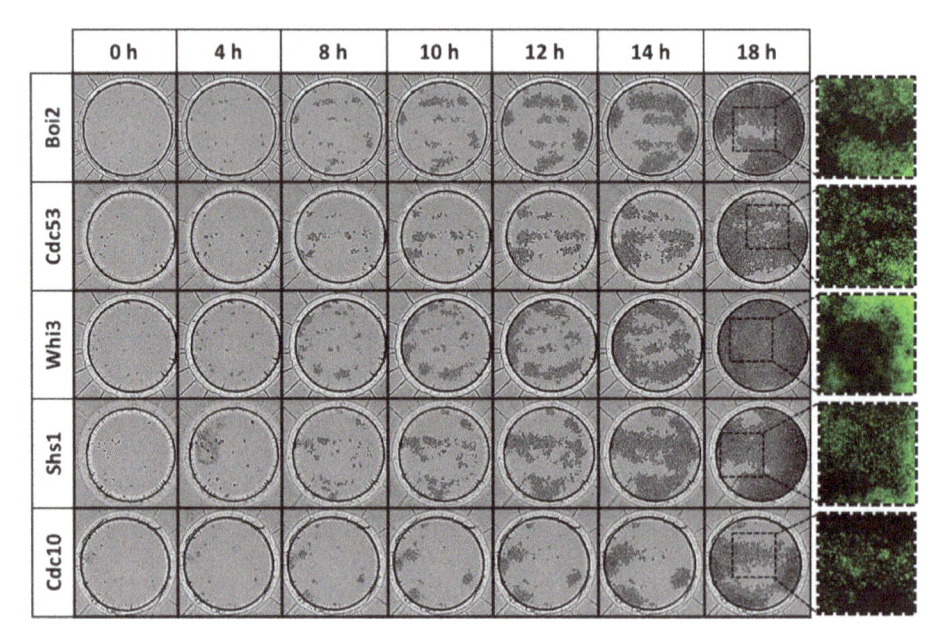

Figure 6. Yeast growth in some selected wells (see Figure S3 for the full overview).

Fermentation 2020, 6, 26

To evaluate the suitability of the developed yeast chips to perform dynamic analyses of protein expression and protein localization in single yeast cells, a small set of 6 clones were selected, i.e. Cdc39, Cla4, Bem1, Shs1, Cdc14 and Cdc28. A time-lapse experiment was performed where these clones were observed at higher resolution (600× magnification) during 3 h (Figure 7).

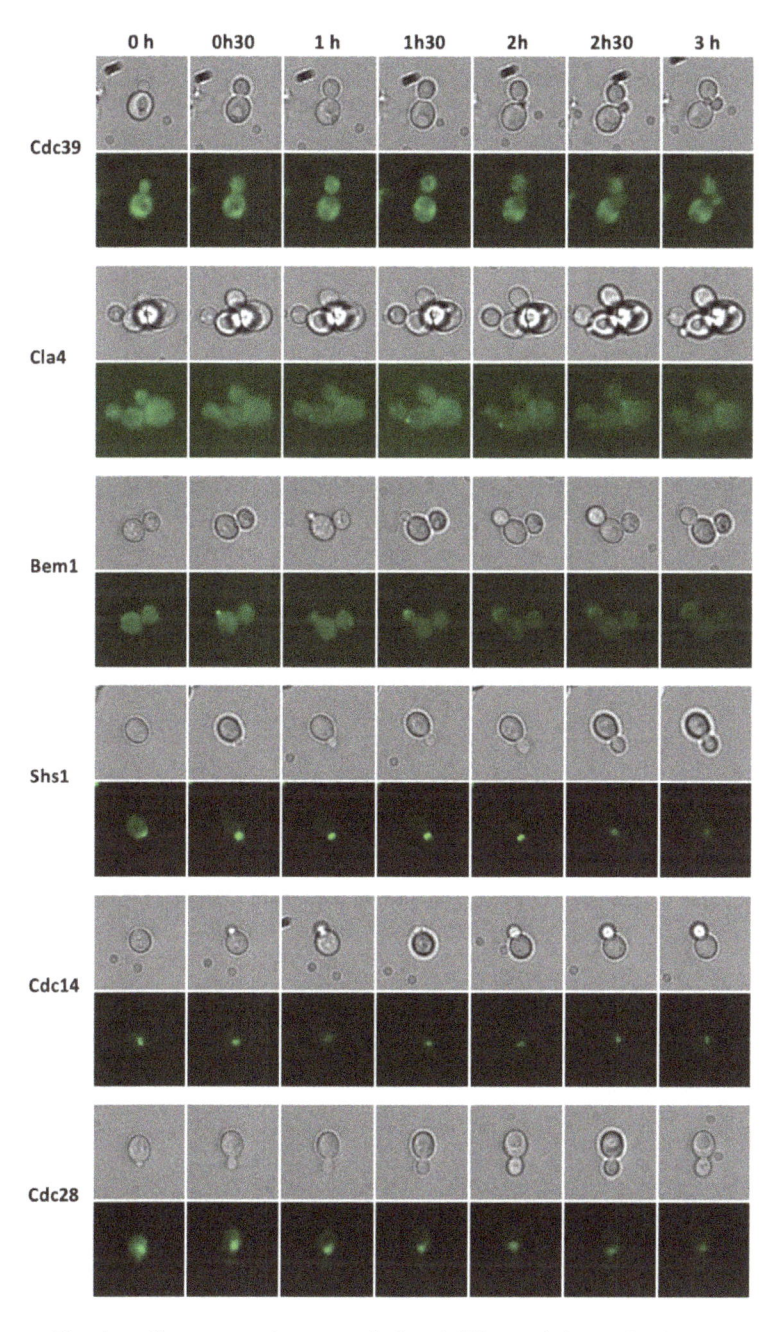

Figure 7. Time-lapse fluorescence microscopy of selected GFP-tagged clones related to the cell cycle.

4. Discussion

To create a living GFP-tagged yeast array, we combined mechanical patterning by constructing an array of microwells with cell printing (robotic cell patterning), which allows the controlled placement of the GFP-tagged clones in the selected wells. The cells were trapped in the wells by sticking them to the glass bottom of the microwells using the lectin Con A. Microfluidic chips where a GFP-tagged yeast clone collection was patterned as an array into microchambers have been previously developed [10,37]. Also, a microfluidic perfusion system where a robotic printed yeast array on agar and sandwiched with a track-etched membrane has been described [38]. These designs and fabrication methods are much more complex and difficult to construct. Due to the open design, robotic patterning of living cells in microwells is much more flexible in creating different filling designs. Additionally, it could be used to create high-density arrays without increasing much the complexity of the microfluidic chip design and construction.

A hybrid SU-8 on glass microfluidic chip containing a living cell array was developed. The direct fabrication of SU-8 microwells on the glass coverslip substrate is a simple, low cost and a high precision method that is suitable to construct disposable biochips [39]. SU-8 is biocompatible and has also the advantage that high density well arrays at high aspect ratios could be created [40]. A microwell array of 3 × 34 microwells containing microwells with a diameter of 300 µm and depth of 50 µm were created. The SU-8 microwell layer stuck well to the glass substrate. The bonding of the SU-8 microwell array to the glass could also withstand temperature shifts to low temperature (refrigerator). This allowed to store the chip filled with yeast cells for a few days before the growth experiment was performed.

Piezo printing was selected as the method to dispense a Con A protein solution and the GFP-tagged yeast collection into the microwells. We performed these dispensing steps with a piezoelectric dispenser (also called drop-on-demand ink-jet printing) [41,42]. Piezo printing of proteins and cells has been used previously for various applications including enzyme printing for glucose biosensors [43], protein arrays [44], netrin-1 adhesive micropattern construction [45], bacterial *Escherichia coli* arrays [46,47], bacterial and yeast cells on cantilever array sensors [48], *S. cerevisiae* cells on an agar layer [38], and mammalian cells [49].

We demonstrated that printing a small array on a flat glass substrate is possible and this method could be used to create cell arrays on glass substrate in a cheap and easy way. However, the droplet size was not proportional to the droplet volume which is a limitation to increase the throughput. We expected that half the number of droplets would result in half the spot volume, half the spot size and that twice the number of spots could be printed in the array. However, this was not the case. Also, significant variations in the droplet locations occurred, which resulted in arrays that were not perfectly arranged, which can result in merging of the spots in case of a small pitch distance (Figure 2e,f).

Printing of cells into microwells compared to on a flat surface has many advantages. Spatial confinement in the microwell results in higher resolution printing: smaller well sizes than droplet spot sizes and smaller pitch distances can be obtained. Droplet position in the wells is more accurate than droplets on a flat surface (Figure 2). Additionally, the liquid–air interface is reduced for microwells compared to droplets on a surface, resulting in a reduced evaporation rate. All these benefits result in higher throughput and stable printing.

The printer setup allowed to aspirate a yeast solution from an MTP (multiwell plate) well and deposit picoliter droplets containing living cells at the target microwell, and to perform this consecutively for all filled MTP wells. A tip dispenser washing protocol based on syringe pump cell solution ejection was successful to avoid cross contamination between different wells since syringe pump ejection occurred at a much higher flow rate than pressure-based ejection and could remove all yeast cells from the piezo tip. A GFP-tagged clone collection in a 3 × 34 cell array was cultivated during 18 h (Figure 6 and S3). This experiment demonstrated that a clone collection of 34 clones in triplicate could be successfully printed at a cell concentration that allows single cell observation during several generations.

As a proof-of-concept experiment to observe changes in GFP-tagged protein expression and cellular location, we selected GFP-tagged proteins that play a role during the yeast growth cycle. The Cdc39 protein is a subunit of the CCR4-NOT1 core complex that has multiple roles in the regulation of mRNA levels [50,51]. A high fluorescence intensity covering the cytoplasm in the mother and daughter cell can be observed (Figure 7) since it is present at a large number of protein molecules per cell (4300) [52]. Cla4p is a Cdc42p-activated signal transducing kinase that is involved in septin ring assembly, vacuole inheritance, cytokinesis, and sterol uptake regulation [53]. It is distributed in the cytoplasm and the bud. Higher intensity spots can be observed at the site where the bud appears (Figure 7). Bem1p is involved in establishing cell polarity and morphogenesis and functions as a scaffold protein for complexes that include Cdc24p, Ste5p, Ste20p, and Rsr1p [54]. A high intensity spot can be observed at the bud site and the growing bud (Figure 7). Shs1p is a component of the septin ring that is required for cytokinesis [55]. Initially, it is present at the cell periphery and moves to the bud site to create the septin ring (Figure 7). Cdc14p is a phosphatase required for the mitotic exit [56]. It is present in the nucleus and the nucleolus (Figure 7). Cdc28p is a cyclin-dependent kinase (CDK) catalytic subunit and master regulator of mitotic and meiotic cell cycles [57,58]. It alternately associates with G1, S, and G2/M phase cyclins. It is observed initially in the cytoplasm and next in the nucleus (Figure 7). These results demonstrate that the microfluidic chip can be used to perform dynamic experiments with subcellular resolution of fluorescently-tagged proteins.

In the future, the number of microwells in the array could be upscaled allowing to perform dynamic proteomics and localizomics experiments. This cell microarray-based systems-biology platform could be used to detect directly chemical disturbances in a small-molecule compound screen of the proteome in contrast to genetic approaches that are based on chemical-genetic interactions and necessitates a multistep indirect approach [59]. The genome-wide tagging of proteins of S. cerevisiae, including with GFP, has already provided a vast resource of such information [9,11,33,60]. Analysis of this high-resolution, high coverage localisation data set in the context of transcriptional, genetic, and protein-protein interaction (PPI) data revealed the combinatorial logic of transcriptional co-regulation and spatial-temporal regulation of proteins, and provided for example a comprehensive view of trafficking and signalling regulatory interactions within and between organelles in eukaryotic cells. As demonstrated here, dynamic movements from one location to another can also be followed as the cell proceeds through the cell cycle. Dynamic movements of proteins have also been described in yeast cells that respond to environmental stresses such as dithiothreitol (DTT) stress [9], hydrogen peroxide stress [9], osmotic stress by potassium chloride [37], and nitrogen starvation [9] or chemical perturbations by rapamycin [11], hydroxyurea [11], or methyl methane sulfonate [10].

5. Conclusions

We developed a perfusion microfluidic chip containing living yeast cell arrays that allows long term cultivation of the yeast cells and high-resolution time-lapse fluorescence microscopy. The creation of the cell array was based on mechanical patterning in SU-8 microwells and piezoelectric filling of the microwells. The microfluidic chip was closed by sticking a top plate that contained the microfluidic channel and inlet and outlet to the microwell substrate. The developed technology and method were validated by a growth experiment of a clone collection of 34 clones (in triplicate) that are linked to the cell cycle. Additionally, a set of six selected GFP-tagged clones were observed by time-lapse fluorescence microscopy at high resolution to observe single cell protein expression and subcellular location of the GFP-tagged proteins.

Future technological challenges lie in the further upscaling of the technology and procedures to construct genome/proteome-wide cell microarrays and analyzing cells dynamically on a whole proteome level. Dynamic proteomics profiling information based on chemical compound (such as e.g. drug compound) perturbation should allow to determine the target(s) and mechanism of action (MoA) [61]. For example, this technology could lead to the discovery of novel antifungal molecules, which are highly desired due to the limited number of available antifungal drugs and the fast emergence of multiresistant pathogens [62,63].

Supplementary Materials: The following are available online at http://www.mdpi.com/2311-5637/6/1/26/s1, Figure S1: Construction of the SU-8 microwell array on the glass substrate, Figure S2: The robotic piezo dispenser, Figure S3: Growth of yeast cells in the microfluidic chip, Table S1: Selected *S. cerevisiae* GFP clones that were used in the growth experiment.

Author Contributions: Conceptualization, C.Y. and R.G.W.; methodology, C.Y., S.T., and R.G.W.; formal analysis, C.Y.; investigation, C.Y.; resources, R.G.W.; writing—original draft preparation, C.Y. and R.G.W.; writing—review and editing, C.Y., S.T., and R.G.W.; supervision, R.G.W.; project administration, R.G.W.; funding acquisition, C.Y. and R.G.W. All authors have read and agreed to the published version of the manuscript.

Funding: This research was funded by The Belgian Federal Science Policy Office (Belspo) and the European Space Agency (ESA) PRODEX program grant number FluoCells and Yeast Bioreactor projects. C.Y. was funded by FWO for the funding of the SB PhD grant. The Research Council of the Vrije Universiteit Brussel (Belgium) and the University of Ghent (Belgium) are acknowledged to support the Alliance Research Group VUB-UGent NanoMicrobiology (NAMI), and the International Joint Research Group (IJRG) VUB-EPFL BioNanotechnology & NanoMedicine (NANO).

Acknowledgments: We acknowledge Rouslan Efremov for the use of lithography equipment and Gangadhar Eluru for allowing us to use his design of the microwell mask. The Hercules Foundation (FWO Flanders) is acknowledged for the funding of the Raith Voyager e-beam equipment (AUGE/13/19).

Conflicts of Interest: The authors declare no conflict of interest.

References

1. Castel, D.; Pitaval, A.; Debily, M.A.; Gidrol, X. Cell microarrays in drug discovery. *Drug Discov. Today* **2006**, *11*, 616–622. [CrossRef] [PubMed]
2. Berthuy, O.I.; Muldur, S.K.; Rossi, F.; Colpo, P.; Blum, L.J.; Marquette, C.A. Multiplex cell microarrays for high-throughput screening. *Lab Chip* **2016**, *16*, 4248–4262. [CrossRef] [PubMed]
3. Hong, H.; Koom, W.; Koh, W.G. Cell Microarray Technologies for High-Throughput Cell-Based Biosensors. *Sensors* **2017**, *17*, 1293. [CrossRef] [PubMed]
4. Willaert, R.; Goossens, K. Microfluidic Bioreactors for Cellular Microarrays. *Fermentation* **2015**, *1*, 38–78. [CrossRef]
5. Jonczyk, R.; Kurth, T.; Lavrentieva, A.; Walter, J.-G.; Scheper, T.; Stahl, F. Living Cell Microarrays: An Overview of Concepts. *Microarrays* **2016**, *5*, 11. [CrossRef]
6. Willaert, R.; Sahli, H. On-Chip Living-Cell Microarrays for Network Biology. In *Bioinformatics—Trends and Methodologies*; InTechOpen: London, UK, 2011.
7. Fernandes, T.G.; Diogo, M.M.; Clark, D.S.; Dordick, J.S.; Cabral, J.M.S. High-throughput cellular microarray platforms: Applications in drug discovery, toxicology and stem cell research. *Trends Biotechnol.* **2009**, *27*, 342–349. [CrossRef]
8. Chen, D.; Davis, M. Molecular and functional analysis using live cell microarrays. *Curr. Opin. Chem. Biol.* **2006**, *10*, 28–34. [CrossRef]
9. Breker, M.; Gymrek, M.; Schuldiner, M. A novel single-cell screening platform reveals proteome plasticity during yeast stress responses. *J. Cell Biol.* **2013**, *200*, 839–850. [CrossRef]
10. Dénervaud, N.; Becker, J.; Delgado-Gonzalo, R.; Damay, P.; Rajkumar, A.S.; Unser, M.; Shore, D.; Naef, F.; Maerkl, S.J. A chemostat array enables the spatio-temporal analysis of the yeast proteome. *Proc. Natl. Acad. Sci. USA* **2013**, *110*, 15842–15847. [CrossRef]
11. Chong, Y.T.; Koh, J.L.Y.; Friesen, H.; Duffy, K.; Cox, M.J.; Moses, A.; Moffat, J.; Boone, C.; Andrews, B.J. Yeast proteome dynamics from single cell imaging and automated analysis. *Cell* **2015**, *161*, 1413–1424. [CrossRef]
12. Yuan, H.; Zhang, R.; Shao, B.; Wang, X.; Ouyang, Q.; Hao, N.; Luo, C. Protein expression patterns of the yeast mating response. *Integr. Biol. (UK)* **2016**, *8*, 712–719. [CrossRef]
13. Shao, B.; Yuan, H.; Zhang, R.; Wang, X.; Zhang, S.; Ouyang, Q.; Hao, N.; Luo, C. Reconstructing the regulatory circuit of cell fate determination in yeast mating response. *PLoS Comput. Biol.* **2017**, *13*, e1005671. [CrossRef] [PubMed]
14. Yarmush, M.L.; King, K.R. Living-Cell Microarrays. *Annu. Rev. Biomed. Eng.* **2009**, *11*, 235–257. [CrossRef] [PubMed]

15. Park, M.C.; Hur, J.Y.; Cho, H.S.; Park, S.H.; Suh, K.Y. High-throughput single-cell quantification using simple microwell-based cell docking and programmable time-course live-cell imaging. *Lab Chip* **2011**, *11*, 79–86. [CrossRef] [PubMed]
16. Minc, N.; Boudaoud, A.; Chang, F. Mechanical Forces of Fission Yeast Growth. *Curr. Biol.* **2009**, *19*, 1096–1101. [CrossRef] [PubMed]
17. Falconnet, D.; Niemistö, A.; Taylor, R.J.; Ricicova, M.; Galitski, T.; Shmulevich, I.; Hansen, C.L. High-throughput tracking of single yeast cells in a microfluidic imaging matrix. *Lab Chip* **2011**, *11*, 466–473. [CrossRef]
18. Groisman, A.; Lobo, C.; Cho, H.; Campbell, J.K.; Dufour, Y.S.; Stevens, A.M.; Levchenko, A. A microfluidic chemostat for experiments with bacterial and yeast cells. *Nat. Methods* **2005**, *2*, 685–689. [CrossRef]
19. Ryley, J.; Pereira-Smith, O.M. Microfluidics device for single cell gene expression analysis inSaccharomyces cerevisiae. *Yeast* **2006**, *23*, 1065–1073. [CrossRef]
20. Bell, L.; Seshia, A.; Lando, D.; Laue, E.; Palayret, M.; Lee, S.F.; Klenerman, D. A microfluidic device for the hydrodynamic immobilisation of living fission yeast cells for super-resolution imaging. *Sens. Actuators B Chem.* **2014**, *192*, 36–41. [CrossRef]
21. Carlo, D.D.; Wu, L.Y.; Lee, L.P. Dynamic single cell culture array. *Lab Chip* **2006**, *6*, 1445–1449. [CrossRef]
22. Barbulovic-Nad, I.; Lucente, M.; Sun, Y.; Zhang, M.; Wheeler, A.R.; Bussmann, M. Bio-microarray fabrication techniques—A review. *Crit. Rev. Biotechnol.* **2006**, *26*, 237–259. [CrossRef] [PubMed]
23. Guillemot, F.; Souquet, A.; Catros, S.; Guillotin, B.; Lopez, J.; Faucon, M.; Pippenger, B.; Bareille, R.; Rémy, M.; Bellance, S.; et al. High-throughput laser printing of cells and biomaterials for tissue engineering. *Acta Biomater.* **2010**, *6*, 2494–2500. [CrossRef] [PubMed]
24. Bean, G.J.; Jaeger, P.A.; Bahr, S.; Ideker, T. Development of ultra-high-density screening tools for microbial "omics". *PLoS ONE* **2014**, *9*, e85177. [CrossRef] [PubMed]
25. Schaack, B.; Reboud, J.; Combe, S.; Fouqué, B.; Berger, F.; Boccard, S.; Odile Filhol-Cochet, F.C. A "DropChip" Cell Array for DNA and siRNA Transfection Combined with Drug Screening. *Nanobiotechnology* **2005**, *1*, 183–189. [CrossRef]
26. Ringeisen, B.R.; Othon, C.M.; Barron, J.A.; Young, D.; Spargo, B.J. Jet-based methods to print living cells. *Biotechnol. J.* **2006**, *1*, 930–948. [CrossRef]
27. Roth, E.A.; Xu, T.; Das, M.; Gregory, C.; Hickman, J.J.; Boland, T. Inkjet printing for high-throughput cell patterning. *Biomaterials* **2004**, *25*, 3707–3715. [CrossRef]
28. Ferris, C.J.; Gilmore, K.G.; Wallace, G.G.; In Het Panhuis, M. Biofabrication: An overview of the approaches used for printing of living cells. *Appl. Microbiol. Biotechnol.* **2013**, *97*, 4243–4258. [CrossRef]
29. Demirci, U.; Montesano, G. Cell encapsulating droplet vitrification. *Lab Chip* **2007**, *7*, 1428–1433. [CrossRef]
30. Lee, W.; Debasitis, J.C.; Lee, V.K.; Lee, J.H.; Fischer, K.; Edminster, K.; Park, J.K.; Yoo, S.S. Multi-layered culture of human skin fibroblasts and keratinocytes through three-dimensional freeform fabrication. *Biomaterials* **2009**, *30*, 1587–1595. [CrossRef]
31. Gonzalez-Macia, L.; Morrin, A.; Smyth, M.R.; Killard, A.J. Advanced printing and deposition methodologies for the fabrication of biosensors and biodevices. *Analyst* **2010**, *135*, 845–867. [CrossRef]
32. Li, J.; Rossignol, F.; Macdonald, J. Inkjet printing for biosensor fabrication: Combining chemistry and technology for advanced manufacturing. *Lab Chip* **2015**, *15*, 2538–2558. [CrossRef] [PubMed]
33. Huh, W.K.; Falvo, J.V.; Gerke, L.C.; Carroll, A.S.; Howson, R.W.; Weissman, J.S.; O'Shea, E.K. Global analysis of protein localization in budding yeast. *Nature* **2003**, *425*, 686–691. [CrossRef] [PubMed]
34. Schindelin, J.; Arganda-Carreras, I.; Frise, E.; Kaynig, V.; Longair, M.; Pietzsch, T.; Preibisch, S.; Rueden, C.; Saalfeld, S.; Schmid, B.; et al. Fiji: An open-source platform for biological-image analysis. *Nat. Methods* **2012**, *9*, 676–682. [CrossRef] [PubMed]
35. Preibisch, S.; Saalfeld, S.; Tomancak, P. Globally optimal stitching of tiled 3D microscopic image acquisitions. *Bioinformatics* **2009**, *25*, 1463–1465. [CrossRef]
36. Pemberton, L.F. Preparation of yeast cells for live-cell imaging and indirect immunofluorescence. *Methods Mol. Biol.* **2014**, *1205*, 79–90.
37. Zhang, R.; Yuan, H.; Wang, S.; Ouyang, Q.; Chen, Y.; Hao, N.; Luo, C. High-throughput single-cell analysis for the proteomic dynamics study of the yeast osmotic stress response. *Sci. Rep.* **2017**, *7*, 42200. [CrossRef]

38. Mirzaei, M.; Pla-Roca, M.; Safavieh, R.; Nazarova, E.; Safavieh, M.; Li, H.; Vogel, J.; Juncker, D. Microfluidic perfusion system for culturing and imaging yeast cell microarrays and rapidly exchanging media. *Lab Chip* **2010**, *10*, 2449–2457. [CrossRef]
39. Kang, D.H.; Han, W.B.; Choi, N.; Kim, Y.J.; Kim, T.S. Tightly Sealed 3D Lipid Structure Monolithically Generated on Transparent SU-8 Microwell Arrays for Biosensor Applications. *ACS Appl. Mater. Interfaces* **2018**, *10*, 40401–40410. [CrossRef]
40. Wu, Z.Z.; Zhao, Y.; Kisaalita, W.S. Interfacing SH-SY5Y human neuroblastoma cells with SU-8 microstructures. *Colloids Surf. B Biointerfaces* **2006**, *52*, 14–21. [CrossRef]
41. Calvert, P. Inkjet printing for materials and devices. *Chem. Mater.* **2001**, *13*, 3299–3305. [CrossRef]
42. Derby, B. Bioprinting: Inkjet printing proteins and hybrid cell-containing materials and structures. *J. Mater. Chem.* **2008**, *18*, 5717–5721. [CrossRef]
43. Newman, J.D.; Turner, A.P.F.; Marrazza, G. Ink-jet printing for the fabrication of amperometric glucose biosensors. *Anal. Chim. Acta* **1992**, *262*, 13–17. [CrossRef]
44. Montenegro-Nicolini, M.; Miranda, V.; Morales, J.O. Inkjet Printing of Proteins: An Experimental Approach. *AAPS J.* **2017**, *19*, 234–243. [CrossRef] [PubMed]
45. Matsugaki, A.; Yamazaki, D.; Nakano, T. Selective patterning of netrin-1 as a novel guiding cue for anisotropic dendrogenesis in osteocytes. *Mater. Sci. Eng. C* **2020**, *108*, 110391. [CrossRef]
46. Xu, T.; Petridou, S.; Lee, E.H.; Roth, E.A.; Vyavahare, N.R.; Hickman, J.J.; Boland, T. Construction of high-density bacterial colony arrays and patterns by the ink-jet method. *Biotechnol. Bioeng.* **2004**, *85*, 29–33. [CrossRef] [PubMed]
47. Zheng, Q.; Lu, J.; Chen, H.; Huang, L.; Cai, J.; Xu, Z. Application of inkjet printing technique for biological material delivery and antimicrobial assays. *Anal. Biochem.* **2011**, *410*, 171–176. [CrossRef] [PubMed]
48. Lukacs, G.; Maloney, N.; Hegner, M. Ink-jet printing: Perfect tool for cantilever array sensor preparation for microbial growth detection. *J. Sensors* **2012**, *2012*. [CrossRef]
49. Zhang, J.; Chen, F.; He, Z.; Ma, Y.; Uchiyama, K.; Lin, J.M. A novel approach for precisely controlled multiple cell patterning in microfluidic chips by inkjet printing and the detection of drug metabolism and diffusion. *Analyst* **2016**, *141*, 2940–2947. [CrossRef]
50. Collart, M.A.; Struhl, K. NOT1(CDC39), NOT2(CDC36), NOT3, and NOT4 encode a global-negative regulator of transcription that differentially affects TATA-element utilization. *Genes Dev.* **1994**, *8*, 525–537. [CrossRef]
51. Kruk, J.A.; Dutta, A.; Fu, J.; Gilmour, D.S.; Reese, J.C. The multifunctional Ccr4-Not complex directly promotes transcription elongation. *Genes Dev.* **2011**, *25*, 581–593. [CrossRef]
52. Ghaemmaghami, S.; Huh, W.K.; Bower, K.; Howson, R.W.; Belle, A.; Dephoure, N.; O'Shea, E.K.; Weissman, J.S. Global analysis of protein expression in yeast. *Nature* **2003**, *425*, 737–741. [CrossRef] [PubMed]
53. Benton, B.K.; Tinkelenberg, A.; Gonzalez, I.; Cross, F.R. Cla4p, a Saccharomyces cerevisiae Cdc42p-activated kinase involved in cytokinesis, is activated at mitosis. *Mol. Cell. Biol.* **1997**, *17*, 5067–5076. [CrossRef] [PubMed]
54. Madden, K.; Snyder, M. Cell Polarity and Morphogenesis in Budding Yeast. *Annu. Rev. Microbiol.* **1998**, *52*, 687–744. [CrossRef]
55. Mino, A.; Tanaka, K.; Kamei, T.; Umikawa, M.; Fujiwara, T.; Takai, Y. Shs1p: A novel member of septin that interacts with Spa2p, involved in polarized growth in Saccharomyces cerevisiae. *Biochem. Biophys. Res. Commun.* **1998**, *251*, 732–736. [CrossRef] [PubMed]
56. Visintin, R.; Craig, K.; Hwang, E.S.; Prinz, S.; Tyers, M.; Amon, A. The phosphatase Cdc14 triggers mitotic exit by reversal of Cdk-dependent phosphorylation. *Mol. Cell* **1998**, *2*, 709–718. [CrossRef]
57. Mendenhall, M.D.; Hodge, A.E. Regulation of Cdc28 cyclin-dependent protein kinase activity during the cell cycle of the yeast Saccharomyces cerevisiae. *Microbiol. Mol. Biol. Rev.* **1998**, *62*, 1191–1243. [CrossRef] [PubMed]
58. Chymkowitch, P.; Eldholm, V.; Lorenz, S.; Zimmermann, C.; Lindvall, J.M.; Bjørås, M.; Meza-Zepeda, L.A.; Enserink, J.M. Cdc28 kinase activity regulates the basal transcription machinery at a subset of genes. *Proc. Natl. Acad. Sci. USA* **2012**, *109*, 10450–10455. [CrossRef]
59. Weig, M.; Brown, A.J.P. Genomics and the development of new diagnostics and anti-Candida drugs. *Trends Microbiol.* **2007**, *15*, 310–317. [CrossRef]

60. Howson, R.; Huh, W.K.; Ghaemmaghami, S.; Falvo, J.V.; Bower, K.; Belle, A.; Dephoure, N.; Wykoff, D.D.; Weissman, J.S.; O'Shea, E.K. Construction, verification and experimental use of two epitope-tagged collections of budding yeast strains. *Comp. Funct. Genom.* **2005**, *6*, 2–16. [CrossRef]
61. Cohen, A.A.; Geva-Zatorsky, N.; Eden, E.; Frenkel-Morgenstern, M.; Issaeva, I.; Sigal, A.; Milo, R.; Cohen-Saidon, C.; Liron, Y.; Kam, Z.; et al. Dynamic proteomics of individual cancer cells in response to a drug. *Science* **2008**, *322*, 1511–1516. [CrossRef]
62. Michael, C.A.; Dominey-Howes, D.; Labbate, M. The Antimicrobial Resistance Crisis: Causes, Consequences, and Management. *Front. Public Health* **2014**, *2*, 145. [CrossRef] [PubMed]
63. Willaert, R. Micro- and Nanoscale Approaches in Antifungal Drug Discovery. *Fermentation* **2018**, *4*, 43. [CrossRef]

© 2020 by the authors. Licensee MDPI, Basel, Switzerland. This article is an open access article distributed under the terms and conditions of the Creative Commons Attribution (CC BY) license (http://creativecommons.org/licenses/by/4.0/).

fermentation

Article

Yeast Nanometric Scale Oscillations Highlights Fibronectin Induced Changes in *C. albicans*

Anne-Céline Kohler [1,*], Leonardo Venturelli [1], Abhilash Kannan [2], Dominique Sanglard [2], Giovanni Dietler [1,3], Ronnie Willaert [3,4,5] and Sandor Kasas [1,2,6]

1. Laboratoire de Physique de la Matière Vivante, EPFL, 1015 Lausanne, Switzerland; leonardo.venturelli@epfl.ch (L.V.); giovanni.dietler@epfl.ch (G.D.); sandor.kasas@epfl.ch (S.K.)
2. Institute of Microbiology, Lausanne University Hospital, CH-1011 Lausanne, Switzerland; Abhilash.Kannan@chuv.ch (A.K.); Dominique.Sanglard@chuv.ch (D.S.)
3. International Joint Research Group VUB-EPFL NanoBiotechnology & NanoMedicine (NANO), Vrije Universiteit Brussel, 1050 Brussels, Belgium; Ronnie.Willaert@vub.be
4. Research Group Structural Biology Brussels, Alliance Research Group VUB-UGent NanoMicrobiology (NAMI), 1050 Ixelles, Belgium
5. Department Bioscience Engineering, University Antwerp, 2000 Antwerp, Belgium; Ronnie.Willaert@uantwerpen.be
6. Plateforme de Morphologie UFAM, CUMRL, University of Lausanne, 1015 Lausanne, Switzerland
* Correspondence: anne-celine.kohler@epfl.ch; Tel.: +41-216930464

Received: 18 December 2019; Accepted: 19 February 2020; Published: 21 February 2020

Abstract: Yeast resistance to antifungal drugs is a major public health issue. Fungal adhesion onto the host mucosal surface is still a partially unknown phenomenon that is modulated by several actors among which fibronectin plays an important role. Targeting the yeast adhesion onto the mucosal surface could lead to potentially highly efficient treatments. In this work, we explored the effect of fibronectin on the nanomotion pattern of different *Candida albicans* strains by atomic force microscopy (AFM)-based nanomotion detection and correlated the cellular oscillations to the yeast adhesion onto epithelial cells. Preliminary results demonstrate that strongly adhering strains reduce their nanomotion activity upon fibronectin exposure whereas low adhering Candida remain unaffected. These results open novel avenues to explore cellular reactions upon exposure to stimulating agents and possibly to monitor in a rapid and simple manner adhesive properties of *C. albicans*.

Keywords: *Candida albicans*; adhesion; fibronectin; nanomotion; atomic force microscope (AFM)

1. Introduction

Yeast biotechnology is a recent field where nanotechniques are used to manipulate and analyze yeast cells and cell constituents at the nanoscale [1]. Among the nanotechniques, AFM-related approaches played a major role in unveiling morphological, mechanical and biochemical properties of yeast [2–4]. Recently, our team demonstrated that living cells attached onto a soft cantilever induce nanometric scale oscillations (referred to as nanomotion) that stop as soon as the organism dies [5]. Commercially available atomic force microscopes (AFM) or dedicated devices easily detect these oscillations. Nanomotion detection has been applied to numerous biological samples such as proteins, single organelles, and a plethora of living cells such as prokaryotes (bacteria) and eukaryotes (fungal, vegetal and mammalian cells) [6]. The most straightforward application of the technique is the ultra-rapid antibiotic sensitivity test (AST). AST can be performed within an hour as compared to long-lasting traditional AST methods, which depend on the replication rate of the bacteria [7–9]. The test consists in attaching the organism of interest onto an AFM cantilever and monitoring its oscillations as a function of time upon addition of antibiotics in the analysis chamber. It is worth noting that the nanometric scale oscillations do not only reflect the living or death state of the organisms but also its activity [5,10].

Fungal infections are a major public health issue nowadays; it is estimated that every year fungi infect about 1.2 billion people [11]. *C. albicans* is a common fungal pathogen that belongs to the human microbiome of healthy individuals [12]. This commensal relationship is a complex interplay of candidial and human factors. However, impairment of the host immunity or the normal host microbiota can lead to *C. albicans* infection (candidiasis) [13]. *C. albicans* is the predominant cause of virtually all types of candidiasis [14]. The first step of the infection is the adhesion of *C. albicans* onto the host. This step is an essential determinant of pathogenesis, as it allows *C. albicans* to attach to host cells and to form biofilms or to disseminate in the host blood vessels. The biofilm increases yeast cell resistance to antifungal therapeutics and protects it from the host immune system [15]. *C. albicans* has developed multiple ways to colonize and infect host cells and tissues. One such mechanism is the specific ligand–receptor interaction through a whole range of adhesins displayed on the yeast cell wall [16–18]. These cell wall proteins are capable of recognizing protein ligands [16], glycolipids [19–22] and carbohydrates [23–29] on the host cells. Fibronectin is an important protein ligand of the host extracellular matrix (ECM) that plays an essential role in *C. albicans* adhesion [30]. Furthermore, targeting fibronectin has shown to alter *C. albicans* biofilm formation [31]. Therefore, a better understanding of the yeast–fibronectin interaction could lead to novel therapeutic options to fight candidiasis.

In this work, we applied nanomotion analysis to monitor the oscillatory activity of *C. albicans* upon exposure to fibronectin. We used an AFM-based nanomotion detector to follow the evolution of cellular oscillations in the absence and the presence of fibronectin on strongly and poorly adherent *C. albicans* cells. Interestingly, these two isolates reacted very differently to the interaction with fibronectin. These preliminary results demonstrate the potential of nanomotion analysis to monitor ligand–receptor interactions in a label free manner.

2. Materials and Methods

2.1. Yeast Strains

The *C. albicans* isolate 101 and CEC 3675 were kindly provided by Salomé Leibundgut and Christophe D'Enfert laboratories [32], respectively. The yeasts were cultured in yeast-extracted peptone-dextrose (YPD) medium (1% m/v yeast extract (Difco Laboratories, Fisher Scientific, Hampton, NH, USA), 2% m/v peptone (Difco Laboratories, Fisher Scientific, Hampton, NH, USA) and 2% m/v glucose (Sigma, St. Louis, MI, USA)) overnight at 30 °C with shaking (160 rpm).

2.2. Experimental Procedures

Rectangular tipless cantilevers (qp-CONT, NanoandMore GmbH, Wetzlar, Germany), with a nominal spring constant of 0.1 N/m and an average resonant peak in liquids of 8 kHz, were coated with 2 mg/mL of concanavalin A (Con A) (Sigma, St. Louis, MI, USA) for 30 min at room temperature. After removing the excess of Con A, the yeast cells were placed in contact with the cantilever for 1 h at room temperature to allow them to attach to its surface. Poorly attached *C. albicans* cells were removed by washing gently with YPD medium. Finally, the *C. albicans* covered cantilever was inserted into the analysis chamber containing 2 mL of filtered (0.2 µm syringe filter, Merck Millipore, Burlington, MA, USA) YPD medium. The measurements were performed at room temperature in YPD medium and in YPD medium containing 25 µg/mL of fibronectin (Sigma, USA). Fibronectin was directly added inside the chip reservoir. For the experiments performed with antifungals, caspofungin (Sigma, USA) was diluted in the YPD present in the analysis chamber to reach a final concentration of 100 µg/mL.

2.3. Nanomotion Detector

The cantilever oscillations were collected in real time using an in-house developed nanomotion detection device. The system relies on a laser-based signal transduction as typically used in commercial AFMs. A typical experiment lasted for 2 h. The control experiments were carried out for at least 4 h.

2.4. Software and Nanomotion Analysis

The cantilever oscillations were recorded and saved at 20 kHz using a USB-4431 DAQ card (National Instruments, Austin, TX, USA). The data acquisition program was developed in LabView. A dedicated Python program was used to process the recorded data and to display the deflection of the cantilever as a function of time. The software first removes the low frequency cantilever displacement signal by calculating a first order fit of the raw signal (deflection of the cantilever) by taking 20 seconds-long window frames. The obtained fit is then subtracted from the raw signal to remove thermally induced cantilever deflection. The thermal drift essentially occurs at the beginning of the experiment and during the fluid exchange procedures. The thermal drift free signal is further processed to obtain its variance in 10 seconds-window frames.

2.5. Viability Assay

Cells were placed inside a commercially available microfluidic chip (Ibidi, Planegg, Germany), and stained with calcofluor white (Sigma, USA), according to the manufacturer's instructions. To detect dead cells, propidium iodide (PI, Sigma, USA) was added to the YPD medium and the fluorescence of the yeast cells was recorded using an Axiovert microscope (Zeiss, Oberkochen, Germany).

2.6. Adhesion Assay

Adherence of *C. albicans* to TR146 cells was measured using the protocols previously described [33,34] with slight modifications (Figure S1). TR146 cells grown as monolayers in 6-well plates were incubated with 100 *C. albicans* cells for 20 min at 37 °C. The supernatant was carefully removed and spread on YPD agar plates to determine the number of non-adherent fungal cells. The adherent fungal cells that were left behind in the 6-well plates were rinsed with PBS and were overlaid with melted Wort agar at 40 °C. The plates were incubated at 30 °C for 36 h to count the colonies. Adherence was determined as the ratio of the number of colonies grown on Wort agar to the number of colonies grown on Wort agar and the number of colonies grown from the culture supernatant.

2.7. Statistical Analysis

Statistical analysis of nanomotion experiments were performed with the Python package Scipy. We performed the non-parametric Mann–Whitney U test for the three independent replicates. We used standard student t-test to process the adhesion assay on three independent replicates using the Graphpad Prism software.

3. Results

To assess a putative differential reaction of strongly and weakly interacting *C. albicans* to fibronectin, we quantified the adhesion of two different isolates, 101 and CEC3675, on oral keratinocytes (TR 146). As shown in Figure 1 isolate 101 was measured to have a significantly higher adhesion compared to isolate CEC3675.

To investigate the *C. albicans*–fibronectin interaction we used an in house nanomotion detector depicted in Figure 2A. The set up consists in an analysis chamber filled with liquid (in our case YPD) containing the cantilever to which yeast cells are attached (Figure 2B). The cantilever oscillations were recorded (Figure 2C) and processed to display the signal variance as a function of time (Figure 2D).

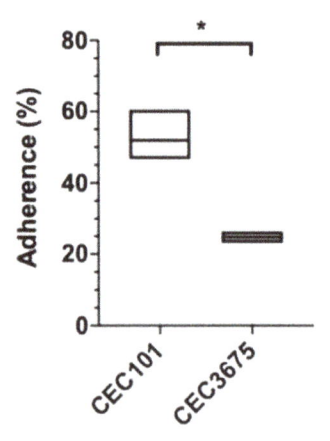

Figure 1. *C. albicans* isolates 101 and CEC3675 adhere differently to oral keratinocytes. Percentage of adherence of both isolates. Statistical analysis ($n = 3$) was done using standard t-test. The asterisk represents $p < 0.05$.

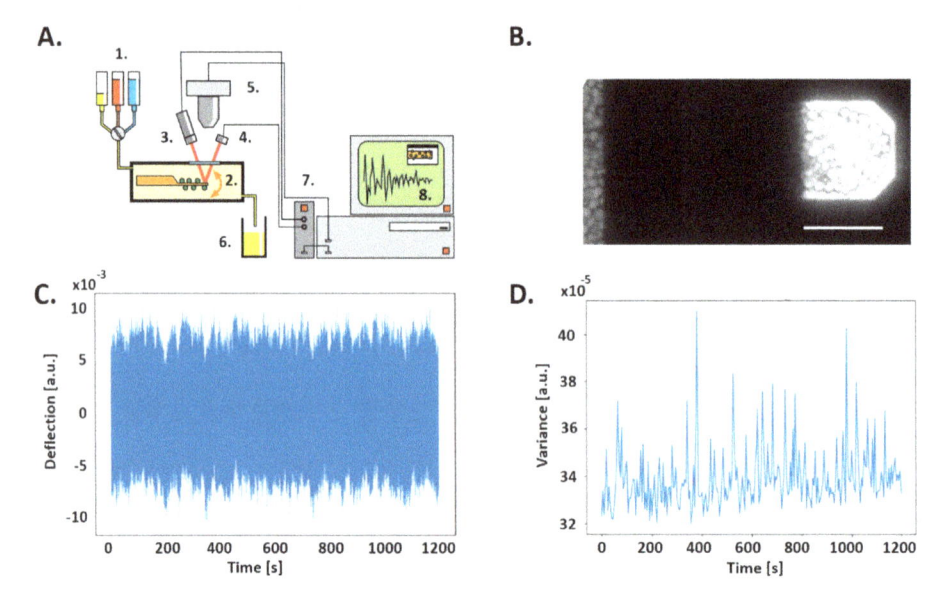

Figure 2. Nano-mechanical sensor system. (**A**) Representative image of a cantilever with attached *C albicans* cells. Scale bar 40 µm. (**B**) Schematic of the experimental system and data collection. (1) Liquids to be injected into the analysis chamber. In our case YPD, YPD containing fibronectin, and YPD containing caspofungin. (2) Analysis chamber with the AFM cantilever and *C. albicans* attached onto its surface (green circles). (3) Super luminescent diode. (4) Four-segment photodiode. (5) Optical microscopy with camera. (6) Liquid waste. (7) In-house dedicated electronics and National Instruments data acquisition card. (8) Desktop computer. (**C**). The collected raw data are processed; and (**D**). analyzed using the variance of the signal.

Using this system, we monitored the nanomotion pattern of *C. albicans* isolates 101 and CEC3675 in the absence and presence of fibronectin (Figure 3). Before addition of fibronectin, both isolates behaved similarly (Figure 3B). However, in the presence of fibronectin, nanomotion activity (variance) of isolate 101 drastically decreased (from 0.9 ± 0.5 to 0.3 ± 0.1) (Figure 3). In contrast, isolate CEC3675 did not present a significant decrease. To confirm that the drop of signal was not due to a change in the temperature, nor convective currents that can appear upon addition of a liquid in the analysis chamber, we performed control experiments, simultaneously, with another nanomotion detector. These experiments consisted in injecting the same quantity of medium, instead of fibronectin, into the analysis chamber. The obtained results showed no significant difference in the nanomotion pattern, for both isolates, upon addition of YPD media (Figure S2). Additionally, we assessed the number of cells present on the cantilever before and after the experiment to determine if the reduced signal was caused by cells being detached from the cantilever. The analysis of the images taken by the optical microscope located above the nanomotion detector (as depicted in the schematic in Figure 2A) confirmed that no cells detached from the cantilever throughout the experiments (Figure S3).

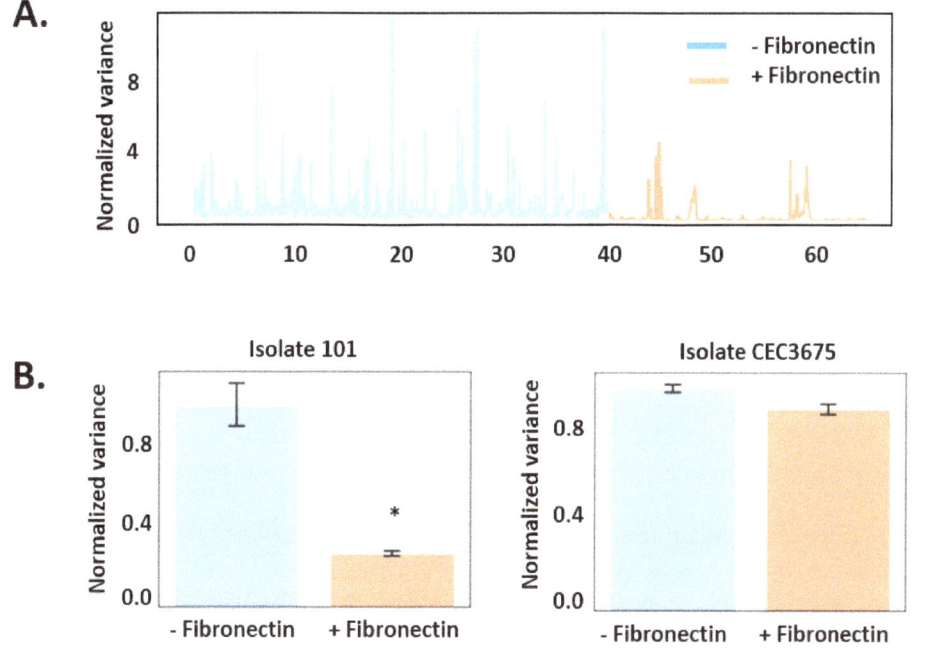

Figure 3. *C. albicans* isolate 101 and CEC 3675 react differently to fibronectin. (**A**). Representative graph of the normalized variance of isolate 101 in YPD (blue) and in YPD with fibronectin (orange). The decrease of the normalized variance is clearly visible between the two conditions. (**B**). The mean of the normalized variance (experiment in triplicate) represented as a bar plot for isolate 101 compared to isolate CEC 3675. Error bars are the confidence of intervals. Statistical analyses were done using Mann–Whitney U test, the asterisk represents $p < 0.05$.

To further exclude another cause of the decrease of the nanomotion signal for isolate 101, such as premature cell death, we monitored *C. albicans* viability by nanomotion and fluorescence microscopy in the absence and presence of fibronectin. Eventually the cells were killed by the antifungal caspofungin. As shown in Figure 3A, the variance of the nanomotion signal drastically dropped after the drug injection. The fluorescent viability test did not show any effect of fibronectin on the cellular viability as it can be noticed in Figure 4B. Similarly, fibronectin also did not have any effect on the viability of isolate CEC3675 (Figure S4).

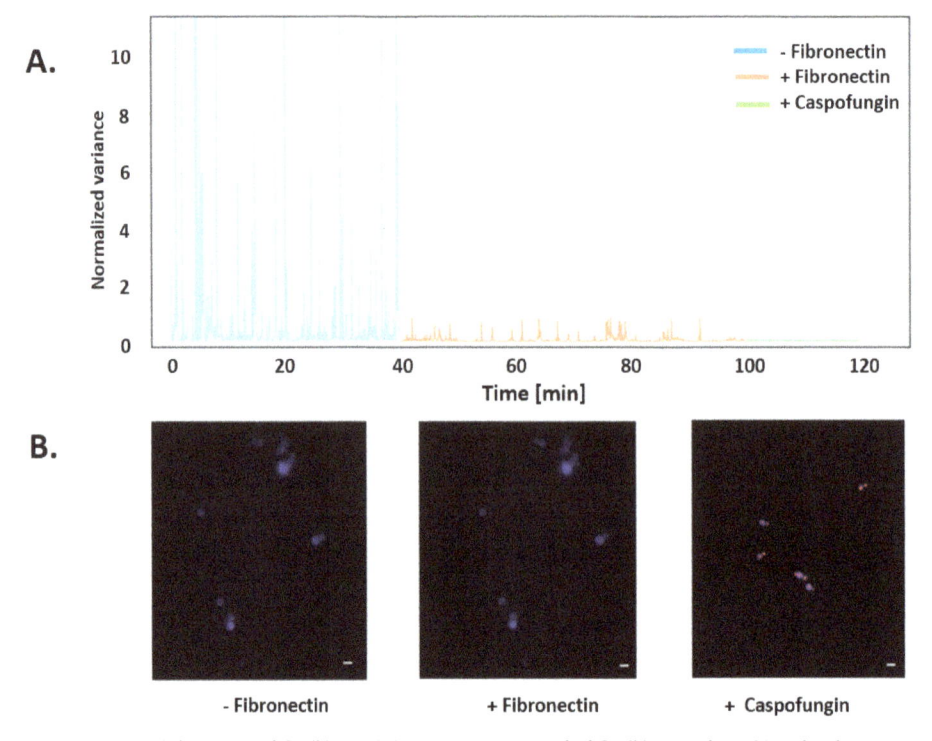

Figure 4. Viability assay of *C. albicans*. (**A**). Nanomotion signal of *C. albicans* isolate 101 in the absence (blue curve) and presence of fibronectin (orange curve), and after killing the cells by the antifungal caspofungin (green curve). (**B**). Representative fluorescence images of *C. albicans* isolate 101 in the absence (left panel) and presence of fibronectin (middle panel), and after killing (right panel). Scale bar 5 µm.

4. Discussion

C. albicans infection is a multistep process, consisting in the binding of *C. albicans* on epithelial cells. In a first adhesion step, the *C. albicans* adhesins of the agglutinin-like sequence (Als) family bind to ECM proteins of the host such as fibronectin [35,36], laminin and collagen. The attachment of the yeast cell to the host is followed by the penetration and transmigration of hypha into host cells, which then leads to vascular dissemination as soon the hypha reaches blood vessels. In this study we only explored the interaction of fibronectin with the yeast form. Adhesins playing a role in the planktonic *C. albicans* adhesion are the Als family members Als1 [27] and Als5 [37], Eap1 [38–40], Csh1 (cell surface hydrophobicity) [41,42], Ihd1 [43,44] and members of the SAP family [45–47]. It has been shown that Als1, Als5, and Csh1 interacts with fibronectin; Sap9 and Sap10 can interact with the ECM proteins collagen and vimentin. It has not yet been demonstrated that fibronectin is a ligand for Sap9/10, Eap1 and Ihd1.

Here, we used nanomotion detection to monitor the oscillation pattern of planktonic *C. albicans* cells upon exposure to fibronectin. Two different clinical isolates that showed a different adhesive phenotype, were used. The isolate 101 adhered significantly stronger to the host epithelial cells compared to isolate CEC3675. Nanomotion experiments showed that fibronectin affects isolate 101 significantly more than CEC3675. This drop of the nanomotion signal indicates a modification of the cellular activity upon fibronectin—*C. albicans* interaction. These results suggest that the initiation of adhesion related signaling in the yeast cell upon fibronectin attachment is mediated by the interaction with adhesins. Potential adhesion candidates that have been shown to interact with fibronectin are

Als1, Als3 and Csh1. The complete elucidation of the molecular mechanisms involved in the process are still unclear and deserve further research. We plan to investigate which specific adhesin(s) is (are) involved in the observed activity reduction. Additionally, the effect of other ligands such as laminin, collagen IV, fibrinogen and gelatin [28,48–50] should also be investigated.

This work demonstrated the ability of nanomotion detection to monitor in real time and in a label-free manner cellular activity changes induced by interacting ligands. Activity changes induced by increasing glucose concentration were observed for *Escherichia coli* in a previous study [5]. This technique opens novel avenues to detect cellular activation or inhibition induced by ligand–receptor interactions.

Supplementary Materials: The following are available online at http://www.mdpi.com/2311-5637/6/1/28/s1, Figure S1: Schematic representation of the adhesion assay protocol, Figure S2: Effect of the injection of YPD medium in the analysis chamber, Figure S3: Density of yeast cells on the cantilever, Figure S4: Viability assay of isolate CEC3675.

Author Contributions: Conceptualization, A.-C.K.; methodology, A.-C.K. and L.V.; software, A.-C.K.; validation, A.-C.K., L.V. and A.K.; formal analysis A.-C.K., investigation, A.-C.K., L.V., A.K.; resources, S.K., writing—original draft preparation, S.K. and A.-C.K.; writing—review and editing all the authors; visualization, A.-C.K., A.K.; supervision, A.-C.K., S.K.; project administration, S.K., G.D.; funding acquisition, S.K., G.D., D.S., R.W. All authors have read and agreed to the published version of the manuscript.

Funding: This research was funded by Schweizerischer Nationalfonds zur Förderung der Wissenschaftlichen Forschung 200021-144321, CRSII5_173863 and 407240_167137, the Gebert Rüf Stiftung GRS-024/14, NASA NNH16ZDA001N-CLDTCH, EPFL and ESA PRODEX project Yeast Bioreactor.

Acknowledgments: The authors thanks C. d'Enfert for providing the *Candida albicans* strains and S. Leibundgut for highly constructive discussions. The technical assistance of Danielle Brandalise is acknowledged. The Belgian Federal Science Policy Office (Belspo) and the European Space Agency (ESA) PRODEX program supported this work. The Research Council of the Vrije Universiteit Brussel (Belgium) and the University of Ghent (Belgium) are acknowledged to support the Alliance Research Group VUB-UGent NanoMicrobiology (NAMI), and the International Joint Research Group (IJRG) VUB-EPFL BioNanotechnology & NanoMedicine (NANO).

Conflicts of Interest: The authors declare no conflict of interest.

References

1. Willaert, R.; Kasas, S.; Devreese, B.; Dietler, G. Yeast Nanobiotechnology. *Fermentation* **2016**, *2*, 18. [CrossRef]
2. Formosa-Dague, C.; Duval, R.E.; Dague, E. Cell biology of microbes and pharmacology of antimicrobial drugs explored by Atomic Force Microscopy. *Semin. Cell Dev. Biol.* **2018**, *73*, 165–176. [CrossRef]
3. Alsteens, D.; Müller, D.J.; Dufrêne, Y.F. Multiparametric Atomic Force Microscopy Imaging of Biomolecular and Cellular Systems. *Acc. Chem. Res.* **2017**, *50*, 924–931. [CrossRef] [PubMed]
4. Kasas, S.; Stupar, P.; Dietler, G. AFM contribution to unveil pro- and eukaryotic cell mechanical properties. *Semin. Cell Dev. Biol.* **2018**, *73*, 177–187. [CrossRef] [PubMed]
5. Longo, G.; Alonso-Sarduy, L.; Rio, L.M.; Bizzini, A.; Trampuz, A.; Notz, J.; Dietler, G.; Kasas, S. Rapid detection of bacterial resistance to antibiotics using AFM cantilevers as nanomechanical sensors. *Nat. Nanotechnol.* **2013**, *8*, 522–526. [CrossRef]
6. Kohler, A.C.; Venturelli, L.; Longo, G.; Dietler, G.; Kasas, S. Nanomotion detection based on atomic force microscopy cantilevers. *Cell Surf.* **2019**, *5*, 100021. [CrossRef]
7. Villalba, M.I.; Stupar, P.; Chomicki, W.; Bertacchi, M.; Dietler, G.; Arnal, L.; Vela, M.E.; Yantorno, O.; Kasas, S. Nanomotion Detection Method for Testing Antibiotic Resistance and Susceptibility of Slow-Growing Bacteria. *Small* **2018**, *14*, 1702671. [CrossRef]
8. Stupar, P.; Opota, O.; Longo, G.; Prod'hom, G.; Dietler, G.; Greub, G.; Kasas, S. Nanomechanical sensor applied to blood culture pellets: A fast approach to determine the antibiotic susceptibility against agents of bloodstream infections. *Clin. Microbiol. Infect.* **2017**, *23*, 400–405. [CrossRef]
9. Mustazzolu, A.; Venturelli, L.; Dinarelli, S.; Brown, K.; Floto, R.A.; Dietler, G.; Fattorini, L.; Kasas, S.; Girasole, M.; Longo, G. A rapid unravelling of mycobacterial activity and of their susceptibility to antibiotics. *Antimicrob. Agents Chemother.* **2019**, *63*, 02194-18. [CrossRef]
10. Kasas, S.; Ruggeri, F.S.; Benadiba, C.; Maillard, C.; Stupar, P.; Tournu, H.; Dietler, G.; Longo, G. Detecting nanoscale vibrations as signature of life. *Proc. Natl. Acad. Sci. USA* **2015**, *112*, 378–381. [CrossRef]

11. Brown, G.D.; Denning, D.W.; Levitz, S.M. Tackling human fungal infections. *Science* **2012**, *336*, 647. [CrossRef] [PubMed]
12. Ghannoum, M.A.; Jurevic, R.J.; Mukherjee, P.K.; Cui, F.; Sikaroodi, M.; Naqvi, A.; Gillevet, P.M. Characterization of the Oral Fungal Microbiome (Mycobiome) in Healthy Individuals. *PLoS Pathog.* **2010**, *6*, e1000713. [CrossRef] [PubMed]
13. Rupp, S. Interactions of the fungal pathogen Candida albicans with the host. *Future Microbiol.* **2007**, *2*, 141–151. [CrossRef] [PubMed]
14. Filler, S.G.; Sheppard, D.C. Fungal Invasion of Normally Non-Phagocytic Host Cells. *PLoS Pathog.* **2006**, *2*, e129. [CrossRef] [PubMed]
15. Nobile, C.J.; Johnson, A.D. *Candida albicans* Biofilms and Human Disease. *Annu. Rev. Microbiol.* **2015**, *69*, 71–92. [CrossRef] [PubMed]
16. Chaffin, W.L.; López-Ribot, J.L.; Casanova, M.; Gozalbo, D.; Martínez, J.P. Cell Wall and Secreted Proteins ofCandida albicans: Identification, Function, and Expression. *Microbiol. Mol. Biol. Rev.* **1998**, *62*, 130–180. [CrossRef]
17. Sundstrom, P. Adhesion in Candida spp. *Cell. Microbiol.* **2002**, *4*, 461–469. [CrossRef]
18. Filler, S.G. Candida-host cell receptor-ligand interactions. *Curr. Opin. Microbiol.* **2006**, *9*, 333–339. [CrossRef]
19. Ghannoum, M.A.; Burns, G.R.; Abu Elteen, K.; Radwan, S.S. Experimental evidence for the role of lipids in adherence of Candida spp. to human buccal epithelial cells. *Infect. Immun.* **1986**, *54*, 189–193. [CrossRef]
20. Jimenez-Lucho, V.; Ginsburg, V.; Krivan, H.C. Cryptococcus neoformans, Candida albicans, and other fungi bind specifically to the glycosphingolipid lactosylceramide (Gal beta 1-4Glc beta 1-1Cer), a possible adhesion receptor for yeasts. *Infect. Immun.* **1990**, *58*, 2085–2090. [CrossRef]
21. Yu, L.; Lee, K.K.; Sheth, H.B.; Lane-Bell, P.; Srivastava, G.; Hindsgaul, O.; Paranchych, W.; Hodges, R.S.; Irvin, R.T. Fimbria-mediated adherence of Candida albicans to glycosphingolipid receptors on human buccal epithelial cells. *Infect. Immun.* **1994**, *62*, 2843–2848. [CrossRef] [PubMed]
22. Cameron, B.J.; Douglas, L.J. Blood group glycolipids as epithelial cell receptors for Candida albicans. *Infect. Immun.* **1996**, *64*, 891–896. [CrossRef] [PubMed]
23. Sandin, R.L.; Rogers, A.L.; Patterson, R.J.; Beneke, E.S. Evidence for mannose-mediated adherence of Candida albicans to human buccal cells in vitro. *Infect. Immun.* **1982**, *35*, 79–85. [CrossRef] [PubMed]
24. Critchley, I.A.; Douglas, L.J. Role of glycosides as epithelial cell receptors for Candida albicans. *J. Gen. Microbiol.* **1987**, *133*, 637–643. [CrossRef]
25. Macura, A.B.; Tondyra, E. Influence of some carbohydrates and concanavalin A on the adherence of Candida albicans in vitro to buccal epithelial cells. *Zentralbl. Bakteriol.* **1989**, *272*, 196–201. [CrossRef]
26. Brassart, D.; Woltz, A.; Golliard, M.; Neeser, J.R. In vitro inhibition of adhesion of Candida albicans clinical isolates to human buccal epithelial cells by Fuc alpha 1—2Gal beta-bearing complex carbohydrates. *Infect. Immun.* **1991**, *59*, 1605–1613. [CrossRef]
27. Donohue, D.S.; Ielasi, F.S.; Goossens, K.V.Y.; Willaert, R.G. The N-terminal part of Als1 protein from Candida albicans specifically binds fucose-containing glycans. *Mol. Microbiol.* **2011**, *80*, 1667–1679. [CrossRef]
28. Ielasi, F.S.; Alioscha-Perez, M.; Donohue, D.; Claes, S.; Sahli, H.; Schols, D.; Willaert, R.G. Lectin-glycan interaction network-based identification of host receptors of microbial pathogenic adhesins. *MBio* **2016**, *7*. [CrossRef]
29. Everest-Dass, A.V.; Kolarich, D.; Pascovici, D.; Packer, N.H. Blood group antigen expression is involved in C. albicans interaction with buccal epithelial cells. *Glycoconj. J.* **2017**, *34*, 31–50. [CrossRef] [PubMed]
30. Calderone, R.A.; Scheld, W.M. Role of fibronectin in the pathogenesis of candidal infections. *Rev. Infect. Dis.* **1987**, *9*, S400–S403. [CrossRef]
31. Nett, J.E.; Cabezas-Olcoz, J.; Marchillo, K.; Mosher, D.F.; Andes, D.R. Targeting fibronectin to disrupt in vivo Candida albicans biofilms. *Antimicrob. Agents Chemother.* **2016**, *60*, 3152–3155. [CrossRef] [PubMed]
32. Ropars, J.; Maufrais, C.; Diogo, D.; Marcet-Houben, M.; Perin, A.; Sertour, N.; Mosca, K.; Permal, E.; Laval, G.; Bouchier, C.; et al. Gene flow contributes to diversification of the major fungal pathogen Candida albicans. *Nat. Commun.* **2018**, *9*, 2253. [CrossRef] [PubMed]
33. Murciano, C.; Moyes, D.L.; Runglall, M.; Tobouti, P.; Islam, A.; Hoyer, L.L.; Naglik, J.R. Evaluation of the Role of Candida albicans Agglutinin-Like Sequence (Als) Proteins in Human Oral Epithelial Cell Interactions. *PLoS ONE* **2012**, *7*, e33362. [CrossRef] [PubMed]

34. Schönherr, F.A.; Sparber, F.; Kirchner, F.R.; Guiducci, E.; Trautwein-Weidner, K.; Gladiator, A.; Sertour, N.; Hetzel, U.; Le, G.T.T.; Pavelka, N.; et al. The intraspecies diversity of C. albicans triggers qualitatively and temporally distinct host responses that determine the balance between commensalism and pathogenicity. *Mucosal Immunol.* **2017**, *10*, 1335–1350. [CrossRef] [PubMed]
35. Skerl, K.G.; Calderone, R.A.; Segal, E.; Sreevalsan, T.; Scheld, W.M. In vitro binding of Candida albicans yeast cells to human fibronectin. *Can. J. Microbiol.* **1984**, *30*, 221–227. [CrossRef] [PubMed]
36. Douglas, L.J. Adhesin—Receptor interactions in the attachment of Candida albicans to host epithelial cells. *Can. J. Bot.* **1995**, *73*, 1147–1153. [CrossRef]
37. Alsteens, D.; Beaussart, A.; Derclaye, S.; El-Kirat-Chatel, S.; Park, H.R.; Lipke, P.N.; Dufrêne, Y.F. Single-cell force spectroscopy of Als-mediated fungal adhesion. *Anal. Methods* **2013**, *5*, 3657–3662. [CrossRef] [PubMed]
38. Li, F.; Palecek, S.P. EAP1, a Candida albicans Gene Involved in Binding Human Epithelial Cells. *Eukaryot. Cell* **2003**, *2*, 1266–1273. [CrossRef] [PubMed]
39. Li, F.; Svarovsky, M.J.; Karlsson, A.J.; Wagner, J.P.; Marchillo, K.; Oshel, P.; Andes, D.; Palecek, S.P. Eap1p, an adhesin that mediates Candida albicans biofilm formation in vitro and in vivo. *Eukaryot. Cell* **2007**, *6*, 931–939. [CrossRef] [PubMed]
40. Li, F.; Palecek, S.P. Distinct domains of the Candida albicans adhesin EAP1 p mediate cell-cell and cell-substrate interactions. *Microbiology* **2008**, *154*, 1193–1203. [CrossRef]
41. Hooshdaran, M.Z.; Barker, K.S.; Hilliard, G.M.; Kusch, H.; Morschhäuser, J.; Rogers, P.D. Proteomic analysis of azole resistance in Candida albicans clinical isolates. *Antimicrob. Agents Chemother.* **2004**, *48*, 2733–2735. [CrossRef] [PubMed]
42. Singleton, D.R.; Fidel, P.L.; Wozniak, K.L.; Hazen, K.C. Contribution of cell surface hydrophobicity protein 1 (Csh1p) to virulence of hydrophobic *Candida albicans* serotype A cells. *FEMS Microbiol. Lett.* **2005**, *244*, 373–377. [CrossRef] [PubMed]
43. de Groot, P.W.J.; Hellingwerf, K.J.; Klis, F.M. Genome-wide identification of fungal GPI proteins. *Yeast* **2003**, *20*, 781–796. [CrossRef] [PubMed]
44. McCall, A.D.; Pathirana, R.U.; Prabhakar, A.; Cullen, P.J.; Edgerton, M. Candida albicans biofilm development is governed by cooperative attachment and adhesion maintenance proteins. *NPJ Biofilms Microbiomes* **2019**, *5*, 21. [CrossRef]
45. Watts, H.; Cheah, F.S.; Hube, B.; Sanglard, D.; Gow, N.A. Altered adherence in strains of *Candida albicans* harbouring null mutations in secreted aspartic proteinase genes. *FEMS Microbiol. Lett.* **1998**, *159*, 129–135. [CrossRef]
46. Naglik, J.R.; Challacombe, S.J.; Hube, B. Candida albicans Secreted Aspartyl Proteinases in Virulence and Pathogenesis. *Microbiol. Mol. Biol. Rev.* **2003**, *67*, 400–428. [CrossRef]
47. Kumar, R.; Breindel, C.; Saraswat, D.; Cullen, P.J.; Edgerton, M. Candida albicans Sap6 amyloid regions function in cellular aggregation and zinc binding, and contribute to zinc acquisition. *Sci. Rep.* **2017**, *7*, 1–15. [CrossRef]
48. Klotz, S.A.; Gaur, N.K.; Lake, D.F.; Chan, V.; Rauceo, J.; Lipke, P.N. Degenerate peptide recognition by Candida albicans adhesins Als5p and Als1p. *Infect. Immun.* **2004**, *72*, 2029–2034. [CrossRef]
49. Sheppard, D.C.; Yeaman, M.R.; Welch, W.H.; Phan, Q.T.; Fu, Y.; Ibrahim, A.S.; Filler, S.G.; Zhang, M.; Waring, A.J.; Edwards, J.E. Functional and structural diversity in the Als protein family of Candida albicans. *J. Biol. Chem.* **2004**, *279*, 30480–30489. [CrossRef]
50. Argimón, S.; Wishart, J.A.; Leng, R.; Macaskill, S.; Mavor, A.; Alexandris, T.; Nicholls, S.; Knight, A.W.; Enjalbert, B.; Walmsley, R.; et al. Developmental regulation of an adhesin gene during cellular morphogenesis in the fungal pathogen Candida albicans. *Eukaryot. Cell* **2007**, *6*, 682–692. [CrossRef]

© 2020 by the authors. Licensee MDPI, Basel, Switzerland. This article is an open access article distributed under the terms and conditions of the Creative Commons Attribution (CC BY) license (http://creativecommons.org/licenses/by/4.0/).

fermentation

Article

Screening and Application of *Cyberlindnera* Yeasts to Produce a Fruity, Non-Alcoholic Beer

Konstantin Bellut [1], Maximilian Michel [2], Martin Zarnkow [2], Mathias Hutzler [2], Fritz Jacob [2], Jonas J. Atzler [1], Andrea Hoehnel [1], Kieran M. Lynch [1] and Elke K. Arendt [1,3,*]

1. School of Food and Nutritional Sciences, University College Cork, College Road, Cork, Ireland
2. Research Center Weihenstephan for Brewing and Food Quality, Technische Universität München, Alte Akademie 3, 85354 Freising-Weihenstephan, Germany
3. APC Microbiome Ireland, University College Cork, College Road, Cork, Ireland
* Correspondence: e.arendt@ucc.ie

Received: 14 November 2019; Accepted: 13 December 2019; Published: 17 December 2019

Abstract: Non-alcoholic beer (NAB) is enjoying growing demand and popularity due to consumer lifestyle trends and improved production methods. In recent years in particular, research into the application of non-*Saccharomyces* yeasts to produce NAB via limited fermentation has gained momentum. Non-*Saccharomyces* yeasts are known to produce fruity aromas, owing to a high ester production. This trait could be harnessed to mask the often-criticized wort-like off-flavor of NAB produced via limited fermentation. Six *Cyberlindnera* strains were characterized and screened in wort extract. Four of the six strains produced a pleasant, fruity aroma while exhibiting low ethanol production. The strain *Cyberlindnera subsufficiens* C6.1 was chosen for fermentation optimization via response surface methodology (RSM) and a pilot-scale (60 L) brewing trial with subsequent sensory evaluation. A low fermentation temperature and low pitching rate enhanced the fruitiness and overall acceptance of the NAB. The NAB (0.36% ABV) produced on pilot-scale was significantly more fruity and exhibited a significantly reduced wort-like off-flavor compared to two commercial NABs. This study demonstrated the suitability of *Cyberlindnera subsufficiens* to produce a fruity NAB, which can compete with commercial NABs. The outcome strengthens the position of non-*Saccharomyces* yeasts as a serious and applicable alternative to established methods in NAB brewing.

Keywords: brewing; *Cyberlindnera*; NABLAB; non-alcoholic beer; non-conventional yeast; non-*Saccharomyces* yeast; response surface methodology

1. Introduction

While the overall market growth of beer is slowing down, non-alcoholic and low alcohol beer (NABLAB) is growing in volume and popularity, owed to stricter legislation, lifestyle trends and improved production methods [1]. The increasing interest has fueled research in NABLAB production methods, especially in recent years, aimed at overcoming taste deficits compared to regular beer and consequently improving consumer acceptance. The two major production methods, physical dealcoholization and limited fermentation, both compromise the taste of the beer. Dealcoholized beer is often criticized for its lack of body and aromatic profile, a consequence of the removal of volatile esters and higher alcohols in conjunction with ethanol. Apart from a sweet taste due to residual sugars, one of the main points of criticism of NAB produced by limited fermentation is its wort-like off-flavor caused by aldehydes present in the wort [2]. In regular beer, ethanol significantly increases aldehyde retention, reducing the perceptibility of the wort-like flavor. However, in NAB produced by limited fermentation, the low ethanol content and higher levels of mono- and disaccharides intensify this undesired off-flavor [3].

It is known that esters, which yeast produce as a by-product of alcoholic fermentation, are extremely important for the flavor profile of beer [4,5]. The lack thereof, as well as their overproduction, can significantly compromise the flavor. Aside from strain-specific differences, the process parameters such as the fermentation temperature, pitching rate and wort gravity have been shown to have a significant influence on ester formation [4,6]. In non-alcoholic beers, ester concentrations are lower compared to regular beer, independent of the production method [7,8]. While physical dealcoholization removes esters that were previously produced, limited fermentation adversely affects the production of substantial amounts in the first place.

Non-*Saccharomyces* yeasts are known for their important contribution to the flavor profile of fermented foods and beverages and have therefore been investigated for their targeted application in bioflavoring and, not least, NABLAB brewing [1,9,10]. Species that have been mentioned in the context of NABLAB production belong to the genera *Cyberlindnera*, *Hanseniaspora*, *Lachancea*, *Mrakia*, *Pichia*, *Torulaspora*, *Saccharomycodes*, *Scheffersomyces* and *Zygosaccharomyces* [1,11–16]. In particular, the *Cyberlindnera* species are known for their high ester production, which was shown in studies with *Cyberlindnera saturnus* (formerly *Williopsis saturnus*), *C. mrakii* (formerly *Williopsis saturnus* var. *mrakii*) and *C. subsufficiens* (formerly *Williopsis saturnus* var. *subsufficiens*) [17–20]. Furthermore, it has been proposed to use yeasts with high production of flavor compounds (i.e., esters, higher alcohols) to mask the wort-like flavor of NAB produced by limited fermentation. However, research in that direction is sparse [21,22]. In addition, such yeasts are capable of reducing aldehydes to their correspondent alcohol, which can also enhance the reduction of the often-criticized wort-like off-flavor [23,24].

In this study, six strains of the genus *Cyberlindnera* were investigated to create a fruity NAB. After identification, the strains were characterized for their substrate utilization, flocculation behavior and stress responses. A screening in diluted wort extract was performed to investigate the strains' potential to produce a pronounced fruity flavor without the production of high concentrations of ethanol. Interspecific differences in sugar consumption and the production of volatile fermentation by-products was investigated by means of high-performance liquid chromatography (HPLC) and gas chromatography (GC). The most promising strain was studied further to determine the optimal fermentation conditions to enhance the fruity flavor, which was performed by means of response surface methodology (RSM). Finally, a non-alcoholic beer was produced on pilot-scale (60 L), and its analytical attributes, aroma, and taste compared to two commercial NABs were examined.

2. Materials and Methods

2.1. Materials

All reagents used in this study were at least analytical grade from Sigma-Aldrich (St Louis MO, USA) unless stated otherwise. The wort extract applied in this study was spray-dried wort from 100% barley malt (Spraymalt Light, Muntons plc, Suffolk, UK). For the pilot-scale brewing, pilsner malt and acidulated malt were sourced from Weyermann (Malzfabrik Weyermann, Bamberg, Germany).

2.2. Yeast Strains

Strain Origin and Identification

Strain 837A was isolated from a brewery cellar, NT Cyb originates from a dried fermentation starter for rice wine, strain C6.1 originates from a coconut, and L1 from "Lulo", the fruit of *Solanum quitoense*. The type strains CBS 1707 and CBS 5763 originate from soil samples. For identification, the D1/D2 domain of the 26S rRNA gene was amplified, sequenced and compared to publicly available sequences in the National Center for Biotechnology Information (NCBI) database using the Basic Local Alignment Search Tool (BLAST; https://blast.ncbi.nlm.nih.gov/Blast.cgi).

The DNA of the yeast isolates was extracted using an extraction kit (Yeast DNA Extraction Kit, Thermo Fisher Scientific, Waltham MA, USA). To amplify the D1/D2 domain of the 26S rRNA gene, the primers NL1 (5′-GCATATCAATAAGCGGAGGAAAAG-3′) and NL4 (5′-GGTCCGTGTTTCAAGACGG-3′) were used. Polymerase chain reaction (PCR) was performed using the temperature protocol: 95 °C/2 min; 30 cycles of 95 °C/30 s, 56 °C/15 s; 72 °C/60 s; 72 °C/5 min. Stock cultures were kept in 50% (*v/v*) glycerol at −80 °C.

2.3. Yeast Characterization

2.3.1. Flocculation Assay and Phenolic Off-Flavor (POF) Test

The flocculation test was performed using a slightly modified Helm's assay [25,26]. Essentially, all cells were washed in ethylenediaminetetraacetic acid (EDTA) and the sedimentation period was extended to 10 min. Wort was composed of 75 g/L spray-dried malt extract (Spraymalt Light, Muntons plc, Suffolk, UK) adjusted to 15 International Bitterness Units (IBU) (15 mg/mL iso-α-acids; from 30% stock solution; Barth-Haas Group, Nürnberg, Germany).

The phenolic off-flavor test was performed according to Meier-Dörnberg et al. [27]. In short, yeast strains were spread on yeast and mold agar plates (YM-agar) containing only one of the following precursors: either ferulic acid, cinnamic acid or coumaric acid. After three days of incubation at 25 °C, plates were evaluated by a trained panel by sniffing to detect any of the following aromas: clove-like (4-vinylguajacol), Styrofoam-like (4-vinylstyrene) and medicinal-like (4-vinylphenol). *Saccharomyces cerevisiae* LeoBavaricus—TUM 68®(Research Center Weihenstephan for Brewing and Food Quality, Freising-Weihenstephan, Germany) was used as a positive control.

2.3.2. Substrate Utilization

To analyze substrate utilization by the *Cyberlindnera* strains, the test kit API ID 32C (BioMérieux, Marcy-l'Étoile, France) was used. Preparation of the inoculum and inoculation of the strips were performed according to the manufacturers' instructions. Colonies for the inoculum were grown on yeast extract peptone dextrose (YPD) agar plates for 48 h at 27 °C. After inoculation, API ID 32C strips were incubated for 2 days at 28 °C. The samples were evaluated visually for turbidity in the wells, differentiating positive (+), negative (−), and weak (w) growth.

2.3.3. Stress Tests

Stress tests were performed via the measurement of yeast growth in a microplate, through the repeated measurement of absorbance over a time period of 96 h (Multiskan FC, Thermo Scientific, Waltham, MA, USA). The substrate for the hop sensitivity test was sterile-filtered wort extract (75 g/L Muntons Spraymalt Light) adjusted to 0, 50 and 100 mg/L iso-α-acids (1 mg/L = 1 International Bitterness Unit, IBU), respectively, by using an aliquot of a stock solution of 3% iso-α-acids in 96% (*v/v*) ethanol (Barth-Haas Group, Nürnberg, Germany). For testing ethanol sensitivity, the sterile-filtered wort extract was adjusted to 0%, 2.5%, 5% and 7.5% ABV with an aliquot of 100% (*v/v*) ethanol. For testing pH sensitivity, the sterile-filtered wort extract was adjusted to the following pHs with 2 M HCl: 5.5 (control without addition of HCl): 5.0, 4.0 and 3.0. For inoculation, strains were grown in sterilized wort extract for 24 h at 25 °C under aerobic conditions. The microtiter plate wells were inoculated with a concentration of 10^5 cells/mL. The wells contained 200 µL of the respective wort substrates. Plates were incubated at 25 °C, and absorbance was measured every 30 min at 600 nm without shaking over a time period of 96 h (Multiskan FC, Thermo Scientific, Waltham, MA, USA). Stress tests were performed in triplicate.

2.4. Yeast Screening

2.4.1. Propagation

Single colonies of the respective strains were taken from yeast extract peptone dextrose (YPD) agar plates after 72 h growth at 25 °C and transferred into a 250 mL sterile Duran glass bottle (Lennox

Laboratory Supplies Ltd, Dublin, Ireland) containing 150 mL propagation wort consisting of 75 g/L spray-dried malt (Spraymalt light, Muntons plc, Suffolk, UK) and 30 g/L glucose (Gem Pack Foods Ltd., Dublin, Ireland), sterilized at 121 °C for 15 min. The bottles were covered with sterile cotton and placed in an incubator with orbital shaker (ES-80 shaker-incubator, Grant Instruments (Cambridge) Ltd, Shepreth, UK) and incubated for 24 h at an orbital agitation of 170 rpm at 25 °C (Strain 837A was incubated for 48 h). Cell count was performed using a Thoma Hemocytometer with a depth of 0.1 mm (Blaubrand, Sigma-Aldrich, St. Louis, MO, USA).

2.4.2. Fermentation

Fermentation wort was prepared by dissolving 75 g/L spray-dried malt extract (Munton Spraymalt light) in 1 L of brewing water and sterilizing at 121 °C for 15 min, followed by filtration through a sterile grade 1V Whatman filter (Whatman plc, Maidstone, UK) to remove hot trub formed during sterilization. The analytical attributes of the fermentation wort for the yeast screening trial and RSM trial is shown in Table 1.

Table 1. Attributes of screening wort from wort extract.

Attribute	Unit	Value
Real Extract	°P	6.97 ± 0.00
pH	–	5.20 ± 0.01
Free amino nitrogen (FAN)	mg/L	115 ± 1
Maltotriose	g/L	8.12 ± 0.15
Maltose	g/L	32.37 ± 0.57
Sucrose	g/L	0.83 ± 0.04
Glucose	g/L	5.68 ± 0.91
Fructose	g/L	1.45 ± 0.10

Fermentation trials were carried out in 1 L sterile Duran glass bottles, equipped with an air lock. Per yeast strain, triplicate bottles were filled with 400 mL of wort and left untouched throughout the fermentation. Yeast cells for pitching were washed by centrifugation at 900 g for 5 min and resuspended in sterile water to ensure no carryover of sugars from the propagation wort into the fermentation wort. Pitching rate was 3×10^7 cells/mL. Fermentation temperature was 25 °C. Fermentation was performed until no change in extract could be measured for two consecutive days.

2.5. Scanning Electron Microscopy (SEM)

Yeast cultures for scanning electron microscopy (SEM) were prepared following the protocol for cultured microorganisms by Das Murtey and Ramasamy [28]. Single colonies were taken from a YPD agar plate and grown in YPD broth for 24 h at 25 °C. One milliliter of sample was centrifuged at 900 g for 2 min for pellet formation and resuspended in 5% glutaraldehyde solution prepared in 0.1 M phosphate buffer (pH 7.2) for fixation. After 30 min, the sample was centrifuged, the supernatant was discarded, and the pellet was washed twice in 0.1 M phosphate buffer. Consequently, the pellet was resuspended in 1% osmium tetroxide prepared in 0.1 M phosphate buffer. After 1 h, cells were again washed twice in 0.1 M phosphate buffer. The sample was then dehydrated through an ethanol series of 35%, 50%, 75%, 95%, absolute ethanol, and hexamethyldisilazane (HDMS), with 30 min per step (last two ethanol steps twice), centrifuging and discarding the supernatant at each change. Lastly, the second HDMS was discarded and the sample left drying overnight in a desiccator.

The dehydrated yeast sample was mounted onto plain aluminum stubs using carbon double surface adhesive and coated with a 5 nm gold-palladium (80:20) layer using a Gold Sputter Coater (BIO-RAD Polaron Division, SEM coating system, England), then observed under a constant accelerating voltage of 5 kV under a JEOL scanning electron microscope type 5510 (JEOL, Tokyo, Japan).

2.6. Response Surface Modeling (RSM)

To investigate optimal fermentation conditions for C6.1 to produce a fruity, non-alcoholic beer, response surface methodology (RSM) was performed using DesignExpert 9 software (StatEase, Minneapolis, MN, USA). A two-factorial, face-centered, central composite design with single factorial points and 5 replications of the center point was chosen. The predictor factors were temperature (17, 22, 27 °C), and pitching rate (10, 35, 60 × 10^6 cells/mL).

Spray-dried malt extract (Spraymalt light, Muntons plc, Suffolk, UK) served as the substrate. Wort preparation, propagation and inoculation were carried out as outlined in 2.4.1. The wort used was the same as in the screening (Table 1). Fermentation volume was 150 mL in 250 mL Duran glass bottles equipped with an air lock. Fermentation was performed until no change in extract could be measured for two consecutive days. Table 2 shows the experimental design.

Table 2. Response surface methodology (RSM) experimental design: Two-factorial, face-centered, central composite design with five repetitions of the center point. Factor 1, A: temperature, range 17, 22, 27 °C. Factor 2, B: pitching rate, range 10, 35, 60 × 10^6 cells/mL.

Run	Factor 1 A: Temperature (°C)	Factor 2 B: Pitching Rate (×10^6 cells/mL)
1	22	60
2	22	10
3	17	35
4	27	35
5 *	22	35
6 *	22	35
7	17	60
8 *	22	35
9 *	22	35
10 *	22	35
11	17	10
12	27	10
13	27	60

* Center point.

Models were produced applying backward elimination regression of insignificant model terms with α to exit of 0.1 (detailed report in supplementary Data Sheet S1). For significant models with insignificant lack of fit (LOF), 3D response surface plots were produced. Fermentations for model validation were performed in the same wort with propagation as outlined in 2.4.1 and fermentation as outlined above.

2.7. Pilot-Scale Brewing

2.7.1. Wort Production

Wort for the pilot brew was produced in a 60 L pilot-scale brewing plant consisting of a combined mash-boiling vessel, a lauter tun and whirlpool (FOODING Nahrungsmitteltechnik GmbH, Stuttgart, Germany). The grain bill comprised 6.65 kg Weyermann Pilsner Malt and 0.35 kg Weyermann Acidulated Malt (Malzfabrik Weyermann, Bamberg, Germany). Grains were milled with a two-roller mill ("Derby", Engl Maschinen, Schwebheim, Germany) at a 0.8 mm gap size. The crushed malt was mashed-in with 30 L of brewing water at 50 °C. The following mashing regime was employed: 20 min at 50 °C, 20 min at 62 °C, 10 min at 72 °C and mashing out at 78 °C. The mash was pumped into the lauter tun, and lautering was performed after a 15 min lauter rest, employing four sparging steps of 5 L hot brewing water each. Boil volume was 50 L at a gravity of 1.030 (7.0 °P), and total boiling time was 60 min. Thirty minutes into the boil, 15 g of Magnum hop pellets (14% iso-α-acids) were added for a calculated IBU content of 9. After boiling, gravity was readjusted to 1.030 (7.0 °P) with hot brewing water, and hot trub precipitates and hop residue were removed in the whirlpool with a rest of 20 min. Clear wort was pumped through a heat exchanger and filled into 60 L cylindroconical fermentation vessels at a temperature of 17 °C.

2.7.2. Propagation, Fermentation and Aftercare

A first propagation step was employed as described in 2.4.1. A second propagation step was performed by transferring the small-scale propagated wort into a 5 L carboy filled with 2 L of sterile wort extract at 7 °P and closed with sterile cotton. The second propagation step was conducted for 24 h under constant agitation at ambient temperature (20 ± 2 °C).

Yeast was pitched into the fermenter at a pitching rate of 10^7 cells/mL. Fermentation was carried out in cylindroconical fermentation vessels with a capacity of 60 L, at ambient pressure and at a glycol-controlled fermentation temperature of 17 °C. Samples were withdrawn every day. Fermentation was carried out until no change in extract could be measured for two consecutive days. The beer was then filled into a 50 L keg and carbonated by repeated pressurization with CO_2 to 1 bar at 2 °C. After 5 days, the carbonated beer was filled into 330 mL brown glass bottles with a counter-pressure hand-filler (TOPINCN, Shenzen, China) and capped. Bottles were pasteurized in a pilot retort (APR-95; Surdry, Abadiano, Vizcaya, Spain) with spray water at 65 °C for 10 min resulting in approximately 23 pasteurization units (PU). The successful pasteurization was confirmed by plating the pasteurized NAB on agar plates. Beer bottles were stored at 2 °C in a dark place for further analysis and sensory evaluation.

2.8. Sensory Evaluation

The sensory evaluation of the samples produced during yeast screening and RSM trial were judged by a panel of 12–15 experienced tasters. Samples were given at ambient temperature (20 °C) with a three-digit code. Each panelist evaluated the samples in an individual booth at ambient temperature (20 °C). The tasters were asked to desribe the sample in their own words, followed by evaluation of the intensity of a fruity smell and the overall acceptance of the smell of the sample on a hedonic scale from 0 ("not fruity"/"dislike extremely") to 5 ("extremely fruity"/"like extremely") according to MEBAK Sensory Analysis 3.2.1 "Simple Descriptive Test" and 3.2.2 "Profile Test", respectively.

The non-alcoholic beer samples (C6.1 pilot scale and commercial samples) were tasted and judged by a sensory panel of ten experienced and certified (DLG International Certificate for Sensory Analysis—beer and beer-based mixed drinks; Deutsche Landwirtschafts-Gesellschaft e.V.) panelists. A "Simple Descriptive Test" and "Profile Test" were performed according to MEBAK Sensory Analysis 3.2.1 and 3.2.2, respectively. Attributes for the aroma were "wort-like", "floral", "fruity", "citrus-like" and "tropical". A taste attribute "sweet taste" was also included. Panelists were asked to evaluate the attributes in their intensity on a line-marking scale from 0, "not perceptible", to 5, "strongly perceptible". Before the evaluation of the intensity, a descriptive sensory was performed, where the panelists were asked to describe the aroma of the samples in their own words. Samples were provided in dark glasses with a three-digit code and evaluated at a temperature of 20 °C in order to evaluate the full flavor profile (following DLG guidelines). The commercial samples NAB A and NAB B were non-alcoholic beers produced by limited fermentation [29] and "dialysis technology" [30], respectively. Each panelist tasted the samples in an individual booth at ambient temperature (20 °C). The amount of sample tasted was 50 mL per sample.

2.9. Wort and Beer Analyses

2.9.1. HPLC Analyses

Sugars and ethanol were determined by HPLC Agilent 1260 Infinity (Agilent Technologies, Santa Clara CA, USA) equipped with a refractive index detector (RID) and a Sugar-Pak I 10 µm,

6.5 mm × 300 mm column (Waters, Milford MA, USA), with 50 mg/L Ca-EDTA as mobile phase and a flow rate of 0.5 mL/min at 80 °C. Differentiation of maltose and sucrose was achieved with a Nova-Pak 4 µm, 4.6 mm × 250 mm column (Waters, Milford MA, USA), with acetonitrile/water 78:22 (*v/v*) as mobile phase and a flow rate of 1.0 mL/min. Quantification was achieved by external standards in a calibration range of 0.5 to 30 mM.

2.9.2. GC Analyses

Free vicinal diketones were quantified by a Clarus 500 gas chromatograph (Perkin-Elmer, Waltham MA, USA) with a headspace unit and Elite-5 60 m × 0.25 mm, 0.5 µm column using a 2,3-hexandione internal standard. Fermentation by products (esters, higher alcohols) was quantified using a Clarus 580 (Perkin-Elmer, Waltham MA, USA) gas chromatograph with a headspace unit and INNOWAX cross-linked polyethylene-glycol 60 m × 0.32 mm, 0.5 µm column (Perkin-Elmer, Waltham MA, USA). Vials containing beer samples were equilibrated for 25 min at 60 °C. The samples were injected at 50 °C, rising to 85 °C after one minute by heating at 7 °C/min. A temperature of 85 °C was maintained for one minute and then elevated to 190 °C at a heating rate of 25 °C/min.

2.9.3. Other

Glycerol was determined via enzymatic assay kit (glucokinase method), following the recommended procedure (K-GCROLGK, Megazyme, Bray Co. Wicklow, Ireland). The method is based on the use of ADP-glucokinase and an increase in absorbance on conversion of NAD^+ to NADH, and is performed at ambient temperature at a sample volume of 2 mL.

Free amino nitrogen (FAN) was measured using a ninhydrin-based dying method, where absorbance is measured at 570 nm against a glycine standard (ASBC Method Wort-12 A). The method is performed at a total volume of 10 mL. Following the color reaction at 95 °C, the samples are measured at ambient temperature.

Extract (apparent and real) and ethanol (for fermentation monitoring) were analyzed via density meter DMA 4500M with Alcolyzer Beer ME (Anton-Paar GmbH, Graz, Austria) at 20 °C and a sample volume of 30 mL.

The pH was determined using a digital pH meter (Mettler Toledo LLC, Columbus, OH, USA).

2.10. Statistical Analyses

Screening fermentations and analyses were carried out in triplicate. Statistical analysis was performed using RStudio, Version 1.1.463 with R version 3.5.2 (RStudio Inc, Boston, MA, USA; R Core Team, r-project). One-way analysis of variance (ANOVA) was used to compare means, and Tukey's post hoc test with 95% confidence intervals was applied for the pairwise comparison of means. When available, values are given as the mean ± standard deviation. Statistical analyses during the RSM trials were performed using the DesignExpert 9 software (StatEase, Minneapolis, MN, USA).

3. Results and Discussion

3.1. Yeast Strain Characterization

To identify the species of the yeast strains, amplification of the D1/D2 domain via PCR was performed and sequenced. The obtained sequences were compared to publicly available sequences in the NCBI nucleotide database via BLAST. The results of the strain identification are shown in Table 3.

Table 3. Yeast strain designation, species and origin of yeast strains used in this study.

Strain Designation	Species	Origin	Yeast Bank
837A	Cyberlindnera misumaiensis	Brewery cellar	FZW BLQ [1], Weihenstephan, Germany
NT Cyb	Cyberlindnera fabianii	Dried yeast starter for rice wine	FZW BLQ [1], Weihenstephan, Germany
L1	Cyberlindnera jadinii	Fruit of *Solanum quitoense*, "Lulo"	UCC Culture Collection, Cork, Ireland
C6.1	Cyberlindnera subsufficiens	Coconut	UCC Culture Collection, Cork, Ireland
CBS 1707 [T]	Cyberlindnera mrakii	Soil	Westerdijk Fungal Biodiversity Institute, Utrecht, Netherlands
CBS 5763 [T]	Cyberlindnera subsufficiens	Soil	Westerdijk Fungal Biodiversity Institute, Utrecht, Netherlands

[1] Research Centre Weihenstephan for Brewing and Food Quality, Technische Universität München; [T] Type strain.

The yeast strains were found to belong to the species *Cyberlindnera misumaiensis* (837A), *C. fabianii* (NT Cyb), *C. jadinii* (L1), and *C. subsufficiens* (C6.1). The *Cyberlindnera mrakii* type strain CBS 1707 (former *Williopsis saturnus* var. *mrakii*; synonym NCYC 500) was included in this study as a strain that has previously been investigated for the production of a low alcohol beer with high levels of esters [20]. The *Cyberlindnera subsufficiens* type strain CBS 5763 was included as an example to investigate potential intraspecific differences from C6.1.

3.2. API Substrate Utilization

Before considering non-conventional yeasts for NABLAB brewing, their behavior regarding utilization of important wort sugars like maltose and sucrose should be investigated. An API ID 32C test was performed to investigate the utilization of those sugars and to show general, interspecific differences between the strains. The results of the API test are shown in Table 4.

Table 4. Results of the API ID 32C substrate utilization test of the individual strains. Substrates without brewing relevance, which were negative for all strains, are not shown. "+" positive, "−" negative, "w" weak.

Substrate	837A	NT Cyb	L1	C6.1	CBS 1707	CBS 5763
Cycloheximide (Actidione)	+	−	−	−	−	−
D-Cellobiose	+	+	+	+	+	+
D-Galactose	−	−	w	−	−	−
D-Glucose	+	+	+	+	+	+
D-Maltose	−	+	+	−	+[1]	−[1]
D-Mannitol	+	+	w	w	w	w
D-Melibiose	−	−	−	−	−	−
D-Melezitose	−	+	+	−	+	−
D-Raffinose	−	+	+	+	+	+
D-Sorbitol	+	+	w	+	−	−
D-Sucrose	−	+	+	+	−	+
D-Trehalose	−	+	−	−	+	−
D-Xylose	−	+	+	+	+	+
Esculin Ferric Acid	+	+	+	+	+	+
Glucosamine	−	−	−	w	−	−
Glycerol	+	+	+	+	+	+
Lactic Acid	−	+	+	+	+	+
Levulinic Acid	−	w	w	w	w	+
L-Sorbose	−	−	−	−	−	+
Methyl-αD-Glucopyranoside	−	+	−	−	−	−
N-Acetyl-Glucosamine	−	−	w	−	w	−
Palatinose	−	+	+	−	+	−
Potassium Gluconate	w	w	−	+	w	+

[1] Growth "variable" according to Kurtzman et al. [31].

Maltose utilization was positive for NT Cyb, L1 and CBS 1707, in accordance with the reported literature, although assimilation of maltose by CBS 1707 is classified as "variable" [31]. Sucrose utilization was positive for four of the six strains and negative for 837A and CBS1707. The results suggest that in brewers' wort, where maltose is the most abundant fermentable sugar, only NT Cyb, L1 and CBS 1707 have the capability to achieve high attenuations. However, the API test investigates substrate utilization under aerobic conditions. Sugar consumption during fermentation, under anaerobic conditions, can differ significantly [31], which is also known as the Kluyver effect [32]. Due to the inability of 837A and CBS 1707 to utilize sucrose, lower attenuations in fermentations in wort could be expected.

3.3. Stress Tests

When considering non-*Saccharomyces* yeast strains for brewing purposes, several brewing-relevant parameters such as flocculation behavior, POF production and stress responses should be investigated [33]. The flocculation behavior can give initial indications regarding yeast handling in terms of potential bottom cropping. POF behavior is important because in most beer styles, POF is not desired. Substances like hop-derived iso-α-acids, ethanol content, or the pH value of the wort can have significant influences on yeast activity, manifesting mainly in a prolonged lag time, and even complete growth inhibition [33–35]. With the investigated yeast strains, iso-α-acid concentrations of up to 100 IBU had no significant effect on the yeast growth (data not shown), which is in accordance with previous reports on seven different non-*Saccharomyces* species [34,35]. However, Michel et al. [33] reported a minor prolongation in the lag time of *Torulaspora delbrueckii* strains in concentrations of up to 90 IBU. The results of the investigated characterization attributes are shown in Table 5.

Table 5. Characterization of yeast strains for flocculation behavior, phenolic off-flavor (POF) production and lag time in wort with and without a stressor at different concentrations. "—" no growth.

	Characterization Attributes	Unit	837A	NT Cyb	L1	C6.1	CBS 1707	CBS 5763
	Flocculation	%	78 ± 3	22 ± 2	35 ± 4	32 ± 1	85 ± 2	51 ± 4
	POF	-	negative	negative	negative	negative	negative	negative
Ethanol	0% ABV	h	18	6	9	6	9	9
	2.5% ABV	h	120	12	18	18	12	18
	5% ABV	h	—	24	36	24	48	—
	7.5% ABV	h	—	42	—	—	126	—
pH	5.5	h	18	6	9	6	9	9
	5	h	18	6	9	6	9	9
	4	h	66	6	9	6	9	9
	3	h	—	12	24	18	78	42

CBS 1707 exhibited the strongest flocculation behavior, at 85%, followed by 837A and CBS 5763, at 78% and 51%, respectively. NT Cyb, L1 and C6.1 exhibited very low flocculation of below 35%. All strains were negative for POF behavior. NT Cyb and C6.1 exhibited the fastest growth in wort (without a stress factor), overcoming the lag time after only 6 hours, followed by L1 and the CBS strains after 9 hours. Strain 837A exhibited a long lag phase of 18 hours (Figure 1). Concentrations of 2.5% ABV ethanol in the wort affected the lag time of all investigated strains. 837A was especially susceptible, with a prolonged lag phase of 120 hours. The remainder of the strains showed an extension of the lag phase of 3 to 12 hours. At 5% ABV, growth was fully inhibited for 837A and CBS 5763, while the other strains again exhibited an extension of the lag phase, of up to a maximum of 48 hours in CBS 1707. Complete growth inhibition was observed for L1 and C6.1 at 7.5% ABV, while the lag phase of NT Cyb and CBS 1707 was prolonged to 42 and 126 hours, respectively. All strains except 837A, which showed a significant extension of the lag phase to 66 hours, remained unaffected by a lower pH of 4. Only at pH 3 were lag times affected, while 837A was fully inhibited. Growth at low pH is important when considering the yeast for sour beer production, where the yeast must withstand pH values of below 4 [36]. However, it has been shown that organic acids like lactic acid can have a stronger inhibitory effect on yeasts and other microorganisms than HCl, which is caused by its chemical properties as a weak acid [35,37]. Inhibition by lactic acid could therefore be more pronounced than the HCl inhibition observed in this study. Figure 1 shows the growth of the investigated yeast strains in wort without the addition of a stressor.

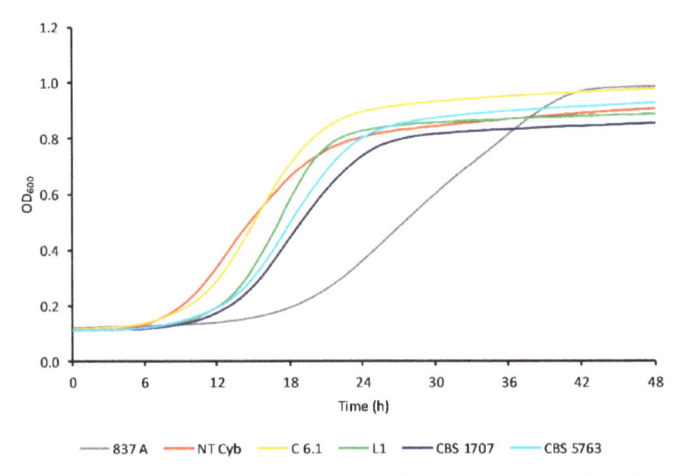

Figure 1. Growth of yeast strains in 7 °P wort extract at 25 °C without a stressor. Growth curves shown are the mean of a triplicate.

3.4. Screening

To investigate interspecific differences in the fermentation of wort, fermentation trials were performed in a diluted wort extract of 7 °P. Previous studies have shown that extract contents of around 7 °P will yield ethanol concentrations of around 0.5% ABV, a popular legal limit for NAB [7], in fermentations with maltose-negative yeast strains [1,14,34,38]. After aerobic propagation for 24 hours, NT Cyb exhibited the highest number of cells, at 2×10^9 cells/mL, more than four-fold the amount of cells compared to L1, C6.1, and the CBS strains with counts between 3.4 and 4.9×10^8 cells/mL (Table 6). Due to a delayed growth (compare Figure 1), 837A had to be propagated for 48 hours, reaching a cell count of 6.1×10^8 cells/mL. For the screening in wort, yeast cells were added at a concentration of 3×10^7 cells/mL, after a gentle washing step in water to prevent carry-over of propagation wort sugars. The results from the yeast screening are shown in Table 6. The fermentations were carried out until no change in extract could be measured for two consecutive days.

Table 6. Results of the screening of the investigated *Cyberlindnera* strains in wort extract.

Non-Alcoholic Beer (NAB) Attributes		Unit	837A	NT Cyb	L1	C6.1	CBS 1707	CBS 5763
Propagation	Cell count (24 h)[1]	× 10⁶ cells/mL	611 ± 34[1]	2055 ± 21	486 ± 27	445 ± 4	338 ± 25	386 ± 48
Fermented wort	Real Extract	°P	6.53 ± 0.03 b	6.40 ± 0.04 ab	6.45 ± 0.05 ab	6.36 ± 0.03 a	6.57 ± 0.10 b	6.35 ± 0.10 a
	Attenuation	%	18 ± 1 ab	23 ± 1 bc	21 ± 2 abc	24 ± 1 c	17 ± 3 a	24 ± 3 c
	Ethanol	% ABV	0.55 ± 0.01 a	0.63 ± 0.01 b	0.66 ± 0.00 bc	0.63 ± 0.00 bc	0.54 ± 0.01 a	0.67 ± 0.02 d
	pH	-	4.41 ± 0.02 ab	4.51 ± 0.02 c	4.44 ± 0.01 bc	4.38 ± 0.03 b	4.37 ± 0.06 ab	4.33 ± 0.01 a
	FAN	mg/L	88 ± 1 b	83 ± 2 ab	80 ± 1 a	81 ± 4 a	78 ± 1 a	84 ± 5 ab
Sugar consumption	Maltotriose	%	3 ± 2	6 ± 1	4 ± 1	5 ± 1	3 ± 1	5 ± 3
	Maltose	%	4 ± 2	4 ± 0	3 ± 1	4 ± 0	4 ± 1	4 ± 4
	Sucrose	%	2 ± 10	100	100	100	2 ± 2	100
	Glucose	%	100	100	100	100	100	100
	Fructose	%	81 ± 1	75 ± 1	100	80 ± 1	73 ± 8	83 ± 2
Fermentation by-products	Glycerol	g/L	0.25 ± 0.05 ab	0.23 ± 0.04 a	0.36 ± 0.03 b	0.30 ± 0.04 ab	0.18 ± 0.05 a	0.21 ± 0.05 a
	Acetaldehyde	mg/L	9.70 ± 2.83 b	8.05 ± 1.48 b	2.60 ± 0.14 a	3.37 ± 0.71 a	3.83 ± 0.45 a	2.57 ± 0.21 a
	Ethyl acetate	mg/L	65.70 ± 14.57 b	22.55 ± 2.90 a	9.27 ± 3.23 a	4.90 ± 0.85 a	8.10 ± 0.28 a	5.17 ± 0.29 a
	Isoamyl acetate	mg/L	0.90 ± 0.14 ab	<LOD	0.15 ± 0.07 a	1.60 ± 0.62 b	1.67 ± 0.12 b	1.03 ± 0.23 ab
	Ethyl formate	mg/L	0.53 ± 0.04 a	0.31 ± 0.06 a	0.57 ± 0.09 a	0.25 ± 0.03 a	2.70 ± 0.57 c	1.45 ± 0.07 b
	Ethyl propionate	mg/L	0.13 ± 0.04 a	0.13 ± 0.01 a	<LOD	<LOD	0.16 ± 0.01 a	0.17 ± 0.03 a
	Isoamyl alcohols	mg/L	11.20 ± 0.14 a	16.40 ± 0.57 b	23.15 ± 0.92 c	11.67 ± 1.74 a	11.93 ± 0.93 a	10.50 ± 0.14 a
	n-Propanol	mg/L	4.03 ± 0.84 ab	3.73 ± 0.21 ab	4.40 ± 0.62 b	3.27 ± 0.15 ab	2.93 ± 0.29 a	3.33 ± 0.15 ab
	Isobutanol	mg/L	7.57 ± 1.24 ab	7.70 ± 0.36 b	8.27 ± 1.38 b	8.03 ± 0.40 b	5.33 ± 0.55 a	7.20 ± 0.20 ab
	Σ Esters	mg/L	67.26 ± 14.79 b	22.99 ± 2.87 a	9.99 ± 3.31 a	6.75 ± 0.61 a	12.62 ± 1.48 a	7.82 ± 0.30 a
	Σ Alcohols	mg/L	22.80 ± 0.14 a	27.83 ± 0.64 b	35.82 ± 1.48 c	22.97 ± 1.97 a	20.20 ± 0.17 a	21.03 ± 0.21 a
Sensory	Aroma	-	Solvent-like, unpleasant	Cabbage-like, unpleasant	Fruity, pleasant	Fruity, pleasant	Fruity, pleasant	Fruity, pleasant

[1] Cell count after 48 h due to delayed growth compared to other strains (compare Figure 1). LOD 'limit of detection'. Different superscripts of values within a row indicate a significant difference ($p \leq 0.05$).

Strains 837A and CBS 1707 exhibited the lowest attenuation of only 18% and 17%, respectively, owing to their inability to utilize sucrose (Table 4), which was confirmed by the lack of sucrose consumption. Liu and Quek [20] also reported the absence of sucrose utilization by CBS 1707. The other strains, which depleted sucrose completely, reached attenuations of 21% to 24%. Consequently, 837A and CBS 1707 also produced, at 0.55% and 0.56% ABV, the lowest amounts of ethanol ($p \leq 0.05$) compared to the remaining strains, where ethanol concentrations ranged from 0.63% to 0.67% ABV. The final pH of the fermented samples ranged from 4.33 (CBS 5763) to 4.51 (NT Cyb). Residual FAN ranged from 78 (CBS 1707) to 88 mg/L (837A). As expected, none of the strains consumed maltotriose. Maltose consumption was also neglectable in all strains, although the species *Cyberlindnera fabianii* (like NT Cyb) has been reported to be able to ferment maltose [31,39]. The observations also underlined that results from the API substrate utilization test (where NT Cyb, L1 and CBS 1707 were positive for maltose) are not necessarily reflected in practice, especially since sugar utilization during respiration and fermentation can differ [31,32,40]. While glucose was depleted by all strains, fructose was only fully depleted by L1. The remaining strains exhibited glucophilic behavior and consumed only 73% to 83% of fructose during fermentation. Regarding fermentation by-products, glycerol concentrations were low, ranging from 0.18 to 0.36 g/L. The strains 837A and NT Cyb accumulated significantly higher amounts of acetaldehyde, at 9.7 and 8.1 mg/L, respectively, compared to 2.6 to 3.8 mg/L in the remaining samples. The sample fermented with *Cyberlindnera misumaiensis* 837A exhibited extremely high values of ethyl acetate, at 65.7 mg/L, twice the flavor threshold concentration in beer [2,41]. Ethyl acetate is described to have a fruity, estery character but also solvent-like, especially in high concentrations. The remaining strains exhibited ethyl acetate production between 4.9 (C6.1) and 22.6 mg/L (NT Cyb). Isoamyl acetate, which is predominantly described as having a fruity, banana-like aroma, has a much lower flavor threshold of only 1.4–1.6 mg/L [2,41]. The strains C6.1 and CBS 1707 produced the highest amounts of isoamyl acetate, at 1.67 and 1.60 mg/L, followed by CBS 5763, 837A and L1, at 1.03, 0.90 and 0.15 mg/L, respectively. NT Cyb did not produce detectable amounts of isoamyl acetate. Concentrations of ethyl formate and ethyl propionate in the fermented samples were low, ranging from undetectable to 2.7 mg/L. Ethyl butyrate and ethyl caproate were not detected in either of the samples (data not shown). The strain L1 produced a significantly higher amount of higher alcohols, at 35.8 mg/L, followed by NT Cyb, at 27.8 mg/L, and the remaining strains at 20–23 mg/L. During sensory evaluation, the high ethyl acetate concentration in the sample fermented with 837A was indeed perceptible and described as an unpleasant, solvent-like aroma. The sample fermented with NT Cyb was described as having an unpleasant, cabbage-like aroma. The remaining samples were characterized by a pleasant, fruity aroma.

The unpleasant, solvent-like aroma in the sample fermented with 837A was attributed to the very high ethyl acetate concentration, well above the flavor threshold. However, the cabbage-like aroma, which is generally associated with sulfides or thiol compounds [41], that was detected in the sample fermented with NT Cyb could not be linked to the volatile by-products that were measured. Interestingly, ethyl acetate concentrations in the remaining samples, characterized by a pleasant, fruity aroma, were low, at only 2.6–3.8 mg/L. However, C6.1, CBS 1707 and CBS 5763 exhibited higher amounts of isoamyl acetate, a desired ester in beer (particularly ales) [42], when compared to the samples with unpleasant aroma. The concentrations of 1.0–1.6 mg/L are within the reported flavor threshold in beer of 0.5–2.0 mg/L [43]. Additionally, it is well known that synergistic effects between esters occur that can push the concentration of perception below their individual flavor thresholds [42,44,45]. Isoamyl acetate could therefore have been a cause of the fruity aroma in the samples fermented with C6.1, CBS 1707 and CBS 5763. However, the sample fermented with L1, which was also characterized by a fruity aroma, only contained a very low isoamyl acetate concentration of 0.15 mg/L. It is noteworthy, however, that the L1 sample contained a significantly higher amount of isoamyl alcohol, at 23.2 mg/L, which is described as having an alcoholic, fruity and banana-like flavor [2]. The results have confirmed that not a high amount of esters, but rather a balanced profile will lead to a pleasant, fruity aroma [5].

Based on the results from the screening, *Cyberlindnera subsufficiens* C6.1 was chosen for optimization of fermentation conditions by means of response surface methodology, followed by an up-scaled brewing trial at 60 L to create a fruity, non-alcoholic beer (≤0.5% ABV). Strains 837A and NT Cyb were eliminated because of their poor flavor characteristics. CBS 1707 was eliminated due to its inability to ferment sucrose, which apart from the lower attenuation, would remain in the wort after fermentation, acting as an additional sweetening agent and potential contamination risk. *Cyberlindnera jadinii* strain L1 was eliminated due to its very low isoamyl acetate production (Table 6) and due to its maltose utilization when oxygen was present (Table 4). The decision between the two similarly performing *Cyberlindnera subsufficiens* strains C6.1 and CBS 5763 was made in favor of C6.1 due to a more pleasant fruitiness. In addition, C6.1 showed increased tolerance towards stress caused by ethanol or low pH (Table 5).

3.5. Response Surface Methodology (RSM)

To find the optimal fermentation conditions for C6.1 for an up-scaled application to produce a fruity, non-alcoholic beer, RSM was performed. Michel et al. [46] applied RSM to optimize the fermentation conditions of a *Torulaspora delbrueckii* strain for brewing purposes. They found that the pitching rate and fermentation temperature were crucial parameters, which influenced the flavor character of the final beer. The optimal fermentation conditions were shown to be at 21 °C with a high pitching rate of 60×10^6 cells/mL. Especially for non-*Saccharomyces* yeasts, the pitching rate can be crucial since most non-*Saccharomyces* species have comparably smaller cell sizes [46]. Figure 2 shows an example of the differing cell size between *Cyberlindnera subsufficiens* strain C6.1 (A) and the brewers' yeast strain *Saccharomyces cerevisiae* WLP001 (B) at identical magnification.

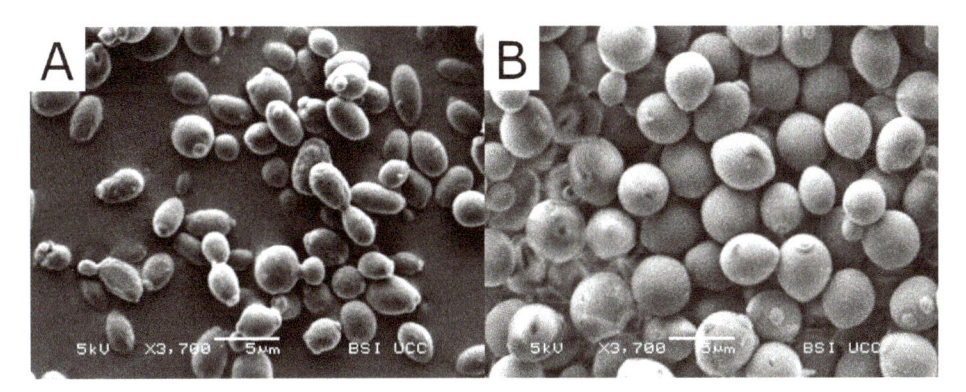

Figure 2. Scanning electron microscopy (SEM) picture of *Cyberlindnera subsufficiens* strain C6.1 (**A**) and the brewers' yeast strain *Saccharomyces cerevisiae* WLP001 (**B**) at a magnification of × 3700. Size of bar: 5 μm.

It is also known that temperature and pitching rate have an influence on ester production, though strain-specific differences also play a role [4,6]. Previously reported fermentation temperatures of *Cyberlindnera subsufficiens* and other *Cyberlindnera* spp. range from 20 to 25 °C [12,17,19,20,47]. Consequently, a two-factorial, face-centered central composite design was chosen with Factor A: fermentation temperature (17, 22, 27 °C), and Factor B: pitching rate (10, 35, 60×10^6 cells/mL). The individual experiment runs are listed in Table 2. The wort extract applied in the RSM trial was the same as that used for the screening, at an extract content of 7 °P (Table 1). Fermentation was conducted until no change in extract could be measured for two consecutive days. With the measured response values, significant models could be produced. The significant response models, with their respective minima and maxima and a summary of the model statistics, are shown in Table 7. Insignificant response models are not shown, and response models with a significant lack of fit will not be discussed in this study but are included in the visualized data for the sake of a complete picture. For a full report on model statistics and response values, refer to the supplementary Data Sheet S1.

Table 7. Analysis of variance (ANOVA) results for response models of the response surface methodology (RSM) trial.

Response	Unit	Minimum	Maximum	Model	P-Value	LOF P-Value
Ethanol	% ABV	0.41	0.60	RQuadratic	2.80×10^{-3} **	0.648
Ethyl acetate	mg/L	3.4	9.3	2FI	3.12×10^{-2} *	0.007 **
Isoamyl acetate	mg/L	0.8	2.2	RQuadratic	1.42×10^{-2} *	0.046 *
Acetaldehyde	mg/L	1.9	3.4	RLinear	1.35×10^{-3} **	0.337
n-Propanol	mg/L	3.2	4.5	2FI	9.03×10^{-3} **	0.029 *
Isobutanol	mg/L	3.2	6.7	RQuadratic	4.30×10^{-9} ***	0.145
Isoamyl alcohols	mg/L	7.3	13.3	Quadratic	2.67×10^{-5} ***	0.270
Σ Esters	mg/L	4.2	11.1	RQuadratic	1.48×10^{-2} *	0.018 *
Σ Alcohols	mg/L	13.7	22.9	RQuadratic	3.28×10^{-8} ***	0.339
Glycerol	g/L	0.17	0.37	RQuadratic	4.85×10^{-5} ***	0.034 *
Acceptance	-	1.08	3.38	Linear	1.31×10^{-2} *	0.377
Fruitiness	-	1.13	3.38	Linear	7.31×10^{-3} **	0.484

Model terminology: "RQuadratic" Reduced Quadratic; "2FI" Two-Factor Interaction; "RLinear" Reduced Linear. "LOF" Lack of Fit. ANOVA significance codes: *** $p \leq 0.001$, ** $p \leq 0.01$, * $p \leq 0.05$.

It was possible to create significant models for 12 responses (Table 7). However, five also exhibited significant lack of fit (LOF), rendering them unusable for predictions. The aim of the RSM was to investigate the optimal fermentation conditions to create a fruity, non-alcoholic beer. The three-dimensional response surface plots of the interactive effects of temperature and pitching rate on the final ethanol content and the fruitiness of the produced NAB are shown in Figures 3 and 4.

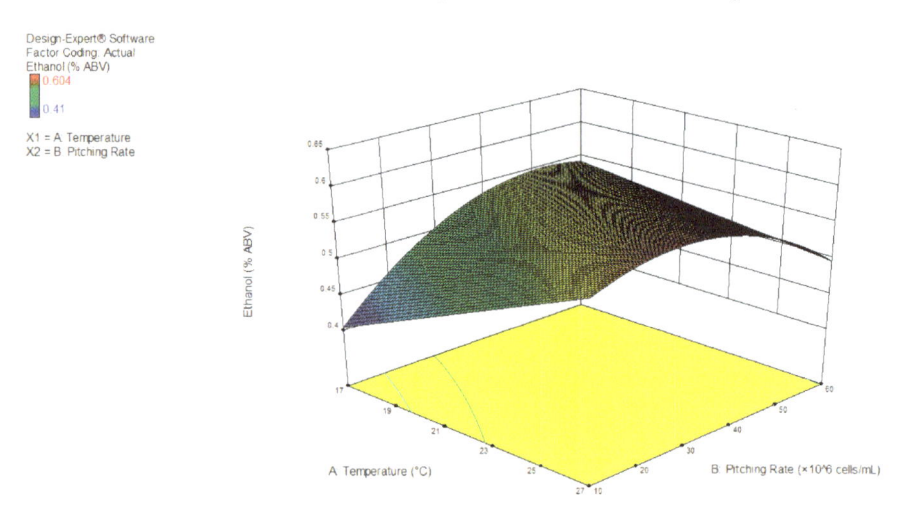

Figure 3. Three-dimensional response surface plot of the interactive effects of temperature and pitching rate on the ethanol content of the produced non-alcoholic beer ($p < 0.01$).

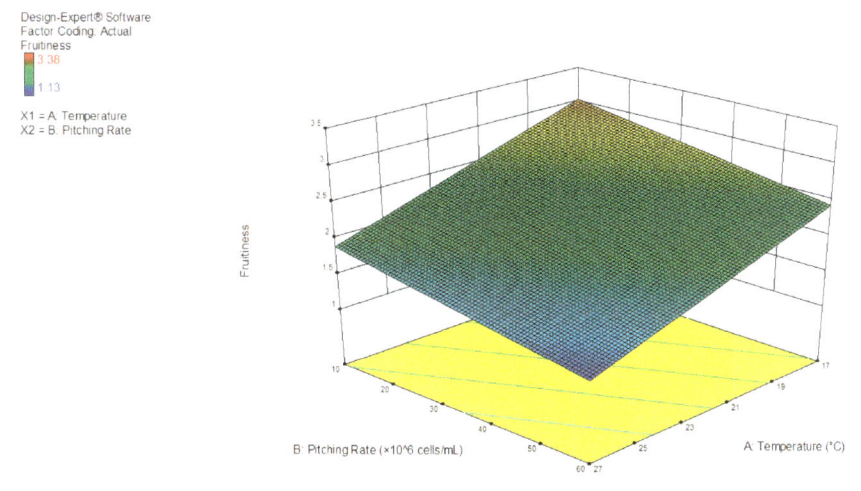

Figure 4. Three-dimensional response surface plot of the effects of temperature and pitching rate on the fruitiness of the produced non-alcoholic beer ($p < 0.01$).

Ethanol content was lowest at a low temperature of 17 °C and low pitching rate (10^7 cells/mL), and it went up with increasing temperature and pitching rate, but lowered again at a high pitching rate combined with a high fermentation temperature (Figure 3). The minium and maximum values were 0.41% and 0.60% ABV. Sugar analysis revealed that at 17 °C and 10^7 cells/mL, about 0.5 g/L of glucose was remaining after fermentation, while it was fully depleted in worts fermented at higher pitching rates and higher temperatures (data not shown). The residual sugar explained the lower final ethanol concentration. Fructose was only fully depleted in the samples that were fermented at 27 °C. At 22 °C, fermented samples exhibited residual fructose concentrations between 0.2 and 0.5 g/L, and at 17 °C, fermented samples showed remaining fructose concentrations between 0.2 and 0.7 g/L. Acetaldehyde concentrations were only dependent on the pitching rate, with increasing amounts of acetaldehyde found at lower pitching rates (Figure A1). This result correlates with other studies that found a decrease in acetaldehyde with increasing pitching rate in wort fermentations with brewers' yeasts [48,49]. However, overdosing yeast ($>5 \times 10^7$ cells/mL) can lead to an increase in acetaldehyde again, as observed by Erten et al. [50]. The temperature did not have a significant effect on the acetaldehyde concentration and was therefore excluded from the model ($p = 0.39$; supplementary Data Sheet S1). However, regarding higher alcohols, the fermentation temperature had a stronger effect, with increasing amounts of higher alcohols found at higher temperatures (Figures 5 and A2), which is consistent with the literature [4,5]. Isoamyl acetate concentrations were generally high and ranged from 0.8 to 2.2 mg/L. Although the model was significant ($p < 0.05$), it was unsuitable for value prediction due to a significant lack of fit ($p = 0.046$).

Interestingly, the production of the esters ethyl acetate and isoamyl acetate did not show a clear correlation to temperature, which underlines that the general rule of thumb, that higher fermentation temperatures lead to increased ester production, is not valid for all yeast strains (Figure 5) [4]. Furthermore, the amount of esters that were quantified in this study did not correlate with the perceived fruitiness of the NAB, which tentatively suggests that the fruity flavor profile was caused by yet unidentified compounds (Figure 5).

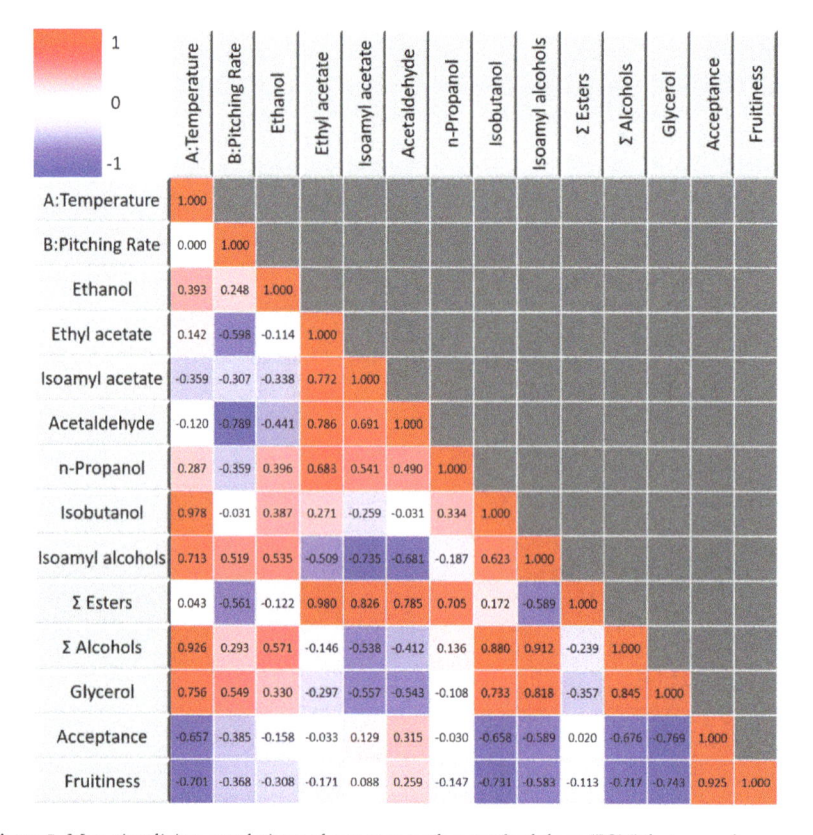

Figure 5. Map visualizing correlations of response surface methodology (RSM) factors and responses based on the Pearson Correlation Coefficient. 1 signifies strong positive correlation, 0 signifies no correlation, and −1 signifies a strong negative correlation.

In terms of fruitiness, a low fermentation temperature paired with a low pitching rate led to the highest perceived fruitiness. Indeed, the highest fruitiness was recorded at 17 °C and 1×10^7 cells/mL and the lowest at 27 °C and 6×10^7 cells/mL, following a linear model. General acceptance showed a strong positive correlation with the fruitiness, indicating that the panel preferred fruity samples (Figures 5 and A3).

Due to the ideal combination of lowest ethanol content and highest fruitiness and acceptance, the fermentation temperature of 17 °C and pitching rate of 1×10^7 cells/mL were chosen as the optimal fermentation conditions for application to produce a fruity, non-alcoholic beer.

A small-scale fermentation at the optimal conditions (17 °C, 10^7 cells/mL) was conducted to validate the RSM model. Table 8 shows the predicted mean including 95% prediction intervals (PI) and the measured ("observed") mean with standard deviation.

Although predicted by a significant model, the observed means for ethanol, acetaldehyde and isobutanol values were not within the 95% prediction interval. Sugar analysis revealed the complete depletion of glucose in the experimental fermentation trial at optimal conditions compared to the RSM model prediction, which explained the increased ethanol production (data not shown). The moderate success in model validation demonstrates the limitations in the application of RSM to optimize fermentations, where small differences in substrate and process conditions can have significant influences on the outcome. Because wort is a very complex substrate, comprising a complex mixture of different sugars, nitrogen sources, minerals and vitamins, among others, any interpretation or the transfer of the RSM results to other substrates (even different wort substrates) should be made with caution. In particular, a different sugar composition will have a significant effect on the responses when applying maltose-negative yeasts. However, the improved fruitiness and therefore higher acceptance of the NAB produced at low temperature and low pitching rate, the main goal from the optimization, was significant and reproducable (Table 8).

Table 8. Response surface methodology (RSM) model validation via predicted value vs. observed value.

Response	95% PI Low	Predicted Mean	95% PI High	Observed Mean	Std. Dev.
Ethanol *	0.33	0.40	0.48	0.53	0.01
Ethyl acetate	0.89	4.74	8.60	6.83	0.59
Isoamyl acetate	0.78	1.63	2.47	2.50	0.10
Acetaldehyde *	2.19	2.97	3.74	1.27	0.29
n-Propanol	2.68	3.28	3.88	3.57	0.06
Isobutanol *	2.91	3.23	3.54	2.80	0.10
Isoamyl alcohols	5.78	7.03	8.29	4.10	0.10
SUM Esters	3.01	7.10	11.19	9.33	0.68
SUM Alcohols *	12.84	13.74	14.64	10.47	0.31
Glycerol	0.13	0.18	0.22	0.27	0.01
Acceptance *	2.12	3.23	4.34	3.75	0.62
Fruitiness *	2.02	3.03	4.05	3.58	0.87

* Significant model with insignificant lack of fit. 'PI' Prediction interval.

3.6. Pilot-Scale Brewing

Despite the limited model validation, the fermentation parameters were successfully optimized to enhance the fruity character of the NAB. Therefore, the pilot-scale brewing trial was conducted with the optimized conditions of 17 °C fermentation temperature and a pitching rate of 10^7 cells/mL.

The grain bill of the wort for the pilot-scale brewing trial consisted of 95% pilsner malt and 5% acidulated malt to lower the starting pH of the wort, to account for the reduced pH drop during fermentations with non-*Saccharomyces* yeasts compared to brewers' yeast. A low beer pH is desired to prevent microbial spoilage and to ensure good liveliness of the beer [51,52]. The analytical attributes of the wort produced at pilot-scale are shown in Table 9.

Table 9. Attributes of the wort produced on pilot-scale.

Wort attributes	Unit	Value
Extract	°P	7.00 ± 0.01
pH		4.86 ± 0.01
FAN	mg/L	107 ± 3
Glucose	g/L	6.01 ± 0.08
Fructose	g/L	0.80 ± 0.01
Sucrose	g/L	2.13 ± 0.03
Maltose	g/L	31.59 ± 0.44
Maltotriose	g/L	9.32 ± 0.13

To assess the suitability of *Cyberlindnera subsufficiens* C6.1 to produce a fruity NAB, it was compared to two commercial NABs. NAB A was a commercial non-alcoholic beer produced by limited fermentation [29], and NAB B was a non-alcoholic beer produced by "dialysis technology" [30]. The NABs were analyzed for their extract, ethanol, FAN and glycerol content as well as their sugar composition and concentration of volatile fermentation by-products. The results are shown in Table 10.

Table 10. Attributes of the non-alcoholic beer (NAB) produced with C6.1 compared to two commercial NABs, NAB A and NAB B.

NAB Attributes	Unit	C6.1 NAB	NAB A	NAB B
Extract (real)	°P	6.60 ± 0.01	6.76 ± 0.07	7.05 ± 0.03
Extract (apparent)	°P	6.46 ± 0.02	6.57 ± 0.06	6.86 ± 0.01
Ethanol	% ABV	0.36 ± 0.00	0.50 ± 0.03	0.49 ± 0.04
pH		4.45 ± 0.01	4.29 ± 0.02	4.29 ± 0.04
FAN	mg/L	96 ± 2	86 ± 6	24 ± 0
Glycerol	g/L	0.30 ± 0.02	0.33 ± 0.01	1.40 ± 0.03
Glucose	g/L	2.77 ± 0.05	2.74 ± 0.04	5.61 ± 0.04
Fructose	g/L	1.65 ± 0.03	1.96 ± 0.03	0.19 ± 0.00
Sucrose	g/L	<LOD	<LOD	<LOD
Maltose	g/L	30.27 ± 0.62	30.11 ± 0.50	17.69 ± 0.24
Maltotriose	g/L	8.67 ± 0.24	8.31 ± 0.21	1.84 ± 0.03
Acetaldehyde	mg/L	10.55	2.40	0.70
Ethyl acetate	mg/L	12.00	<0.10	2.70
Isoamyl acetate	mg/L	0.80	<0.1	0.70
Isoamyl alcohols	mg/L	4.00	4.80	17.40
n-Propanol	mg/L	2.20	<0.5	2.50
Isobutanol	mg/L	3.60	1.00	4.90
Diacetyl	mg/L	<0.01	0.02	0.04
2,3-Pentandione	mg/L	<0.01	<0.01	<0.01
Σ Esters	mg/L	12.8	<0.1	3.4
Σ Alcohols	mg/L	9.8	5.8	24.8

The C6.1 NAB reached final attenuation after 13 days of fermentation at 17 °C, at an ethanol content of 0.36% ABV. At the end of fermentation, 2.77 g/L glucose was remaining in the wort and sucrose was fully depleted. Compared to the initial sugar concentration of the wort (Table 9), fructose concentrations in the final beer were significantly higher, at 1.65 g/L, twice as high as the starting concentration in the wort. Since sucrose was fully depleted, it can be assumed that it was converted to glucose and fructose by the yeast's invertase. The high residual fructose could therefore be attributed to the previously observed glucophilic character of the C6.1 strain in the screening and RSM trial. As a result, fructose was not consumed by the yeast due to the permanent presence of glucose until fermentation came to a halt. As expected, maltose and maltotriose consumption was negligible. Despite the limited fermentation, C6.1 produced a relatively high amount of esters, at 12.8 mg/L, the majority of which was ethyl acetate (12 mg/L). NAB A had an ethanol content of 0.50% ABV. Interestingly, the sugar composition was very similar to that of the C6.1 NAB. Regarding fermentation by-products, however, NAB A exhibited very low concentrations, at about half the amount of higher alcohols and a total lack of the esters ethyl acetate and isoamyl acetate. NAB B had an ethanol content of 0.49% ABV. Owing to its fundamentally different production method, the analyzed attributes were very different from those of the two NABs produced solely by limited fermentation. The low FAN content together with a high glycerol content compared to the other NABs were indicators of a more extensive fermentation, with subsequent removal of ethanol. However, NAB B still exhibited high amounts of monosaccharides, which suggested that the production of the NAB either also entailed a limited fermentation, or the dealcoholized beer was blended with wort (or other means of sugar addition). The increased amounts of higher alcohols in NAB B, at 24.8 mg/L, are uncommon for beers dealcoholized via dialysis, since the process commonly reduces their content in the final NAB by 90%–95% [7]. Despite the addition of acid malt during the wort production for the C6.1 NAB, the final pH after fermentation was, at 4.45, higher compared to 4.29 in the commercial NABs.

Due to the high amounts of residual sugars, proper pasteurization is essential for non-alcoholic beers produced by limited fermentation to avoid microbial spoilage [1,38,53]. After bottling, C6.1 NAB was therefore pasteurized with approximately 23 PU, and the successful pasteurization was confirmed

3.7. Sensory Evaluation

For a holistic evaluation of the C6.1 NAB compared to the two commercial NABs, a sensory trial was conducted with 10 trained and experienced panelists. The panel was asked to describe the flavor of the beer in their own words, followed by an assessment of several intensity attributes. The mean score values of the parameters wort-like, floral, fruity, citrus-like and tropical aroma, as well as sweet taste, of the NABs are shown in Figure 6.

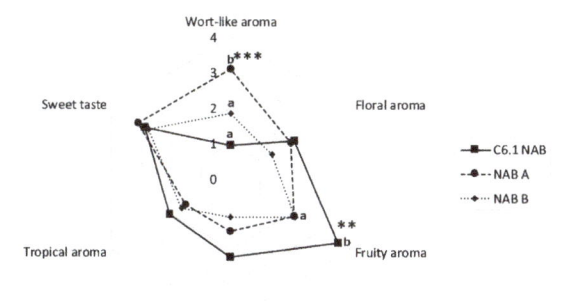

Figure 6. Spider web with the means of the descriptors from the sensory trial of the NAB produced with *Cyberlindnera subsufficiens* C6.1 and the two commercial NABs. Different letters next to data points indicate a significant difference as per Tukey's *post hoc* test. Significance codes: *** $p \leq 0.001$, ** $p \leq 0.01$.

The NAB produced with C6.1 was described as very fruity with aromas of pear, banana, mango and maracuja together with a slightly wort-like character. NAB A was described as malty, wort-like and hoppy, while NAB B was described as wort-like and caramel-like. The C6.1 NAB was indeed evaluated as being significantly more fruity than the commercial NABs ($p \leq 0.01$), at an average of 3.6 out of 5 compared to 2.1 and 2.2 out of 5, scoring also higher in citrus-like and tropical aromas. Consequently, the wort-like aroma, one of the most criticized flaws of NABs produced by limited fermentation [1,2,52], was least pronounced in the NAB produced with C6.1 with an average of 1 out of 5, followed by NAB B with 1.8 out of 5. NAB A exhibited, at an average of 3.2, a significantly more pronounced wort-like aroma ($p \leq 0.001$). A sweet taste, caused by a high amount of residual sugars, is another major point of criticism for NABs produced by limited fermentation [1,2,52]. All NABs scored similarly in sweet taste without significant differences. NAB B scored lower for "floral" compared to the other NABs. However, the difference was not statistically significant. When the panelists were asked for their favorite sample, 40% chose C6.1 NAB, 40% chose NAB A, and 20% chose NAB B. Similarly, Strejc et al. [3] investigated the production of a non-alcoholic beer (0.5% ABV) by a cold contact process (characterized by a low temperature and high pitching rate) with a mutated lager yeast strain (*Saccharomyces pastorianus*). The strain's targeted mutation resulted in an overproduction of isoamyl acetate and isoamyl alcohols. The authors reported that the fruity flavour of the NAB produced with the mutated strain was "partially able to disguise" the typical wort-like off-flavor [21]. However, the isoamyl acetate concentration of the resulting NAB was, at 0.5 mg/L, lower than the concentration in the C6.1 NAB in this study (Table 10). Furthermore, the complex mutation and isolation procedure paired with a potentially limited stability of the mutation limits its applicability in practice. Saerens and Swiegers [22] reported the successful production of a NAB at 1000 L scale with a *Pichia kluyveri* strain, owing to its high production of isoamyl acetate (2–5 mg/L), which reportedly gave the NAB a fruity flavor that was more like that of a regular beer than commercial NABs. In accordance, the results of the sensory indicated that a strong fruity aroma can mask the wort-like off flavor, and that the non-*Saccharomyces* yeasts, which produce a pronounced fruity character, can therefore be a means to produce NAB with improved flavor characteristics.

4. Conclusions

The *Cyberlindnera* genus was found to be a promising non-*Saccharomyces* genus for application in the production of a fruity, non-alcoholic beer. Four of the six investigated species produced a fruity character, despite the limited fermentative capacity, which resulted in a low ethanol concentration. It was shown that through optimization of the fermentation parameters of temperature and pitching rate, the fruity character could be enhanced. Process up-scaling with *Cyberlindnera subsufficiens* strain C6.1 produced a NAB that was significantly more fruity compared to two commercial NABs. Owing to the strong fruity aroma, the often-criticized wort-like aroma could successfully be masked. Yeast handling throughout the process (i.e., propagation, yeast pitching, fermentation) proved to be suitable for pilot-scale brewing, with potential for application at industrial scale. Further studies should investigate if the masking effect was enhanced by a reduction of wort aldehydes via yeast metabolism.

This study demonstrated the suitability of the non-*Saccharomyces* species *Cyberlindnera subsufficiens* for the production of non-alcoholic beer (<0.5% ABV) with novel flavor characteristics that can compete with commercial NABs. The successful pilot-scale (60 L) brewing trial gives prospect to future studies with diverse non-*Saccharomyces* yeasts and strengthens their position as a serious and applicable alternative to established methods in non-alcoholic and low alcohol beer brewing.

Supplementary Materials: The following are available online at http://www.mdpi.com/2311-5637/5/4/103/s1, Data Sheet S1: RSM response values and model statistics.

Author Contributions: Conceptualization, K.B., M.M., M.H., M.Z. and E.K.A.; methodology, K.B. and M.M.; investigation, K.B., J.J.A., and A.H.; resources, M.H., F.J. and E.K.A.; formal analysis, K.B.; writing—original draft preparation, K.B.; writing—review and editing, K.B., M.M., M.Z., M.H., K.M.L. and E.K.A.; visualization, K.B.; supervision, E.K.A.; project administration, E.K.A.; funding acquisition, E.K.A.

Funding: This research was supported by the Baillet Latour Fund within the framework of a scholarship for doctoral students.

Acknowledgments: The authors would like to thank Philippe Lösel for his valuable contribution to this study, Muntons plc for supplying the sprayed wort extract, the Barth-Haas Group for supplying the iso-α-acids extract, and David De Schutter and Luk Daenen for their review.

Conflicts of Interest: The authors declare no conflict of interest.

Appendix A

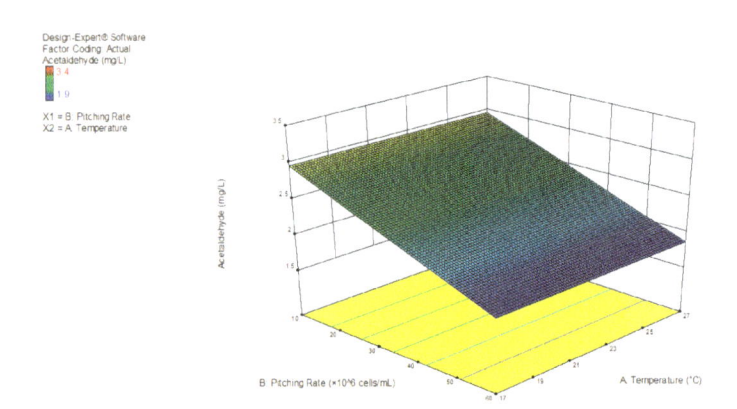

Figure A1. Three-dimensional response surface plot of the effect of pitching rate on the acetaldehyde content of the produced NAB ($p < 0.01$). The factor temperature was excluded from the model due to insignificance ($p = 0.39$; supplementary Data Sheet 1).

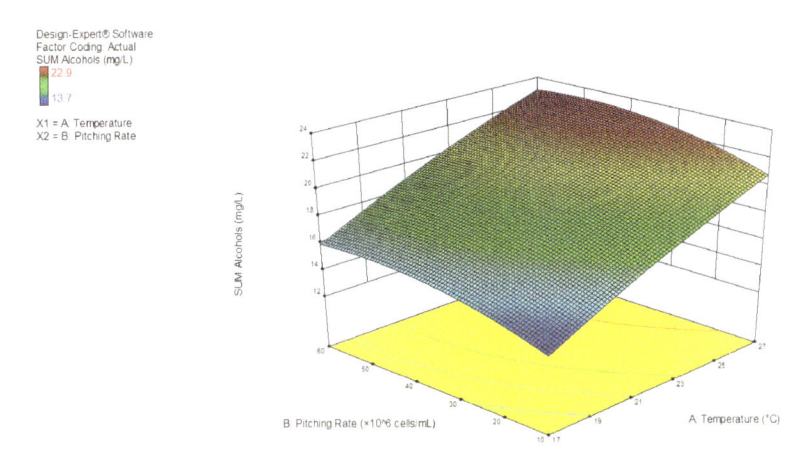

Figure A2. Three-dimensional response surface plot of the interactive effects of temperature and pitching rate on the sum of higher alcohols of the produced NAB ($p < 0.001$).

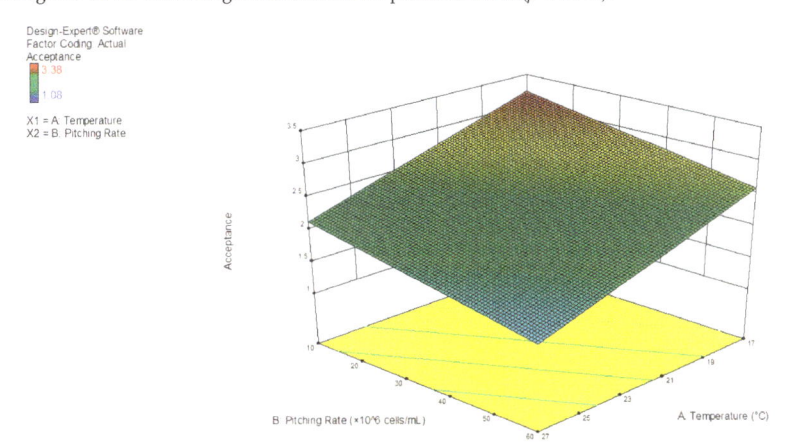

Figure A3. Three-dimensional response surface plot of the effects of temperature and pitching rate on the overall acceptance of the produced NAB ($p < 0.05$).

References

1. Bellut, K.; Arendt, E.K. Chance and Challenge: Non-*Saccharomyces* yeasts in nonalcoholic and low alcohol beer brewing: A Review. *J. Am. Soc. Brew. Chem.* **2019**, *77*, 77–91. [CrossRef]
2. Blanco, C.A.; Andrés-Iglesias, C.; Montero, O. Low-alcohol Beers: Flavor Compounds, Defects, and Improvement Strategies. *Crit. Rev. Food Sci. Nutr.* **2016**, *56*, 1379–1388. [CrossRef] [PubMed]
3. Perpète, P.; Collin, S. Influence of beer ethanol content on the wort flavour perception. *Food Chem.* **2000**, *71*, 379–385. [CrossRef]
4. Verstrepen, K.J.; Derdelinckx, G.; Dufour, J.P.; Winderickx, J.; Thevelein, J.M.; Pretorius, I.S.; Delvaux, F.R. Flavor-active esters: Adding fruitiness to beer. *J. Biosci. Bioeng.* **2003**, *96*, 110–118. [CrossRef]
5. Pires, E.J.; Teixeira, J.A.; Brányik, T.; Vicente, A.A. Yeast: The soul of beer's aroma—A review of flavour-active esters and higher alcohols produced by the brewing yeast. *Appl. Microbiol. Biotechnol.* **2014**, *98*, 1937–1949. [CrossRef]
6. Olaniran, A.O.; Hiralal, L.; Mokoena, M.P.; Pillay, B. Flavour-active volatile compounds in beer: Production, regulation and control. *J. Inst. Brew.* **2017**, *123*, 13–23. [CrossRef]
7. Müller, M.; Bellut, K.; Tippmann, J.; Becker, T. Physical Methods for Dealcoholization of Beverage Matrices and their Impact on Quality Attributes. *Chem. Ing. Tech.* **2016**, *88*. [CrossRef]

8. Riu-Aumatell, M.; Miró, P.; Serra-Cayuela, A.; Buxaderas, S.; López-Tamames, E. Assessment of the aroma profiles of low-alcohol beers using HS-SPME-GC-MS. *Food Res. Int.* **2014**, *57*, 196–202. [CrossRef]
9. Basso, R.F.; Alcarde, A.R.; Portugal, C.B. Could non-*Saccharomyces* yeasts contribute on innovative brewing fermentations? *Food Res. Int.* **2016**, *86*, 112–120. [CrossRef]
10. Holt, S.; Mukherjee, V.; Lievens, B.; Verstrepen, K.J.; Thevelein, J.M. Bioflavoring by non-conventional yeasts in sequential beer fermentations. *Food Microbiol.* **2018**, *72*, 55–66. [CrossRef]
11. Bellut, K.; Michel, M.; Hutzler, M.; Zarnkow, M.; Jacob, F.; De Schutter, D.P.; Daenen, L.; Lynch, K.M.; Zannini, E.; Arendt, E.K. Investigation into the application of *Lachancea fermentati* strain KBI 12.1 in low alcohol beer brewing. *J. Am. Soc. Brew. Chem.* **2019**, *77*, 157–169.
12. Bellut, K.; Michel, M.; Zarnkow, M.; Hutzler, M.; Jacob, F.; Lynch, K.M.; Arendt, E.K. On the suitability of alternative cereals, pseudocereals and pulses in the production of alcohol-reduced beers by non-conventional yeasts. *Eur. Food Res. Technol.* **2019**, *245*, 2549–2564. [CrossRef]
13. De Francesco, G.; Sannino, C.; Sileoni, V.; Marconi, O.; Filippucci, S.; Tasselli, G.; Turchetti, B. *Mrakia gelida* in brewing process: An innovative production of low alcohol beer using a psychrophilic yeast strain. *Food Microbiol.* **2018**, *76*, 354–362. [CrossRef] [PubMed]
14. Narziss, L.; Miedaner, H.; Kern, E.; Leibhard, M. Technology and composition of non-alcoholic beers—Processes using arrested fermentation. *Brauwelt Int.* **1992**, *4*, 396–410.
15. Michel, M.; Meier-Dörnberg, T.; Jacob, F.; Methner, F.J.; Wagner, R.S.; Hutzler, M. Review: Pure non-*Saccharomyces* starter cultures for beer fermentation with a focus on secondary metabolites and practical applications. *J. Inst. Brew.* **2016**, *122*, 569–587. [CrossRef]
16. Li, H.; Liu, Y.; Zhang, W. Method for manufacturing alcohol-free beer through *Candida shehatae*. China Patent CN102220198B, 13 May 2011.
17. Yilmaztekin, M.; Erten, H.; Cabaroglu, T. Production of Isoamyl Acetate from Sugar Beet Molasses by *Williopsis saturnus* var. *saturnus*. *J. Inst. Brew.* **2008**, *114*, 34–38. [CrossRef]
18. Inoue, Y.; Fukuda, K.; Wakai, Y.; Sudsai, T.; Kimura, A. Ester Formation by Yeast *Hansenula mrakii* IFO 0895: Contribuition of Esterase fir Iso-Amyl Acetate Production in Sake Brewing. *LWT Food Sci. Technol.* **1994**, *27*, 189–193. [CrossRef]
19. Aung, M.T.; Lee, P.R.; Yu, B.; Liu, S.Q. Cider fermentation with three *Williopsis saturnus* yeast strains and volatile changes. *Ann. Microbiol.* **2015**, *65*, 921–928. [CrossRef]
20. Liu, S.Q.; Quek, A.Y.H. Evaluation of Beer Fermentation with a Novel Yeast *Williopsis saturnus*. *Food Technol. Biotechnol* **2016**, *54*, 403–412. [CrossRef]
21. Strejc, J.; Siříšťová, L.; Karabín, M.; Almeida e Silva, J.B.; Brányik, T. Production of alcohol-free beer with elevated amounts of flavouring compounds using lager yeast mutants. *J. Inst. Brew.* **2013**, *119*, 149–155. [CrossRef]
22. Saerens, S.; Swiegers, J.H. Production of low-alcohol or alcohol-free beer with *Pichia kluyveri* yeast strains. International Patent WO2014135673A2, 7 March 2014.
23. Gibson, B.; Geertman, J.-M.A.; Hittinger, C.T.; Krogerus, K.; Libkind, D.; Louis, E.J.; Magalhães, F.; Sampaio, J.P. New yeasts—New brews: Modern approaches to brewing yeast design and development. *Fems. Yeast Res.* **2017**, *17*, 1–32. [CrossRef] [PubMed]
24. Saison, D.; De Schutter, D.P.; Vanbeneden, N.; Daenen, L.; Delvaux, F.; Delvaux, F.R. Decrease of aged beer aroma by the reducing activity of brewing yeast. *J. Agric. Food Chem.* **2010**, *58*, 3107–3115. [CrossRef] [PubMed]
25. Bendiak, D.; Van Der Aar, P.; Barbero, F.; Benzing, P.; Berndt, R.; Carrick, K.; Dull, C.; Dunn-Default, S.; Eto, M.; Gonzalez, M.; et al. Yeast Flocculation by Absorbance Method. *J. Am. Soc. Brew. Chem.* **1996**, *54*, 245–248.
26. D'Hautcourt, O.; Smart, K.A. Measurement of Brewing Yeast Flocculation. *J. Am. Soc. Brew. Chem.* **1999**, *57*, 123–128. [CrossRef]
27. Meier-Dörnberg, T.; Hutzler, M.; Michel, M.; Methner, F.-J.; Jacob, F. The Importance of a Comparative Characterization of *Saccharomyces Cerevisiae* and *Saccharomyces Pastorianus* Strains for Brewing. *Fermentation* **2017**, *3*, 41. [CrossRef]
28. Das Murtey, M.; Ramasamy, P. Sample Preparations for Scanning Electron Microscopy—Life Sciences. *Intech Open* **2018**, *2*, 64.

29. Stiegl-Freibier Alkoholfrei. Available online: https://www.stiegl-shop.at/braushop/en/shop/beer/non-alcoholic/stiegl-freibier-alcohol-free/?card=4193 (accessed on 12 October 2019).
30. Baltika 0 Alcohol Free Beer. Available online: https://eng.baltika.ru/products/baltika/baltika-0-alcohol-free-beer/ (accessed on 11 October 2019).
31. Kurtzman, C.P.; Fell, J.W.; Boekhout, T. *The Yeasts, A Taxonomic Study*, 5th ed.; Kurtzman, C.P., Fell, J.W., Boekhout, T., Eds.; Elsevier: London, UK, 2011; ISBN 9780444521491.
32. Fukuhara, H. The Kluyver effect revisited. *Fems. Yeast Res.* **2003**, *3*, 327–331. [CrossRef]
33. Michel, M.; Kopecká, J.; Meier-Dörnberg, T.; Zarnkow, M.; Jacob, F.; Hutzler, M. Screening for new brewing yeasts in the non-*Saccharomyces* sector with *Torulaspora delbrueckii* as model. *Yeast* **2016**, *33*, 129–144. [CrossRef]
34. Bellut, K.; Michel, M.; Zarnkow, M.; Hutzler, M.; Jacob, F.; De Schutter, D.P.; Daenen, L.; Lynch, K.M.; Zannini, E.; Arendt, E.K. Application of Non-*Saccharomyces* Yeasts Isolated from Kombucha in the Production of Alcohol-Free Beer. *Fermentation* **2018**, *4*, 66. [CrossRef]
35. Bellut, K.; Krogerus, K.; Arendt, E.K. *Lachancea fermentati* strains isolated from kombucha: fundamental insights, and practical application in low alcohol beer brewing. manuscript in preparation.
36. Peyer, L.C.; Zarnkow, M.; Jacob, F.; De Schutter, D.P. Sour Brewing: Impact of *Lactobacillus amylovorus* FST2.11 on Technological and Quality Attributes of Acid Beers. *J. Am. Soc. Brew. Chem.* **2017**, *75*, 207–216. [CrossRef]
37. Peyer, L.C.; Bellut, K.; Lynch, K.M.; Zarnkow, M.; Jacob, F.; De Schutter, D.P.; Arendt, E.K. Impact of buffering capacity on the acidification of wort by brewing-relevant lactic acid bacteria. *J. Inst. Brew.* **2017**, *123*, 497–505. [CrossRef]
38. Meier-Dörnberg, T.; Hutzler, M. Alcohol-Free Wheat Beer with Maltose Negative Yeast Strain *Saccharomycodes ludwigii*. In 3rd Young Scientists Symposium. Poster No. P.3.5. Available online: https://www.researchgate.net/publication/307992436_AlcoholFree_Wheat_Beer_with_Maltose_Negative_Yeast_Strain_Saccharomycodes_ludwigii (accessed on 14 November 2019).
39. Nyanga, L.K.; Nout, M.J.R.; Gadaga, T.H.; Theelen, B.; Boekhout, T.; Zwietering, M.H. Yeasts and lactic acid bacteria microbiota from masau (*Ziziphus mauritiana*) fruits and their fermented fruit pulp in Zimbabwe. *Int. J. Food Microbiol.* **2007**, *120*, 159–166. [CrossRef] [PubMed]
40. Goffrini, P.; Ferrero, I.; Donnini, C. Respiration-dependent utilization of sugars in yeasts: A determinant role for sugar transporters. *J. Bacteriol.* **2002**, *184*, 427–432. [CrossRef] [PubMed]
41. Meilgaard, M.C. Flavor chemistry in beer: Part II: Flavor and flavor threshold of 239 aroma volatiles. *Master Brew. Assoc. Am. Tech. Q.* **1975**, *12*, 151–168.
42. Holt, S.; Miks, M.H.; de Carvalho, B.T.; Foulquié-Moreno, M.R.; Thevelein, J.M. The molecular biology of fruity and floral aromas in beer and other alcoholic beverages. *Fems Microbiol. Rev.* **2019**, *43*, 193–222. [CrossRef] [PubMed]
43. American Society of Brewing Chemists ASBC Beer Flavors Database. Available online: http://methods.asbcnet.org/extras/flavors_database.pdf (accessed on 10 October 2019).
44. Meilgaard, M.C. Flavor chemistry in beer: Part I: Flavor interaction between principal volatiles. *Master Brew. Assoc. Am. Tech. Q.* **1975**, *12*, 107–117.
45. Lytra, G.; Tempere, S.; Le Floch, A.; De Revel, G.; Barbe, J.C. Study of sensory interactions among red wine fruity esters in a model solution. *J. Agric. Food Chem.* **2013**, *61*, 8504–8513. [CrossRef]
46. Michel, M.; Meier-Dornberg, T.; Jacob, F.; Schneiderbanger, H.; Haselbeck, K.; Zarnkow, M.; Hutzler, M. Optimization of Beer Fermentation with a Novel Brewing Strain *Torulaspora delbrueckii* Using Response Surface Methodology. *Master Brew. Assoc. Am. Tech. Q.* **2017**, *54*, 23–33.
47. Van Rijswijck, I.M.H.; Wolkers-Rooijackers, J.C.M.; Abee, T.; Smid, E.J. Performance of non-conventional yeasts in co-culture with brewers' yeast for steering ethanol and aroma production. *Microb. Biotechnol.* **2017**, *10*, 1591–1602. [CrossRef]
48. Kucharczyk, K.; Tuszyński, T. The effect of pitching rate on fermentation, maturation and flavour compounds of beer produced on an industrial scale. *J. Inst. Brew.* **2015**, *121*, 349–355. [CrossRef]
49. Jonkova, G.N.; Georgieva, N.V. Effect of some technological factors on the content of esters in beer. *Sci. Study Res.* **2009**, *10*, 271–276.
50. Erten, H.; Tanguler, H.; Cakiroz, H. The effect of pitching rate on fermentation and flavour compounds in high gravity brewing. *J. Inst. Brew.* **2007**, *113*, 75–79. [CrossRef]
51. Vriesekoop, F.; Krahl, M.; Hucker, B.; Menz, G. 125th Anniversary review: Bacteria in brewing: The good, the bad and the ugly. *J. Inst. Brew.* **2012**, *118*, 335–345. [CrossRef]

52. Brányik, T.; Silva, D.P.; Baszczyňski, M.; Lehnert, R.; Almeida, E.; Silva, J.B. A review of methods of low alcohol and alcohol-free beer production. *J. Food Eng.* **2012**, *108*, 493–506. [CrossRef]
53. Rachon, G.; Rice, C.J.; Pawlowsky, K.; Raleigh, C.P. Challenging the assumptions around the pasteurisation requirements of beer spoilage bacteria. *J. Inst. Brew.* **2018**, *124*, 443–449. [CrossRef]

© 2019 by the authors. Licensee MDPI, Basel, Switzerland. This article is an open access article distributed under the terms and conditions of the Creative Commons Attribution (CC BY) license (http://creativecommons.org/licenses/by/4.0/).

Article

Evolution of Aromatic Profile of *Torulaspora delbrueckii* Mixed Fermentation at Microbrewery Plant

Laura Canonico [1,*], Enrico Ciani [2], Edoardo Galli [1], Francesca Comitini [1] and Maurizio Ciani [1,*]

[1] Department of Life and Environmental Sciences, Marche Polytechnic University, Via Brecce Bianche, 60131 Ancona, Italy; edogah@hotmail.it (E.G.); f.comitini@univpm.it (F.C.)
[2] Birra dell'Eremo, Microbrewery, Via Monte Peglia, 5, 06081 Assisi, Italy; enrico@birradelleremo.it
* Correspondence: l.canonico@univpm.it (L.C.); m.ciani@univpm.it (M.C.); Tel.: +39-071-2204150 (L.C.); +39-071-2204987 (M.C.)

Received: 10 December 2019; Accepted: 6 January 2020; Published: 8 January 2020

Abstract: Nowadays, consumers require quality beer with peculiar organoleptic characteristics and fermentation management has a fundamental role in the production of aromatic compounds and in the overall beer quality. A strategy to achieve this goal is the use of non-conventional yeasts. In this context, the use of *Torulaspora delbrueckii* was proposed in the brewing process as a suitable strain to obtain a product with a distinctive aromatic taste. In the present work, *Saccharomyces cerevisiae*/*T. delbrueckii* mixed fermentation was investigated at a microbrewery plant monitoring the evolution of the main aromatic compounds. The results indicated a suitable behavior of this non-conventional yeast in a production plant. Indeed, the duration of the process was very closed to that exhibited by *S. cerevisiae* pure fermentation. Moreover, mixed fermentation showed an increase of some aromatic compounds as ethyl hexanoate, α-terpineol, and β-phenyl ethanol. The enhancement of aromatic compounds was confirmed by the sensory evaluation carried out by trained testers. Indeed, the beers produced by mixed fermentation showed an emphasized note of fruity/citric and fruity/esters notes and did not show aroma defects.

Keywords: *Torulaspora delbrueckii*; craft beer; microbrewery plant; mixed fermentation; aroma profile

1. Introduction

In the last years, there has been a worldwide growth in microbreweries, which leads to competition in the beer market to find new beers and also those that are characterized by peculiar aroma taste. To achieve this, the brewers paid attention to the ingredients which are water, malts, hops, and yeast [1–4]. In particular, the brewers focused their attention on the yeast strains to use in brewing fermentation which are selected not only for their good fermentation efficiency but also for their characteristic aroma and flavors.

In this regard, several recent investigations were focused on the selection of non-conventional yeasts [5–8]. Non-*Saccharomyces* yeasts represent a large source of biodiversity to produce new beer styles. In the last years, different non-*Saccharomyces* yeasts were proposed in brewing, such as *Brettanomyces bruxellensis*, *Torulaspora delbrueckii*, *Candida shehatae*, *Candida tropicalis*, *Zygosaccharomyces rouxii*, *Lachancea thermotolerans*, *Saccharomycodes ludwigii*, and *Pichia kluyveri* [9–13]. *T. delbrueckii* is one of the most well-known non-*Saccharomyces* yeasts and it can be found in wild environments such as plants and soils as well as in wine or in fermented food processes. In the brewing process, *T. delbrueckii* received particular attention due to its ability to ferment maltose, produce ester compounds, and biotransform the monoterpenoid flavor compounds of hops [12,14–16]. In particular, *T. delbrueckii* can improve the amount of different fruity aromas, such as β-phenyl ethanol ("rose" flavors), n-propanol, iso-butanol, amyl alcohol ("solvent brandy" aroma), and ethyl acetate [17–19].

Canonico et al. [6,12] evaluated the use of *T delbrueckii* for beer production, both pure and in mixed cultures with different *S. cerevisiae* starter strains. *T. delbrueckii* in mixed fermentation with different *S. cerevisiae* starter strains showed different behavior and resulting in beers with distinctive flavors. Generally, the main aromatic compounds that were affected by *T. delbrueckii* are some fruity esters. Furthermore, in mixed fermentation, *T. delbrueckii* provided higher levels of higher alcohols, in contrast to data obtained in winemaking, where higher alcohols had lower levels. Moreover, beers obtained with *T. delbrueckii* pure cultures were characterized by a distinctive analytical, aromatic profile, and a low alcohol content (2.66% *v/v*) [12].

Michel et al. [16] investigated different *T. delbrueckii* strains coming from different habitats. One strain was able to produce a fruity and floral aroma (β-phenyl ethanol) and amyl alcohols. Furthermore, two strains were found to be suitable for producing low-alcohol beer owing to their inability to ferment maltose and maltotriose but still produced good flavor. However, investigation into the use of non-conventional yeasts in the brewing process has been performed at a laboratory scale or at a pilot scale while validation trials are lacking at the industrial level, which would give a more accurate assessment of their brewing ability. For this reason and based on the results of previous investigations [6,12] in this study, the contribution of *T. delbreuckii* in mixed fermentation with *S. cerevisiae* starter strain at inoculum ratio 1:20 was assessed at the microbrewery plant. The effect of this non-conventional yeast in mixed fermentation on the evolution of biomass and aroma profile as well as on the final beer composition was evaluated. The sensorial profile of the final beers was also tested.

2. Materials and Methods

2.1. Yeast Strains

T. delbrueckii DiSVA 254 comes from the Yeast Collection of the Department of Life and Environmental Sciences (DiSVA) of the Polytechnic University of Marche (Italy). *T. delbrueckii* strain DiSVA 254 and *S. cerevisiae* commercial strain US-05 (Fermentis, Lesaffre, Marcq En Baroeul, France) were used in mixed fermentation at inoculum ratio 20:1 as reported in a previous study [12]. The US-05 was rehydrated following the manufacturer's instructions and was plated on YPD agar medium at 25 °C, by spreading 0.1 mL yeast suspension onto the surface of the medium.

The yeast strains were maintained on yeast extract (10 g/L), peptone (20 g/L), dextrose (20 g/L), (YPD) agar (18 g/L) at 4 °C, for short-term storage, and in YPD liquid with 80% (*w/v*) glycerol at −80 °C for long-term storage.

2.2. Wort Production and Fermentation Condition

The wort used for the trials was produced at Birra dell'Eremo Microbrewery (Assisi, Italy) from a batch of 1500 L in duplicate fermentations. The wort was made with pilsner malt (100%), the Cascade hop variety, and produced according to the scheme reported by Canonico et al. [6]. The main analytical characters of this wort were pH 5.5, specific gravity 12.3° GPlato, and 20 IBU. The fermentation process was carried out in 2 different batches of 1500 L at 20 °C.

2.3. Growth Kinetics

The biomass evolution was monitored during the fermentation process using viable cell counts on WL Nutrient Agar (Oxoid, Hampshire, UK) and Lysine Agar (Oxoid, Hampshire, UK). Lysine Agar is a medium unable to support the growth of *S. cerevisiae* [20] for the differentiation of *T. delbrueckii* yeast from *S. cerevisiae* US-05 starter strain.

2.4. Bottle Conditioning

At the end of the fermentation process, the beers obtained were transferred into 500-mL bottles, adding 5.5 g/L of sucrose. The secondary fermentation in the bottle was carried out at 18–20 °C for 7–10 days.

2.5. Analytical Procedures

The contents of acetaldehyde, ethyl acetate, higher alcohols (n-propanol, isobutanol, amyl alcohol, isoamyl alcohol) were determined by direct injection into a gas–liquid chromatography system. The volatile compounds were determined by the solid-phase microextraction (HS-SPME) method. Five ml of each sample was placed in a vial containing 1g NaCl closed with a septum-type cap. HS-SPME was carried out under magnetic stirring for 10 min at 25 °C. After this period, an amount of 3-octanol as the internal standard (1.6 mg/L) was added and the solution was heated to 40 °C and extracted with a fiber Divinylbenzene/Carboxen/Polydimethylsiloxane (DVB/CAR/PDMS) for 30 min by insertion into the vial headspace. The compounds were desorbed by inserting the fiber into the Shimadzu gas chromatograph GC injector for 5 min. A glass capillary column was used: 0.25 µm Supelcowax 10 (length, 60 m; internal diameter, 0.32 mm). The fiber was inserted in split–splitless mode: 60 s splitless; the temperature of injection, 220 °C; the temperature of detector, 250 °C; carrier gas, with nitrogen; flow rate, 2.5 mL/min. The temperature program was 50 °C for 5 min, 3 °C/min to 220 °C, and then 220 °C for 20 min. The compounds were identified and quantified by comparisons with external calibration curves for each compound.

2.6. Sensorial Analysis

At the end of the fermentation process, the beers obtained were transferred into 330-mL bottles, adding 5.5 g/L sucrose. The secondary fermentation in the bottle was carried out at 18–20 °C for 7–10 days. After this period, the beers were stored at 4 °C underwent sensory analysis using a scale from 1 to 10 (Analytica EBC, 1997). This was carried out by a group of 14 trained testers, that evaluated the main aromatic notes regarding the olfactory and gustatory perception and structural features. The data were elaborated with statistical analyses to obtained information about the contribution of each descriptor on the organoleptic quality of beer.

2.7. Statistical Analysis

Analysis of variance (ANOVA) was applied to the main characteristics of the beers. The means were analyzed using the STATISTICA 7 software. The significant differences were determined by the means of Duncan tests, and the results were considered significant if the associated p-Values were < 0.05. The results of the sensory analysis were also subjected to Fisher ANOVA, to determine the significant differences with a p-Value < 0.05.

3. Results

3.1. Yeast Species Evolution

The growth kinetics of *T. delbrueckii* in mixed fermentations and *S. cerevisiae* pure culture were reported in Figure 1.

The growth kinetics of the *S. cerevisiae* US-05 pure cultures achieved ca. 10^7 CFU/mL at 3 days of fermentation and decreased at 10^6 CFU/mL until the end of fermentation. Regarding the mixed fermentation, *S. cerevisiae* reached cell concentrations <106 CFU/mL at 3 days of fermentation and decreased at 10^5 CFU/mL, while *T. delbrueckii*, started at a concentration >10^6 CFU/mL, achieved the maximum cell concentration at 3 days of fermentation (107 CFU/mL), and decreased at the end of fermentation (10^6 CFU/mL). The results for mixed fermentation indicated that *T. delbrueckii* at

20-folds higher than *S. cerevisiae* dominated the fermentation process and highlighted a high level of competitiveness of *T. delbrueckii* towards *S. cerevisiae* commercial strain.

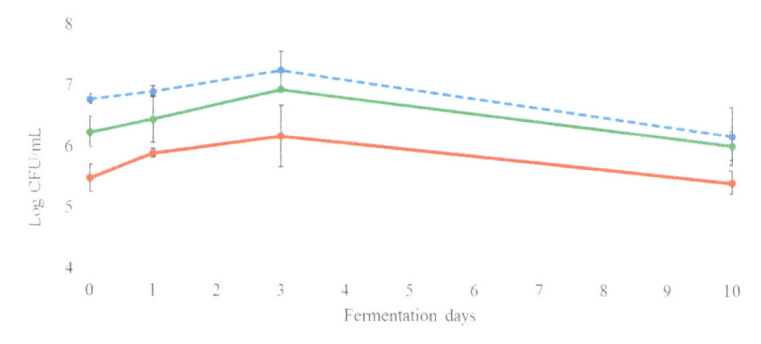

Figure 1. Growth kinetics of pure and mixed fermentation. Pure culture of *S. cerevisiae* (——), *S. cerevisiae* (——), and *T. delbrueckii* (- - -) individually for the mixed fermentation.

3.2. Main Analytical Profile

The analytical compositions of the beers are reported in Table 1.

Table 1. The main analytical characteristics of the beer produced by pure and mixed fermentations.

Fermentation	Wort Gravity Attenuation (°P)	Ethanol % v/v	Residual Sucrose g/L	Residual Glucose g/L	Residual Maltose g/L
S. cerevisiae pure culture	2.85 ± 0.00	4.75 ± 0.03	0.00 ± 0.00	0.00 ± 0.00	0.06 ± 0.02
S. cerevisiae/T. delbrueckii	2.79 ± 0.17	4.68 ± 0.05	0.24 ± 0.02	0.02 ± 0.01	1.36 ± 0.14

Data are means ± standard deviation. The initial composition of the sugars in the wort was sucrose 5.9 g/L; glucose 8.2 g/L; maltose 61.76 g/L. The wort gravity at the start was 12.3 °P.

Both trials finished the process on the 10th day of fermentation highlighting that *T. delbrueckii* in the condition used at the microbrewery plant did not influence the time of the fermentation process.

S. cerevisiae/T. delbrueckii mixed fermentation and *S. cerevisiae* pure fermentation, produce beer with a comparable amount of ethanol content and final values of °P. Regarding the residual sugar, both fermentation trials consumed all sucrose and glucose content, while beer brewed by *S. cerevisiae/T. delbrueckii* mixed fermentation exhibited a slightly higher amount of maltose.

3.3. By-Products and Volatile Compounds

The main volatile compound by-products are reported in Table 2.

For the main volatile compound by-products, *S. cerevisiae* US-05/*T. delbrueckii* mixed fermentations showed different profiles to those produced by *S. cerevisiae* US-05 pure fermentation. In particular, the evolution of the main aroma compounds during the fermentation process showed that β-phenyl ethanol significantly increases in mixed fermentation in all steps of the fermentation process if compared with *S. cerevisiae* starter strain pure culture. Differently, there were no significant differences between the trials for amyl and isoamyl alcohol content with the exception of *S. cerevisiae* pure culture trials, which exhibited a lower amount of these two alcohols at the first step of fermentation (after one day).

Table 2. Evolution of the main volatile compounds (mg/L) in the beer produced by pure (*S. cerevisiae*) and mixed fermentation (*S. cerevisiae/T. delbrueckii*) during the fermentation process. Data are means ± standard deviation. Data with different superscript letters ([a,b,c,d]) within each column are significantly different (Duncan tests; $p < 0.05$). ND: not detected.

Fermentation	Alcohols				Carbonyl Compound			Esters				Terpene
	n-Propanol	Isobutanol	Amyl Alcohol	Isoamyl Alcohol	β-Phenyl Ethanol	Acetaldehyde	Ethyl Acetate	Ethyl Butyrate	Phenylethyl Acetate	Ethyl Hexanoate	Ethyl Octanoate	Isoamyl Acetate	α-Terpineol
S. cerevisiae 1 day fermentation	ND	4.51 ± 0.30 [c]	2.56 ± 0.25 [b]	17.85 ± 0.06 [b]	3.5 ± 0.02 [e]	15.72 ± 0.43 [a]	ND	0.144 ± 0.012 [b]	ND	ND	0.002 ± 0.001 [b]	1.70 ± 0.75 [ab]	ND
S. cerevisiae 3 days of fermentation	20.86 ± 0.29 [c]	13.93 ± 0.31 [b]	6.42 ± 0.60 [a]	45.82 ± 0.52 [a]	15.77 ± 1.08 [c]	11.44 ± 1.03 [b]	3.83 ± 0.45 [b]	0.738 ± 0.123 [a]	ND	ND	0.043 ± 0.001 [a]	2.79 ± 0.32 [ab]	ND
S. cerevisiae 10 days fermentation	28.09 ± 0.16 [a]	27.54 ± 0.24 [a]	7.28 ± 0.13 [a]	46.72 ± 0.48 [a]	17.39 ± 2.1 [c]	4.44 ± 0.98 [d]	7.42 ± 0.35 [a]	0.754 ± 0.048 [a]	ND	0.05 ± 0.03 [b]	0.044 ± 0.003 [a]	4.00 ± 0.24 [a]	ND
S. cerevisiae/T. delbrueckii 1 day of fermentation	24.47 ± 0.40 [b]	3.25 ± 0.15 [d]	6.85 ± 0.17 [a]	46.27 ± 0.19 [a]	2.15 ± 0.29 [d]	7.94 ± 1.36 [c]	5.62 ± 0.97 [a]	0.15 ± 0.03 [b]	0.060 ± 0.001 [c]	0.054 ± 0.006 [b]	0.001 ± 0.001 [b]	0.32 ± 0.0 [b]	0.082 ± 0.01 [b]
S. cerevisiae/T. delbrueckii 3 days of fermentation	26.28 ± 0.22 [ab]	27.44 ± 0.33 [a]	7.06 ± 0.15 [a]	46.49 ± 0.25 [a]	36.37 ± 1.02 [b]	6.19 ± 0.88 [c]	6.52 ± 0.84 [a]	0.367 ± 0.12 [ab]	0.405 ± 0.023 [b]	0.143 ± 0.08 [a]	0.002 ± 0.002 [b]	3.85 ± 0.14 [a]	0.123 ± 0.03 [b]
S. cerevisiae/T. delbrueckii 10 days of fermentation	25.38 ± 0.17 [b]	27.52 ± 0.17 [a]	6.96 ± 0.12 [a]	46.38 ± 0.89 [a]	42.71 ± 0.98 [a]	7.07 ± 0.85 [c]	6.07 ± 0.73 [a]	0.770 ± 0.03 [a]	0.987 ± 0.124 [a]	0.157 ± 0.09 [a]	0.02 ± 0.002 [a]	3.56 ± 0.25 [a]	0.163 ± 0.03 [a]

Acetaldehyde content showed a different trend: pure culture trials showed a progressive reduction of this carbonyl compound during the fermentation, while the *S. cerevisiae/T. delbrueckii* mixed fermentation exhibited the same acetaldehyde content during the process. Ethyl acetate and ethyl hexanoate were detected only in mixed fermentation until the beginning of fermentation. The same trend was also exhibited by α-terpineol. Moreover, regarding ethyl hexanoate and α-terpineol, there was a significant increase at the end of fermentation. For isoamyl acetate content, the results did not show a significant difference between the two fermentations.

3.4. Sensory Analysis

The beers obtained by pure and mixed fermentations underwent sensory analysis, and the results were illustrated in Figure 2.

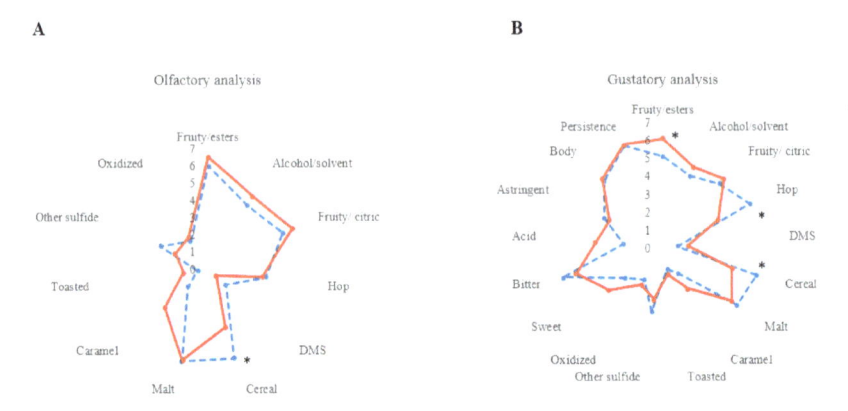

Figure 2. Sensory analysis of beer produced in the microbrewery plant by the *T. delbrueckii* mixed fermentation. From pure cultures of *S. cerevisiae* (‐•‐) and mixed cultures of *S. cerevisiae/T. delbrueckii* (▬). (**A**) Olfactory analysis; (**B**) gustatory analysis. DMS, dimethyl sulfide. * = Significantly different (Fisher ANOVA).

All of the beers analyzed showed significant differences for their main aromatic notes regarding the olfactory and gustatory analysis. In particular, for the main sensorial descriptors, the data showed that the beer obtained with the mixed fermentation was significantly different from that of the *S. cerevisiae* US-05 starter strain for a variety of the sensorial characteristics. Regarding olfactory analysis (Figure 2A), beers brewed with *S. cerevisiae/T. delbrueckii* mixed fermentation showed a bouquet with notes that emphasized the fruity/esters, fruity/citric, and caramel. Moreover, the perception of DMS (dimethyl sulfide) and other sulfide compounds shows they are less well perceived than beers obtained by *S. cerevisiae* pure culture. However, the only significant difference between the two beers was exhibited by the cereal note, which resulted in the emphasis of the product brewed with the *S. cerevisiae* starter strain.

Regarding gustatory analysis (Figure 2B), the beers obtained by mixed fermentation are characterized by the significant perception of fruity/esters notes. The beers obtained with *S. cerevisiae* pure culture were significantly characterized by hop and cereal notes.

In addition, the beers produced by *T. delbrueckii* mixed fermentation were characterized by a pale yellow color, clarity, and persistent and compact foam, which are very important features in the assessment of the quality of a beer (data not shown).

4. Discussion

The use of non-conventional yeasts in the brewing process was recently proposed with the aim to produce beers with distinctive aromatics note or to develop a new technology to increase the typicity of specialty beers such as low-calorie beer, low alcohol beer, novel flavored beer, and gluten-free beer [8,12,13,16,21–24]. In previous studies, the use of *T. delbrueckii* (strain DiSVA 254) in mixed fermentation with *S. cerevisiae* starter strains was investigated at a laboratory scale [6,12]. The results indicated a promising behavior of this yeast for use at microbrewery plants. Indeed, the interactions between *S. cerevisiae* and *T. delbrueckii* produced beers characterized by a distinctive aromatic profile (fruity/citric notes, fruity/esters notes, and full-bodied attributes) [12]. For these reasons, the application of *S. cerevisiae/T. delbrueckii* was assessed at the microbrewery plant evaluating the evolution of the volatile compounds during the fermentation process. The first relevant aspect for its application at the industrial level was the duration of fermentation. Similarly, to the laboratory-scale trials, the brewing process carried out with *T. delbrueckii* mixed fermentation showed a comparable fermentation time to that exhibited by the *S. cerevisiae* starter strain showing good competitiveness with *S. cerevisiae* in the co-culture. This aspect is crucial for its application in a microbrewery where for economic reasons the fermentation process should not exceed 10–15 days.

Regarding the evolution of aroma compounds, generally higher alcohols did not show a significant difference between mixed and pure fermentations. Regarding β-phenyl ethanol content, known for the rose and floral aroma with an odor threshold of 10 mg/L [18], the results showed an increase in mixed fermentation exhibiting a different trend by a previous study [6,12]. An increase of β-phenyl ethanol was observed by Toh et al. [25] and Drosou et al. [26] highlighted that the production of this compound was determined by *T. delbrueckii* and *S. cerevisiae* strains used in fermentation but also by the fermentation condition.

Esters compounds produced by esterification between alcohol and short- or long-chain fatty acids are important compounds that can affect the aroma of the beer [27]. Phenyl ethyl acetate is known for the floral, sweet, honey, and fruity aroma with a threshold of 3.8 mg/L [16] was significantly affected by the presence of *T. delbrueckii* as previously reported [6,12,26].

This study, confirming a previous study [12], showed the significant increase of ethyl hexanoate, fruity esters associated with apple flavor [27], when *T. delbrueckii* was used in mixed fermentation while a different trend was observed with different *T. delbrueckii* and *S. cerevisiae* strains [25,26]. α-terpineol, the terpene responsible for balsamic/fruit notes, was detected only in mixed fermentation and highlighted that these aroma compounds were related to *T. delbrueckii*.

Regarding the evolution of the main aroma compounds during fermentation, the *S. cerevisiae* pure culture and mixed fermentation exhibited a different trend. In particular, the evolution of acetaldehyde content is related to a different metabolic pathway of *S. cerevisiae* and *T. delbrueckii*. Indeed, in *S. cerevisiae* fermentation the acetaldehyde content decreased during the fermentation process, while in mixed fermentation the content of this carbonyl compound remains similar from beginning to end. The same trend was also exhibited for the main alcohol compounds.

Few works are present in the literature regarding the application of *T. delbrueckii* in the brewing process and there are no data about its use at the industrial level. These results confirming the fermentation behavior of *T. delbrueckii* in mixed fermentation, emphasize and reinforce its possible use at the industrial level allowing one to obtain beers with characteristics different from those obtained with *S. cerevisiae* starter strains and with a sensory profile appreciated by tasters.

Author Contributions: L.C., E.C., E.G., F.C., and M.C. contributed equally to this manuscript. All authors participated in the design and discussion of the research. L.C. carried out the experimental part of the work. E.C. carried out the fermentation in the microbrewery. L.C., E.C., E.G., F.C., and M.C. carried out the analysis of the data and wrote the manuscript. All authors have read and agreed to the published version of the manuscript.

Funding: This research received no external funding.

Acknowledgments: The authors wish to thank the Birra dell'Eremo Microbrewery (Assisi, Italy) for making the microbrewery available and for supporting the technical experimental design. Moreover, thanks go to the UNIONBIRRAI association (Milano, Italy) and the trained testers belonging to UNINOBIRRAI BEER TASTERS (UBT Marche region) for helping the authors to complete the study for an industrial application.

Conflicts of Interest: The authors declare no conflict of interest.

References

1. Bernstein, J.M. *Imbibe Magazine*; Portland, OR, USA, 2010; Volume 2017. Available online: https://www.bjcp.org/ (accessed on 20 November 2019).
2. Brungard, M. Calcium and magnesium in brewing water. *New Brew.* **2014**, *31*, 80–88.
3. So, A. Developing new barely varities: A work in progress. *New Brew.* **2014**, *31*, 60–68.
4. Osburn, K.; Ahmad, N.N.; Bochman, M.L. Bio-prospecting, selection, and analysis of wild yeasts for ethanol fermentation. *Zymurgy* **2016**, *39*, 81–88.
5. van Rijswijck, I.M.; Wolkers-Rooijackers, J.C.; Abee, T.; Smid, E.J. Performance of non-conventional yeasts in co-culture with brewers' yeast for steering ethanol and aroma production. *Microb. Biotechnol.* **2017**, *10*, 1591–1602. [CrossRef] [PubMed]
6. Canonico, L.; Comitini, F.; Ciani, M. *Torulaspora delbrueckii* contribution in mixed brewing fermentations with different *Saccharomyces cerevisiae* strains. *Int. J. Food Microbiol.* **2017**, *259*, 7–13. [CrossRef] [PubMed]
7. Osburn, K.; Amaral, J.; Metcalf, S.R.; Nickens, D.M.; Rogers, C.M.; Sausen, C.; Caputo, R.; Miller, J.; Li, H.; Tennessen, J.M.; et al. Primary souring: A novel bacteria-free method for sour beer production. *Food Microbial.* **2018**, *70*, 76–84. [CrossRef]
8. Canonico, L.; Galli, E.; Ciani, E.; Comitini, F.; Ciani, M. Exploitation of Three Non-Conventional Yeast Species in the Brewing Process. *Microorganisms* **2019**, *7*, 11. [CrossRef]
9. Sarens, S.; Swiegers, J.H. Enhancement of Beer Flavor by a Combination of *Pichia* Yeast and Different Hop Varieties. U.S. Patent Application No. 14/241,761, 21 August 2014.
10. De Francesco, G.; Turchetti, B.; Sileoni, V.; Marconi, O.; Perretti, G. Screening of new strains of *Saccharomycodes ludwigii* and *Zygosaccharomyces rouxii* to produce low-alcohol beer. *J. Inst. Brew.* **2015**, *121*, 113–121. [CrossRef]
11. Domizio, P.; House, J.F.; Joseph, C.M.L.; Bisson, L.F.; Bamforth, C.W. *Lachancea thermotolerans* as an alternative yeast for the production of beer. *J. Inst. Brew.* **2016**, *122*, 599–604. [CrossRef]
12. Canonico, L.; Agarbati, A.; Comitini, F.; Ciani, M. *Torulaspora delbrueckii* in the brewing process: A new approach to enhance bioflavour and to reduce ethanol content. *Food Microbial.* **2016**, *56*, 45–51. [CrossRef]
13. Holt, S.; Mukherjee, V.; Lievens, B.; Verstrepen, K.J.; Thevelein, J.M. Bioflavoring by non-conventional yeasts in sequential beer fermentations. *Food Microbial.* **2018**, *72*, 55–66. [CrossRef] [PubMed]
14. King, A.; Dickinson, J.R. Biotransformation of monoterpene alcohols by *Saccharomyces cerevisiae*, *Torulaspora delbrueckii* and *Kluyveromyces lactis*. *Yeast* **2000**, *16*, 499–506. [CrossRef]
15. Tataridis, P.; Kanelis, A.; Logotetis, S.; Nerancis, E. Use of non-*Saccharomyces Torulaspora delbrueckii* yeast strains in winemaking and brewing. *Zb. Matitse Srp. Prir. Nauke* **2013**, *124*, 415–426. [CrossRef]
16. Michel, M.; Kopecká, J.; Meier-Dörnberg, T.; Zarnkow, M.; Jacob, F.; Hutzler, M. Screening for new brewing yeasts in the non-*Saccharomyces* sector with *Torulaspora delbrueckii* as model. *Yeast* **2016**, *33*, 129–144. [CrossRef]
17. Pires, E.J.; Teixeira, J.A.; Brányik, T.; Vincente, A.A. Yeast: The soul of beer's aroma and a review of flavour-active esters and higher alcohols produced by the brewing yeast. *Appl. Microbiol. Biotechnol.* **2014**, *98*, 1937–1949. [CrossRef]
18. Etschmann, M.; Huth, I.; Walisko, R.; Schuster, J.; Krull, R.; Holtmann, D.; Wittmann, C.; Schrader, J. Improving 2-phenylethanol and 6-pentyl-α-pyrone production with fungi by microparticle-enhanced cultivation (MPEC). *Yeast* **2015**, *32*, 145–157. [CrossRef]
19. Basso, R.F.; Alcarde, A.R.; Portugal, C.B. Could non-*Saccharomyces* yeasts contribute on innovative brewing fermentations? *Food Res. Int.* **2016**, *86*, 112–120. [CrossRef]
20. Lin, Y. Detection of wild yeasts in brewery. Efficiency of differential media. *J. Inst. Brew.* **1975**, *81*, 410–417. [CrossRef]
21. Petruzzi, L.; Carbo, M.; Sinigaglia, M.; Bevilacqua, A. Brewers' yeast in controlled and uncontrolled fermentation, with a focus on novel, non-conventional and superior strains. *Food Rev. Int.* **2016**, *32*, 341–363. [CrossRef]
22. Michel, M.; Meier-Dörnberg, T.; Jacob, F.; Schneiderbanger, H.; Haselbeck, K.; Zarnkow, M.; Hutzler, M. Optimization of beer fermentation with a novel brewing strain *Torulaspora delbrueckii* using response surface methodology. *MBAA TQ* **2017**. [CrossRef]
23. Senkarcinova, B.; Dias, I.A.G.; Nespor, J.; Branyik, T. Probiotic alcohol-free beer made with *Saccharomyces cerevisiae* var. boulardii. *LWT* **2019**, *100*, 362–367. [CrossRef]

24. Sannino, C.; Mezzasoma, A.; Buzzini, P.; Turchetti, B. Non-conventional Yeasts for Producing Alternative Beers. In *Non-Conventional Yeasts: From Basic Research to Application*; Springer: Cham, Germany, 2019; pp. 361–388.
25. Toh, D.W.K.; Chua, J.Y.; Liu, S.Q. Impact of simultaneous fermentation with *Saccharomyces cerevisiae* and *Torulaspora delbrueckii* on volatile and non-volatile constituents in beer. LWT **2018**, *91*, 26–33. [CrossRef]
26. Drosou, F.; Tataridis, P.; Oreopoulou, V.; Dourtoglou, V. Use of wine Non-*Saccharomyces* yeasts in brewing. In Proceedings of the 36th European Brwery Convention, Antwerpen, Belgium, 2–6 June 2019.
27. Nykänen, L.; Suomalainen, H. *Aroma of Beer, Wine and Distilled Alcoholic Beverages*; Springer Science & Business Media, Reidel: Dordrecht, The Netherlands, 1983.

© 2020 by the authors. Licensee MDPI, Basel, Switzerland. This article is an open access article distributed under the terms and conditions of the Creative Commons Attribution (CC BY) license (http://creativecommons.org/licenses/by/4.0/).

Communication

Adaptive Evolution of Industrial Brewer's Yeast Strains towards a Snowflake Phenotype

Yeseren Kayacan [1,2,†], Thijs Van Mieghem [1,2,†], Filip Delvaux [3], Freddy R. Delvaux [3] and Ronnie Willaert [1,2,4,*]

[1] Alliance Research Group VUB-UGent NanoMicrobiology (NAMI), Research Group Structural Biology, 1050 Brussels, Belgium; Yeseren.Kayacan@vub.be (Y.K.); thijs.van.mieghem@hotmail.com (T.V.M.)
[2] International Joint Research Group VUB-EPFL NanoBiotechnology & NanoMedicine (NANO), Vrije Universiteit Brussel, 1050 Brussels, Belgium
[3] Biercentrum Delvaux, 3040 Neerijse, Belgium; filip@biercentrum.be (F.D.); Freddy.Delvaux@biw.kuleuven.be (F.R.D.)
[4] Department Bioscience Engineering, University Antwerp, 2020 Antwerp, Belgium
* Correspondence: Ronnie.Willaert@vub.be; Tel.: +32-2629-1846
† These authors contributed equally to this work.

Received: 10 January 2020; Accepted: 3 February 2020; Published: 5 February 2020

Abstract: Flocculation or cell aggregation is a well-appreciated characteristic of industrial brewer's strains, since it allows removal of the cells from the beer in a cost-efficient and environmentally-friendly manner. However, many industrial strains are non-flocculent and genetic interference to increase the flocculation characteristics are not appreciated by the consumers. We applied adaptive laboratory evolution (ALE) to three non-flocculent, industrial *Saccharomyces cerevisiae* brewer's strains using small continuous bioreactors (ministats) to obtain an aggregative phenotype, i.e., the "snowflake" phenotype. These aggregates could increase yeast sedimentation considerably. We evaluated the performance of these evolved strains and their produced flavor during lab scale beer fermentations. The small aggregates did not result in a premature sedimentation during the fermentation and did not result in major flavor changes of the produced beer. These results show that ALE could be used to increase the sedimentation behavior of non-flocculent brewer's strains.

Keywords: *Saccharomyces cerevisiae*; industrial brewer's strains; adaptive laboratory evolution (ALE); snowflake phenotype; beer fermentation

---

1. Introduction

Bulk sedimentation of yeast cells during fermentation is a crucial part of the brewing process. At the end of the fermentation, single yeast cells aggregate and form macroscopic "flocs" [1,2]. These clumps of cells then rapidly sediment from the beer and can be harvested from the bottom (lager fermentation) or float and can be harvested from the top (open ale fermentation) at the end of the primary fermentation. This phenomenon allows the brewer to separate the yeast from the beer in an effective, cost-efficient, and environmentally-friendly way, leaving only the clear and almost cell-free product. The neatly harvested yeast can also be "repitched" into the next fermentation. The timing of sedimentation is of considerable importance to the process. Sedimentation should not take place prematurely and cause stuck fermentation leading to beers with low quality flavor profiles. Complete sedimentation at the end of fermentation is preferred by the brewer, which provides the opportunity for a neat separation of the yeast cells from the beer [3].

Flocculation is the reversible, asexual self-adhesion of yeast cells which leads to their sedimentation [4]. Although single yeast cells do sediment, the large clumps formed by flocculating cells sediment at a much higher rate. The flocculation capacity of yeast is highly strain-dependent, influenced mainly by the genetic background since expressed cell wall flocculins that are lectins, effectuate cell-cell binding [5–7]. However, several environmental factors can affect flocculation too. Calcium availability, pH, temperature, ethanol concentration and oxygen concentration are some of the physiological factors that influence flocculation while physical factors such as cell surface hydrophobicity and favorable hydrodynamic conditions can also affect formation of flocs [8–10]. The sedimentation rate is dependent on the size, shape and density of these flocs.

Optimization of the brewer yeast towards a more flocculating phenotype can lead to a more efficient beer production and a higher final beer quality. Recent advances in DNA sequencing, high-throughput technologies and genetic manipulation methods have led to the molecular and genomic characterization of the brewer's yeast. However, the exponential increase in knowledge generated in the field of functional genomics of yeast can only facilitate strain improvement efforts to some degree. Procedures to obtain approval for modified GMO yeasts are complicated and consumer acceptance for a GMO-produced beer is lacking [11,12]. These hurdles have guided researchers to look elsewhere to generate strains with desired properties.

One attractive more "natural" approach to enhance the attributes of microorganisms is the adaptive laboratory evolution (ALE) approach [13]. In ALE, microorganisms are cultivated under clearly defined conditions for long periods of time, allowing metabolic engineering of microorganisms utilizing genetic variation and selection for beneficial mutations [14,15]. Already a more-and-more used tool in microbial strain improvement, ALE has been applied for improving yeast strains such as for the utilization of alternative sugars by *S. cerevisiae* [16], increasing tolerance of *S. cerevisiae* to environmental conditions [17], for increasing the fermentation capacity of a lager *S. pastorianus* brewing strain under hyperosmotic conditions [18], modifying the production of flavor compounds by *S. pastorianus* for alcohol-free beer production [19], the adaptation of lager strains to very high-gravity brewing conditions [20,21], and for enhancing the fermentation rate with decreased formation of acetate and greater production of fermentative aroma of *S. cerevisiae* wine strains [22,23].

Microbial cells can be cultivated in parallel serial cultures for ALE but varying population densities, fluctuating growth rate, nutrient supply, and environmental conditions characterize this batch cultivation. Continuous (chemostat) cultures, however, ensure more stable conditions such as constant growth rate, tightly controlled nutrient supply and stable pH and oxygen availability [24–26].

Previously, we performed ALE of a *S. cerevisiae* strain in a 3D-printed continuous mini tower fermentor using gravity as a selective pressure to obtain a snowflake phenotype [27]. In this work, we've used the ALE approach for the continuous cultivation of three non-flocculating industrial *S. cerevisiae* brewing strains in miniature chemostats (ministats) [28]. This simple and low-cost setup was used to carry out adaptive evolution experiments where gravity is also the selective pressure on the planktonic cells, which are continuously removed while aggregating cells' sediment are retained. Stable aggregating cells were observed during continuous cultivation, showing a "snowflake" phenotype of unseparated daughter and mother cells. Finally, we have also demonstrated the beer fermentation performance of these evolved strains in lab scale tall tubes fermentors.

2. Materials and Methods

2.1. Yeast Strains and Media

The industrial *Saccharomyces cerevisiae* brewer's strains BCD1, BCD2, BCD3, and BCD4 were provided by Biercentrum Delvaux (Neerijse, Belgium). The lab strains BY4742 [29], BY4742 [*FLO1*], and BY4742::*FLO8* [5] and the strong flocculating industrial strain BCD4 were used as control strains in the flocculation assay. All strains were precultured in YPD (Yeast extract–peptone–dextrose) medium (1% m/v yeast extract, 2% m/v peptone, 4% m/v glucose) overnight at 30 °C. For the continuous ALE in ministats, a high-glucose medium (100 g/L D-glucose, 4 g/L (NH$_4$)SO$_4$, 1.5 g/L KH$_2$PO$_4$, 1 g/L MgSO$_4$·7H$_2$O, and 5 g/L yeast extract) was used. The yeast cells and aggregates were visualized by microscopy (Nikon Eclipse Ti2, Tokyo, Japan).

2.2. Flocculation Assay

The assay described by D'Hautcourt and Smart [30] was used with minor modifications. Cells were cultivated for 24 h in YPD, harvested by centrifugation and resuspended in EDTA buffer (50 mM EDTA, pH 7) to reach an OD_{600nm} value of 10. A sample of 50 µL was taken at 0.5 mL below the meniscus, and the sample was diluted 20 times in a 1.5 mL cuvette with EDTA buffer (50 mM EDTA, pH 7). The tubes were centrifuged (4000 rpm, 3 min), and the supernatant was discarded. The cells were resuspended in 1 mL flocculation buffer A (3 mM $CaSO_4$). The last step was repeated, but the cells were resuspended in flocculation buffer B (3 mM $CaSO_4$, 83 mM CH_3COONa, 4% v/v ethanol, pH 4.5). The tubes were shaken at 100 rpm for 10 min. Prior to taking 50 µL samples 0.5 mL below the meniscus, 3 min of sedimentation in a vertical position took place. The sample was diluted 20 times with EDTA buffer in a 1.5 mL cuvette. The absorbance of both suspensions in the cuvettes was determined, and the related flocculation percentage was calculated:

$$\text{Flocculation percentage (\%)} = \frac{OD_{EDTA} - OD_{Flocculation\ buffer}}{OD_{EDTA}} \times 100$$

For the evolved strains, OD_{EDTA} corresponded to the OD_{600nm} value of the non-evolved reference strain.

2.3. Experimental Setup with Ministats

The continuous fermentation of industrial strains was carried out in a ministat set-up (Figure S1) [28]. Briefly, 15 mL test tubes were kept in an analog heat block (VWR®, Bridgeport, NJ, USA) at 30 °C and fed with high-glucose growth medium using a peristaltic pump (Type ISM833A, Ismatec®, Zurich, Switzerland). This high-glucose growth medium was previously also used in the adaptive evolution of *S. cerevisiae* strains towards a snowflake phenotype using a mini tower fermentor [27]. Medium was supplied at a flow rate of 30 µl/min. Air from a 4-port aquarium pump was fed to the medium in the test tube through an air filter (0.20 µm) and the needle was pushed to the bottom of the tube, in order to agitate the solution. During the experiment, a volume of 7–10 mL was maintained. The ministats were inoculated with 1 mL of an overnight culture. The pH, cell concentration (OD at 600 nm) and the glucose concentration (estimated using a refractometer (Brouwland, Belgium)) were measured during the ALE experiments.

2.4. Wort Fermentations in Tall Tubes

Laboratory-scale tall tubes, made from glass (75 cm high and 8-cm diameter), were used to assess beer fermentation with the evolved strains (Figure S2a). The tall tubes were filled with 2 L of wort with a density of 11 °P, which was provided by Biercentrum Delvaux (Neerijse, Belgium), and autoclaved before inoculation. The evolved strains from the ALE experiment and the original brewer's strains BCD1, BCD2, and BCD3 were added to the tall tubes at a cell concentration of 10×10^6 cells/mL. Fermentations were carried out in duplicates and sampled daily.

Alcolyzer Plus Beer Analyzing System (Anton Paar®, Graz, Austria) and headspace gas chromatography (GC) (Autosystem XL, Perkin Elmer®, Waltham, MA, USA) were used for the analyses of the fermentation process. For the GC analysis, the samples were filtered through a filter paper (Grade MN 713 $\frac{1}{4}$, Macherey-Nagel®, Düren, Germany). During the experiment, the apparent extract (% m/m) and the ethanol content (% v/v) were measured. The apparent extract (Ea) is a direct measurement of the dissolved solids in brewer's wort, gauged according to specific gravity. During fermentation, the fermentable carbohydrates (glucose, maltose, and maltotriose) are consumed by the yeast and the progress of the fermentation is monitored by measuring the disappearance of these solids [31].

Concentrations of the volatile compounds (acetaldehyde, ethyl acetate, diacetyl, propanol, 2,3-pentanedione, isobutanol, isoamyl acetate (3-methyl-1-butylacetate), isoamyl alcohol (3-methyl-

1-butanol), and ethyl caproate) in the beer samples were determined by headspace gas chromatography (HS-GC FID/ECD) as previously described [32,33]. Shortly, collected samples were cooled on ice and after centrifugation, 5 mL of the cooled supernatant was transferred to a vial. The vials were analyzed with a calibrated Autosystem XL gas chromatograph with a headspace autosampler (HS40; Perkin Elmer, Wellesley, MA, USA), equipped with a Chrompack-Wax 52 CB column (length 50 m, 0.32 mm internal diameter, 1.2 µm layer thickness; Varian, Palo Alto, CA, USA). Samples were heated for 16 min at 60 °C in the headspace autosampler before injection (needle temperature 70 °C). Helium was used as the carrier gas. The oven temperature was kept at 50 °C for 7.5 min, increased to 110 °C at 25 °C/min, and was held at that temperature for 3.5 min. Detection of esters, and higher alcohols was established with a flame ionization detector (FID); diacetyl was detected with an electron capture detector (ECD). The FID and ECD temperatures were kept constant at 250 °C and 200 °C, respectively.

3. Results

3.1. Adaptive Evolution in Ministats

Adaptive evolution experiments were performed with the three selected non-flocculating industrial strains using continuous cultivation in the ministats. The flocculation behavior before evolution was assessed and compared to the non-flocculating haploid lab strain BY4742, the strongly flocculating BY4742 [*FLO1*] (constitutively overexpressed *FLO1*), the naturally flocculating BY4742::*FLO8* (functional Flo8p) and the strongly flocculating industrial strain BCD4 (Figure 1a). The flocculation percentages of the three industrial strains were low (<21% ± 2%), ranking them below the natural flocculating BY4742::*FLO8* reference lab strain (46% ± 4%). The strongly flocculating industrial brewer's strain BCD4 shows the same flocculation capacity as the BY4742 lab strain with constitutively overexpressed *FLO1*.

For the ALE experiments, cultivation with the three strains was initiated with a dilution rate of approximately 0.2 h^{-1} using a medium feeding flow rate of 30 µL/min. The cultivation was monitored daily by measuring the glucose concentration, pH, and cell density (Figure 2). Between days 5–7, steady-state was reached and over time, and the dilution rate was increased gradually to 0.35 h^{-1} (BCD1) and 0.45 h^{-1} (BCD2, BCD3) to avoid wash-out and to select for larger aggregates. The continuous cultivation was stopped after 45 days, and the yeast populations were examined by microscopy (Figure 3). The evolved BCD2 aggregates were smaller than the BCD1 and BCD3 aggregates. The nature of the cell clusters was determined by resuspending the cells in EDTA-buffer, which chelates Ca^{+2} ions and disrupts yeast cell clusters if they are formed via flocculin-dependent adhesion. The aggregates persisted for all three strains. This indicated that the clusters are not the result of flocculin interactions and are likely due to failure in separation of the mother and daughter cells, described previously as the "snowflake" phenotype [27,34–36]. The evolved strains were subsequently cultivated in batch cultures and the aggregating phenotype was found to be stable.

The flocculation assay was repeated with the evolved strains to estimate and compare their sedimentation velocity to that of the BCD1, BCD2, and BCD3 strains before evolution (Figure 1). All three evolved strains showed an increase in "flocculation" percentage. Even though none of the strains evolved towards a real flocculating phenotype, their multicellular aggregates contributed to a significant larger sedimentation velocity.

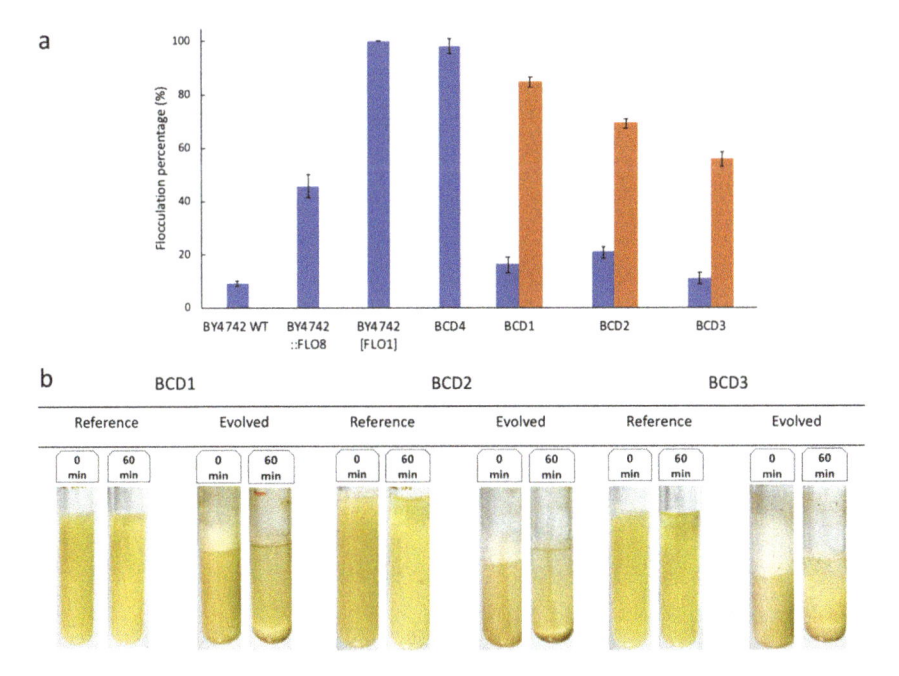

Figure 1. Flocculation and sedimentation behavior of laboratory and industrial strains. (**a**) Flocculation percentages determined for the control strains (BY4742, BY4742::*FLO8*, BY4742 (*FLO1*), the industrial BCD4 strain) and for the three industrial non-flocculating strains BCD1, BCD2, and BCD3 before (■) and after (■) ALE in ministats. All measurements were performed in triplicates. (**b**) The sedimentation behavior of the three industrial brewer's strains before and after adaptive laboratory evolution (ALE) in the ministats.

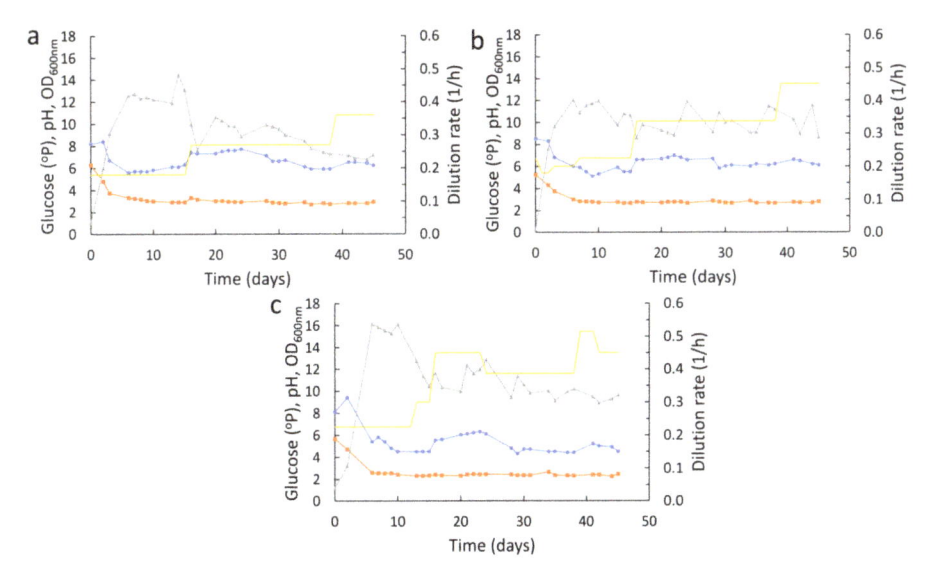

Figure 2. Cultivation of the industrial strains in the ministats: (**a**) BCD1, (**b**) BCD2, and (**c**) BCD3. Glucose content (●), pH (■), OD$_{600\,nm}$ (▲), and the dilution rate (1/h) (—).

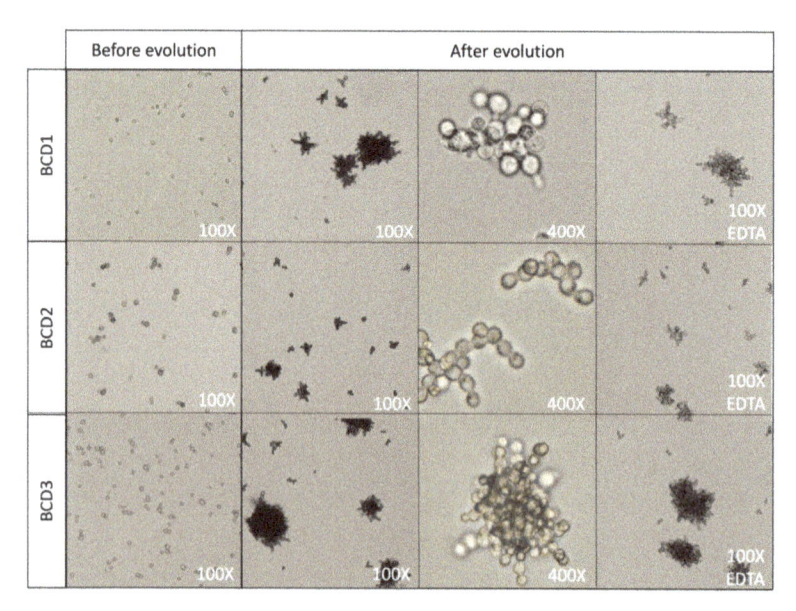

Figure 3. Microscopic observations of the industrial strains before and after the evolution in the ministats. Snowflake clumps are observed for each strain and could not be disrupted by treatment with 50 mM EDTA.

3.2. Performance of Evolved Strains During Beer Fermentation

The performance and behavior of the evolved yeast strains, compared to their reference strains, were evaluated during wort fermentations in tall tubes (Figure S2a). The fermentations were monitored by sampling and measuring cell concentration and beer characteristics such as apparent extract (E_a), ethanol content (Table S1), and the concentration of several flavor compounds (Figure S3 and S4).

The reference strain BCD1 showed a faster fermentation capacity than the evolved strain: The fermentation was almost completed after only 1 day as observed from the evolution of the apparent extract and ethanol concentration (Table S1) as well as from the suspended cell concentration (Figure S2b). The fermentation capacity of the evolved and reference BCD2 strains were similar (Table S1). In contrast, the evolved BCD3 strain showed a faster fermentation than the reference strain (Table S1). The evolution of the suspended cell concentrations during the fermentations for the evolved and reference strains is shown in Figure S1b. There are no large differences observable in the number of suspended cells between the evolved and the reference strains, except for the evolved BCD1 at the second day of fermentation where a much lower cell concentration of the evolved strain was present.

Flavor compounds in the beer were quantified by headspace gas chromatography (Figure S3 and S4). In general, no major influence of the evolved yeast strains on the development of the flavor profile was observed. Some remarkable observations include an increased content of vicinal diketones diacetyl (up to 2.1 ppm) and 2,3-pentanedione (up to 0.9 ppm) during the initial stages of the fermentation for the evolved BCD1 strain, which was decreased by the third day for both compounds to 0.1 ppm. Also, the aliphatic higher alcohol isobutanol concentration increased to 178 ppm (compared to 53 ppm for the reference strain) at the third day of the fermentation. The isobutanol concentration of the evolved BCD2 strain fermentation was doubled at the second day compared to the reference strain, but was still below the isobutanol flavor threshold of 100–200 ppm [37]. By the third day, the evolved BCD3 gave a lower concentration of acetaldehyde than its reference, but a higher concentration of the higher alcohol propanol and isobutanol, and the esters ethylacetate and isoamylacetate.

4. Discussion

Miniature, low-cost chemostats (ministats) were used for adaptive laboratory evolution (ALE) of three industrial *S. cerevisiae* brewer's strains towards a more favorable, aggregating phenotype. The three strains—BCD1, BCD2, and BCD3—were characterized with a low flocculation ability and were continuously cultivated with high-glucose medium for 45 days. Small clusters of cells were observed around day 15, corresponding to approximately 110 generations, which is comparable to other *S. cerevisiae* ALE experiments [27,38,39]. The yeast cell clusters were not disrupted by treatment by EDTA, which indicated that the cell-cell interactions are not based on flocculins. Microscopy showed that the multicellular clusters look like "snowflakes". This "snowflake" phenotype was previously described as the result of the failed separation of daughter cells from mother cells. This phenotype is caused by a frameshift mutation in the transcription factor *ACE2*, which is responsible for the activation of the *CTS1* gene encoding the chitinase necessary to break down the septum between the mother and the daughter cells [34,36,40]. The clusters of cells formed in this way are unlike flocs in that they consist entirely of genetically identical cells and surrounding cells can not adhere to the cluster [36].

The performance of the evolved strains was compared to the reference strains in beer fermentations using tall tubes fermentors. During the fermentation only small aggregates of the evolved strains were observed. These aggregates were kept in suspension during the convective mixing by the CO_2 release by the fermenting yeast cells as was clear from the evolution of the suspended cell concentration (Figure S2b). Apparently, shear stress by convective mixing will break up large aggregates and the presence of these small aggregates will not lead to premature sedimentation during fermentation. The CO_2 production stops at the end of the fermentation and convective currents are reduced significantly. At this moment, the sedimentation of the snowflake aggregates will be significantly faster (Figure 1b) than the reference strains.

To assess the effect of adapted evolution of the 3 strains on the beer flavor, the evolution of a few flavor compounds was determined during the first 3 days of the fermentation. The evolved BCD1 strain showed an increased production of the vicinal diketones and the higher alcohol isobutanol. The synthesis of these compounds is linked to the isoleucine–leucine–valine (ILV) pathway. Although the flavor threshold of diacetyl (0.1–0.15 ppm [41]) was exceeded and the 2,3-pentanedione concentration was close to the flavor threshold (1.0–1.5 ppm [41]) during the first 2 days of fermentation, these concentrations were reduced significantly below the flavor threshold at the third day. Also, the isobutanol content in the beer fermented by the evolved BCD1 strain was much higher than for the reference strain, but did not exceed the flavor threshold of 100–200 ppm [37]. After 7 days, the green beer from the evolved and the reference BCD1 strain both tasted fruity (isoamyl acetate and acetaldehyde). Although the isobutanol concentration was larger in the beer from the evolved BCD2 strain, no difference was tasted. In both beers, the apple flavor (acetaldehyde) could be recognized. The beer from the evolved BCD3 strain contained a higher concentration of the ester isoamyl acetate (banana aroma) and ethyl acetate ester (fruity, solvent-like aroma), which presence could be tasted in the green beer.

Supplementary Materials: The following are available online at http://www.mdpi.com/2311-5637/6/1/20/s1, Figure S1: The experimental setup of the ministats. Figure S2: Beer fermentations in tall tube fermenters, Figure S3 and Figure S4: Comparison of the evolution of some flavor compounds during the tall tubes' fermentations; Table S1: Evolution of the apparent extract and ethanol content during the tall tube fermentations.

Author Contributions: Conceptualization, R.W. and F.R.D.; methodology, R.W., T.V.M.; formal analysis, T.V.M.; investigation, T.V.M., Y.K., F.D., F.R.D., and R.W..; resources, R.W.; writing—original draft preparation, Y.K., T.V.M.; writing—review and editing, Y.K., R.W., F.R., and F.R.D.; supervision, R.W., F.R.D., and F.D.; project administration, R.W.; funding acquisition, R.W. All authors have read and agree to the published version of the manuscript.

Funding: This research was funded by ESA-Belspo, grant PRODEX "Yeast Bioreactor".

Acknowledgments: The Belgian Federal Science Policy Office (Belspo) and the European Space Agency (ESA) PRODEX program supported this work. The Research Council of the Vrije Universiteit Brussel (Belgium) and the University of Ghent (Belgium) are acknowledged to support the Alliance Research Group VUB-UGent NanoMicrobiology (NAMI), and the International Joint Research Group (IJRG) VUB-EPFL BioNanotechnology & NanoMedicine (NANO).

Conflicts of Interest: The authors declare no conflict of interest.

References

1. Bowden, C.P.; Leaver, G.; Melling, J.; Norton, M.G.; Whittington, P.N. Recent and novel developments in the recovery of cells from fermentation broths. In *Separations for Biotechnology*; Verral, M.S., Hudson, M.J., Eds.; Ellis Horwood: Chichester, UK, 1987; pp. 49–61.
2. Stratford, M. Yeast flocculation: A new perspective. *Adv. Microb. Physiol.* **1992**, *33*, 1–71.
3. Vidgren, V.; Londesborough, J. 125th anniversary review: Yeast flocculation and sedimentation in brewing. *J. Inst. Brew.* **2011**, *117*, 475–487. [CrossRef]
4. Goossens, K.; Willaert, R. Flocculation protein structure and cell-cell adhesion mechanism in *Saccharomyces cerevisiae*. *Biotechnol. Lett.* **2010**, *32*, 1571–1585. [CrossRef]
5. Van Mulders, S.E.; Christianen, E.; Saerens, S.M.G.; Daenen, L.; Verbelen, P.J.; Willaert, R.; Verstrepen, K.J.; Delvaux, F.R. Phenotypic diversity of Flo protein family-mediated adhesion in *Saccharomyces cerevisiae*. *FEMS Yeast Res.* **2009**, *9*, 178–190. [CrossRef] [PubMed]
6. Willaert, R.G. Adhesins of yeasts: Protein structure and interactions. *J. Fungi* **2018**, *4*, 119. [CrossRef] [PubMed]
7. Brückner, S.; Mösch, H.-U. Choosing the right lifestyle: Adhesion and development in *Saccharomyces cerevisiae*. *FEMS Microbiol. Rev.* **2011**, *36*, 25–58. [CrossRef] [PubMed]
8. Verstrepen, K.J.; Derdelinckx, G.; Verachtert, H.; Delvaux, F.R. Yeast flocculation: What brewers should know. *Appl. Microbiol. Biotechnol.* **2003**, *61*, 197–205. [CrossRef] [PubMed]
9. Gibson, B.R.; Lawrence, S.J.; Leclaire, J.P.R.; Powell, C.D.; Smart, K.A. Yeast Responses to Stresses Associated with Industrial Brewery Handling. *FEMS Microbiol. Rev.* **2007**, *31*, 535–569. [CrossRef]
10. Soares, E.V. Flocculation in *Saccharomyces cerevisiae*: A review. *J. Appl. Microbiol.* **2010**, *110*, 1–18. [CrossRef]
11. Dequin, S. The potential of genetic engineering for improving brewing, wine-making and baking yeasts. *Appl. Microbiol. Biotechnol.* **2001**, *56*, 577–588. [CrossRef]
12. Saerens, S.M.G.; Duong, C.T.; Nevoigt, E. Genetic improvement of brewer's yeast: Current state, perspectives and limits. *Appl. Microbiol. Biotechnol.* **2010**, *86*, 1195–1212. [CrossRef] [PubMed]
13. Conrad, T.M.; Lewis, N.E.; Palsson, B.Ø. Microbial laboratory evolution in the era of genome-scale science. *Mol. Syst. Biol.* **2011**, *7*, 509. [CrossRef] [PubMed]
14. Portnoy, V.A.; Bezdan, D.; Zengler, K. Adaptive laboratory evolution-harnessing the power of biology for metabolic engineering. *Curr. Opin. Biotechnol.* **2011**, *22*, 590–594. [CrossRef] [PubMed]
15. Dragosits, M.; Mattanovich, D. Adaptive laboratory evolution—Principles and applications for biotechnology. *Microb. Cell Fact.* **2013**, *12*, 64. [CrossRef]
16. Sonderegger, M.; Sauer, U. Evolutionary engineering of *Saccharomyces cerevisiae* for anaerobic growth on xylose. *Appl. Environ. Microbiol.* **2003**, *69*, 1990–1998. [CrossRef]
17. Dhar, R.; Sägesser, R.; Weikert, C.; Wagner, A. Yeast adapts to a changing stressful environment by evolving cross-protection and anticipatory gene regulation. *Mol. Biol. Evol.* **2013**, *30*, 573–588. [CrossRef]
18. Ekberg, J.; Rautio, J.; Mattinen, L.; Vidgren, V.; Londesborough, J.; Gibson, B.R. Adaptive evolution of the lager brewing yeast *Saccharomyces pastorianus* for improved growth under hyperosmotic conditions and its influence on fermentation performance. *FEMS Yeast Res.* **2013**, *13*, 335–349. [CrossRef]
19. Strejc, J.; Siříšťová, L.; Karabín, M.; Almeida e Silva, J.B.; Brányik, T. Production of alcohol-free beer with elevated amounts of flavouring compounds using lager yeast mutants. *J. Inst. Brew.* **2013**, *119*, 149–155. [CrossRef]
20. Huuskonen, A.; Markkula, T.; Vidgren, V.; Lima, L.; Mulder, L.; Geurts, W.; Walsh, M.; Londesborough, J. Selection from Industrial Lager Yeast Strains of Variants with Improved Fermentation Performance in Very-High-Gravity Worts. *Appl. Environ. Microbiol.* **2010**, *76*, 1563–1573. [CrossRef]
21. Yu, Z.; Zhao, H.; Li, H.; Zhang, Q.; Lei, H.; Zhao, M. Selection of *Saccharomyces pastorianus* variants with improved fermentation performance under very high gravity wort conditions. *Biotechnol. Lett.* **2012**, *34*, 365–370. [CrossRef]
22. Cadière, A.; Ortiz-Julien, A.; Camarasa, C.; Dequin, S. Evolutionary engineered *Saccharomyces cerevisiae* wine yeast strains with increased in vivo flux through the pentose phosphate pathway. *Metab. Eng.* **2011**, *13*, 263–271. [CrossRef] [PubMed]

23. Cadière, A.; Aguera, E.; Caillé, S.; Ortiz-Julien, A.; Dequin, S. Pilot-scale evaluation the enological traits of a novel, aromatic wine yeast strain obtained by adaptive evolution. *Food Microbiol.* **2012**, *32*, 332–337. [CrossRef] [PubMed]
24. Weikert, C.; Sauer, U.; Bailey, J.E. Use of a glycerol-limited, long-term chemostat for isolation of Escherichia coli mutants with improved physiological properties. *Microbiology* **1997**, *143*, 1567–1574. [CrossRef] [PubMed]
25. Koppram, R.; Albers, E.; Olsson, L. Evolutionary engineering strategies to enhance tolerance of xylose utilizing recombinant yeast to inhibitors derived from spruce biomass. *Biotechnol. Biofuels* **2012**, *5*, 32. [CrossRef] [PubMed]
26. Wang, L.; Spira, B.; Zhou, Z.; Feng, L.; Maharjan, R.P.; Li, X.; Li, F.; McKenzie, C.; Reeves, P.R.; Ferenci, T. Divergence involving global regulatory gene mutations in an Escherichia coli population evolving under phosphate limitation. *Genome Biol. Evol.* **2010**, *2*, 478–487. [CrossRef]
27. Conjaerts, A.; Willaert, R. Gravity-Driven Adaptive Evolution of an Industrial Brewer's Yeast Strain towards a Snowflake Phenotype in a 3D-Printed Mini Tower Fermentor. *Fermentation* **2017**, *3*, 4. [CrossRef]
28. Miller, A.W.; Befort, C.; Kerr, E.O.; Dunham, M.J. Design and use of multiplexed chemostat arrays. *J. Vis. Exp.* **2013**, e50262. [CrossRef]
29. Brachmann, C.B.; Davies, A.; Cost, G.J.; Caputo, E.; Li, J.; Hieter, P.; Boeke, J.D. Designer Deletion Strains derived from *Saccharomyces cerevisiae* S288C: A Useful set of Strains and Plasmids for PCR-mediated Gene Disruption and Other Applications. *Yeast* **1998**, *14*, 115–132. [CrossRef]
30. D'Hautcourt, O.; Smart, K.A. Measurement of brewing yeast flocculation. *J. Am. Soc. Brew. Chem.* **1999**, *57*, 123–128. [CrossRef]
31. Shimizu, H.; Mizuno, S.; Hiroshima, T.; Shioya, S. Effect of carbon and nitrogen additions on consumption activity of apparent extract of yeast cells in a brewing process. *J. Am. Soc. Brew. Chem.* **2002**, *60*, 163–169. [CrossRef]
32. Verbelen, P.J.; Dekoninck, T.M.L.; Saerens, S.M.G.; Van Mulders, S.E.; Thevelein, J.M.; Delvaux, F.R. Impact of pitching rate on yeast fermentation performance and beer flavour. *Appl. Microbiol. Biotechnol.* **2009**, *82*, 155–167. [CrossRef] [PubMed]
33. Dekoninck, T.M.L.; Mertens, T.; Delvaux, F.; Delvaux, F.R. Influence of beer characteristics on yeast refermentation performance during bottle conditioning of belgian beers. *J. Am. Soc. Brew. Chem.* **2013**, *71*, 23–34. [CrossRef]
34. Ratcliff, W.C.; Fankhauser, J.D.; Rogers, D.W.; Greig, D.; Travisano, M. Origins of multicellular evolvability in snowflake yeast. *Nat. Commun.* **2015**, *6*, 6102. [CrossRef] [PubMed]
35. Ratcliff, W.C.; Denison, R.F.; Borrello, M.; Travisano, M. Experimental evolution of multicellularity. *Proc. Natl. Acad. Sci. USA* **2012**, *109*, 1595–1600. [CrossRef]
36. Oud, B.; Guadalupe-Medina, V.; Nijkamp, J.F.; De Ridder, D.; Pronk, J.T.; Van Maris, A.J.A.; Daran, J.M. Genome duplication and mutations in ACE2 cause multicellular, fast-sedimenting phenotypes in evolved *Saccharomyces cerevisiae*. *Proc. Natl. Acad. Sci. USA* **2013**, *110*, E4223–E4231. [CrossRef]
37. Djordjević, V.; Willaert, R.; Gibson, B.; Nedović, V. Immobilized Yeast Cells and Secondary Metabolites. In *Fungal Metabolites*; Springer International Publishing: Berlin/Heidelberg, Germany, 2016; pp. 1–40.
38. Ratcliff, W.C.; Pentz, J.T.; Travisano, M. Tempo and mode of multicellular adaptation in experimentally evolved *Saccharomyces cerevisiae*. *Evolution* **2013**, *67*, 1573–1581. [CrossRef]
39. Hope, E.A.; Amorosi, C.J.; Miller, A.W.; Dang, K.; Heil, C.S.; Dunham, M.J. Experimental evolution reveals favored adaptive routes to cell aggregation in yeast. *Genetics* **2017**, *206*, 1153–1167. [CrossRef]
40. King, L.; Butler, G. Ace2p, a regulator of CTS1 (chitinase) expression, affects pseudohyphal production in *Saccharomyces cerevisiae*. *Curr. Genet.* **1998**, *34*, 183–191. [CrossRef]
41. Meilgaard, M.C. Flavor chemistry of beer: Part ii: Flavour and threshold of 239 aroma volatiles. *MBAA Tech. Q.* **1975**, *12*, 151–168.

© 2020 by the authors. Licensee MDPI, Basel, Switzerland. This article is an open access article distributed under the terms and conditions of the Creative Commons Attribution (CC BY) license (http://creativecommons.org/licenses/by/4.0/).

fermentation

Article

Characterization of Old Wine Yeasts Kept for Decades under a Zero-Emission Maintenance Regime

Katrin Matti [1], Beatrice Bernardi [1,2], Silvia Brezina [1], Heike Semmler [1], Christian von Wallbrunn [1], Doris Rauhut [1] and Jürgen Wendland [1,*]

- [1] Department of Microbiology and Biochemistry, Hochschule Geisenheim University, Von-Lade-Strasse 1, D-65366 Geisenheim, Germany; Katrin.Matti@hs-gm.de (K.M.); Beatrice.Bernardi@hs-gm.de (B.B.); Silvia.Brezina@hs-gm.de (S.B.); Heike.Semmler@hs-gm.de (H.S.); Christian.Wallbrunn@hs-gm.de (C.v.W.); Doris.Rauhut@hs-gm.de (D.R.)
- [2] Research Group of Microbiology (MICR)—Functional Yeast Genomics, Vrije Universiteit Brussel, Pleinlaan 2, BE-1050 Brussels, Belgium
- * Correspondence: juergen.wendland@hs-gm.de; Tel.: +49-6722-502-332

Received: 14 December 2019; Accepted: 9 January 2020; Published: 11 January 2020

Abstract: All laboratories dealing with microbes have to develop a strain maintenance regime. While lyophilization based on freeze-drying may be feasible for large stock centers, laboratories around the world rely on cryopreservation and freezing of stocks at −80 °C. Keeping stocks at these low temperatures requires investments of several thousand kW/h per year. We have kept yeast stocks for several decades at room temperature on agar slants in glass reagent tubes covered with vaspar and sealed with cotton plugs. They were part of the Geisenheim Yeast Breeding Center stock collection that was started in the 19th century, well before −80 °C refrigeration technology was invented. Of these stocks, 60 tubes were analyzed and around one-third of them could be regrown. The strains were typed by sequencing of rDNA PCR fragments. Based on BlastN analyses, twelve of the strains could be assigned to *Saccharomyces cerevisiae*, two to *S. kudriavzevii*, and the others to *Meyerozyma* and *Candida*. The strains were used in white wine fermentations and compared to standard wine yeasts Uvaferm/GHM (Geisenheim) and Lalvin EC1118. Even with added nitrogen, the strains exhibited diverse fermentation curves. Post-fermentation aroma analyses and the determination of residual sugar and organic acid concentrations indicated that some strains harbor interesting flavor characteristics, surpassing current standard yeast strains. Thus, old strain collections bear treasures for direct use either in wine fermentations or for incorporation in yeast breeding programs aimed at improving modern wine yeasts. Furthermore, this provides evidence that low-cost/long-term culture maintenance at zero-emission levels is feasible.

Keywords: strain collection; aroma profiling; gas chromatography; wine yeast; *Saccharomyces*; fermentation; volatile aroma compounds

1. Introduction

At the end of the 19th century, Emil Christian Hansen at the Carlsberg Laboratory in Copenhagen, Denmark, established the first pure culture lager yeast strain, Unterhefe No. 1 [1]. This strain then became known as *Saccharomyces carlsbergensis*. The finding that one yeast strain was sufficient to generate a fermented beverage of high quality started a new era and lead to new developments in the beer and dairy industry. It was soon recognized by Julius Wortmann at the Geisenheim Research Center in Germany that Hansen's findings were also applicable to wine making [2]. This started efforts in collecting wine yeast strains from different vineyards and wineries in the Rheingau area. These strains were characterized for their fermentation capacity and flavor attributes. At the "Geisenheimer

Hefe-Reinzuchtstation" (Geisenheim Yeast Breeding Center), founded in 1894, these strains were produced as liquid starter cultures and dispatched to the wineries upon request.

With the isolation of pure yeast cultures came the responsibility to maintain stocks of these cultures. Wine making requires yeast starter cultures only once a year just after the grape harvest. By contrast, yeasts for beer production are in constant use throughout the year. Even before the isolation of pure cultures, there was an interest in generating dry yeast cakes for longer term storage. The history of both the patents and literature in this field has been covered in depth by a recent excellent review [3]. A solid supply of dehydrated yeast became a necessity for long distance shipments, which came around 1940 when the Fleischman Co. produced active dry yeast. This yeast required 'reactivation', i.e., rehydration prior to use. In the 1970s, Lesaffre introduced an instant dry yeast which could be used directly without reactivation. In microbiological laboratories, however, bacterial and yeast cultures are nowadays generally preserved by storing at −80 °C which, since the 1970s, has become technically feasible on a larger scale [4].

Thus, for decades after the 1880s, yeast cultures had to be kept by other means. Two techniques used were water stocks and yeast slants covered with vaspar (a mix of paraffin and Vaseline) [5,6]. The method of storing yeasts in distilled water at room temperature, as proposed by Castellani, is not only a cheap way of preserving cultures but is also a very effective way of culturing a collection over many years without the need for constant propagation [7]. This method is particularly useful for yeasts [8]. Storage of fungal cells in distilled water can be extended for 20 years [9]. Therefore, it was stated that storing yeast cultures in distilled water may reach similar efficiencies as freezing at −80 °C [10]. The use of a paraffin or vaspar overlay is also a very cheap way of yeast culturing, although the viability may be reduced when compared to the other methods.

There are only a few long-term studies describing yeast viability, one of which used the traditional method of yeasts grown on slants and covered by paraffin oil and found cells to be viable after a seven-year incubation period [11]. At the Geisenheim Yeast Breeding Center, we have a large collection of wine yeasts and non-conventional yeasts dating back to the 1890s. Samples were routinely stored with a vaspar overlay. Of course, over time, the strain collection was transferred to either storage in liquid nitrogen or in freezers at −80 °C. Nevertheless, we still stored a few samples for over 30 years at room temperature in the old way. In this study, we examined 60 tubes containing these decade-old samples. The yeast were restreaked, and those strains that could be regrown were subjected to fermentation studies and volatile aroma analyses. Our results show that strain collections can safely be stored at room temperature. Such a strain maintenance regime could contribute to energy conservation and the reduction of CO_2 emissions.

2. Materials and Methods

2.1. Strains and Media

The yeast strains used in this study are shown in Table 1, including standard wine yeast strains used for comparison. Yeast strains were subcultured in YPD (1% yeast extract, 2% peptone, 2% glucose).

Table 1. Strains used in this study.

Original Label	Sequence-Based Assignment
Zell 1895	*Saccharomyces cerevisiae*
Valdepenas Criptana 1909	*Saccharomyces cerevisiae*
Brettanomyces claussenii IHG Berlin 1959	*Meyerozyma guilliermondii*
Ungstein 1892	*Saccharomyces cerevisiae*
Riesling Krim 1896	*Saccharomyces cerevisiae*
Olewig II 1896	*Meyerozyma guilliermondii*
Dürkheim 1892	*Meyerozyma guilliermondii*
Rüdesheimer Hinterhaus 1893	*Saccharomyces cerevisiae*
Alpiarca II 1896	*Saccharomyces cerevisiae*
Candida tropicalis	*Candida sanyaensis*
Heimersheimer Ruth 1895	*Saccharomyces cerevisiae*
Steinberg 1893	*Saccharomyces cerevisiae*
Rüdesheimer Berg	*Saccharomyces cerevisiae*
Würzburg (Stein)	*Saccharomyces kudriavzevii*
Winningen 1892	*Saccharomyces kudriavzevii*
Scy 1892	*Saccharomyces cerevisiae*
Bordeaux 1892	*Saccharomyces cerevisiae*
Geisenheimer (Mäuerchen) 1893	*Saccharomyces cerevisiae*

Bold: to indicate true *S. cerevisiae* strains. These strains will be important for winemaking. The others are non-conventional yeasts.

2.2. Molecular Analysis of Yeast Strains

Typing of the strains was determined by performing ITS-PCR (ITS, internal transcribed spacer) using standard ITS1F-fungal specific-(5'-CTTGGTCATTTAGAGGAAGTAA-3') and ITS4-universal-(5'-TCCTCCGCTTATTGATATGC-3') primers and sequencing of the PCR products as previously described [12]. Sequencing was conducted by Starseq, Mainz, Germany.

2.3. Fermentation Conditions

Lab-scale fermentations were carried out in duplicate with a standard pasteurized white wine must with a sugar concentration of 72 °Oechsle. The must was supplemented by the addition of Fermaid E (inactivated yeast product; according to the supplier's instructions; Lallemand, Vienna, Austria). Cells were inoculated at a density of OD_{600} = 0.5. The fermentation temperature was set to 18 °C and cultures were incubated with constant stirring at 300 rpm.

2.4. Analytical Methods

At the end of the fermentations, several compounds including fructose, glucose, ethanol, and organic acids were analyzed by high-performance liquid chromatography (HPLC) with an Agilent 1100 Series (Agilent Technologies, Waldbronn, Germany). Quantitative analyses were done as described in [13]. The HPLC equipment was equipped with a variable wavelength detector (UV/VIS) and a refractive index detector (RID). The column for separation is an Allure Organic Acids (Restek GmbH, Bad Homburg, Germany) with a 5 μm particle size, 60 A pore size and dimensions of 250 mm x i.d. 4.6 mm. A water-based solution of sulfuric acid (0.0139% *v/v*) and ethanol (0.5% *v/v*) was used as eluent. Gas chromatography was conducted using a GC 7890A (Agilent, Santa Clara, CA, USA), coupled with a MSD 5977B mass spectrometer (Agilent, Santa Clara, CA, USA). The determination of aroma compounds followed the analytical approach described in Belda et al. [14] according to the method of Camara et al. [15].

3. Results

3.1. Regrowing of Dormant Strains Kept under Vaspar for Decades

At the Geisenheim Yeast Breeding Center (GYBC), we stored two racks containing 60 reagent tubes with yeast strains. The tubes contained slants on which yeast strains were spread and grown and then overlaid with vaspar (Figure 1). This was the standard procedure for maintaining strains at the GYBC. The collections were generated before the introduction of −80 °C refrigeration. These samples represent an even older stock as younger samples (20+ years of age) were kept in reagent tubes with screw cap closures. The apparent age of the old samples is, therefore, estimated to be over 30 years but less than 60 years, as one isolate carried a label with the year 1959 indicative of the year of isolation (Table 1).

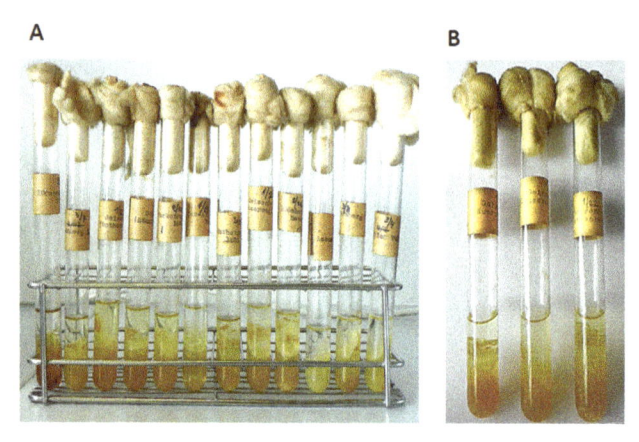

Figure 1. (**A**,**B**). Old samples from the Geisenheimer Yeast Breeding Center. Samples were generated as agar slants with yeasts grown and covered in vaspar. Tubes were plugged by a cotton ball. Samples were labeled according to the location of the isolate and the year of isolation (see Table 1).

We wanted to find out if these strains were still alive and, once propagated, what their fermentation behavior would be like. To this end, we either took samples with an inoculation loop and restreaked them on full medium YPD or inoculated them in liquid YPD. Astonishingly, about one-third (18 out of 60) of the strains could be regrown and cultivated under these conditions.

The strain labels often indicated the area where these strains had been isolated and did not necessarily identified the species. Thus, we went on to type the strains by PCR amplification of a region of the ribosomal DNA using a standard primer pair designed for fungal species (ITS1 and ITS4). These primers are located at the end of the 18S and start of the 28S rDNA and thus amplify the internal transcribed spacer (ITS) region including the 5.8S rDNA. This region is highly variable and allows strain determination to the species level. PCR products were sequenced, and the sequences compared to the NCBI non-redundant database using BlastN. The sequence comparisons indicated that most strains could indeed be assigned to *Saccharomyces cerevisiae*. Two of the strains were found to be *S. kudriavzevii* while three strains matched *Meyerozyma guilliermondii* and one strain could be assigned to a newly described species of *Candida sanyaensis* (Table 1).

3.2. Fermentation Performance

We went on to study the fermentation characteristics of the *Saccharomyces* strains. To this end, strains were used to ferment a standard white wine must of 72 °Oechsle to which additional amino nitrogen was added via an inactivated yeast product (Fermaid E). Fermentations were carried out at 18 °C with stirring over a period of 12 days and fermentation rates were followed by daily measurements of CO_2 release. Strains were compared to the standard wine strain Lalvin EC1118 (Figure 2). It turned out that half of the strains generated a weight loss slightly larger than Lalvin EC1118, while the other half performed less well than Lalvin EC1118 in this respect. The largest weight loss was found with *S. cerevisiae* strain Steinberg 1893 and the *S. kudriavzevii* strain Würzburg (Stein). Lalvin EC1118 required a short lag phase of one day to enter alcoholic fermentation. Several of the tested yeast strains exhibited an extended lag phase of 3–4 days, particularly the strains that later showed the greatest weight loss and also Geisenheimer Mäuerchen, Winningen 1892, and Bordeaux 1892. Thus, even given the extended lag phase, these strains managed a complete fermentation within the 12-day fermentation window.

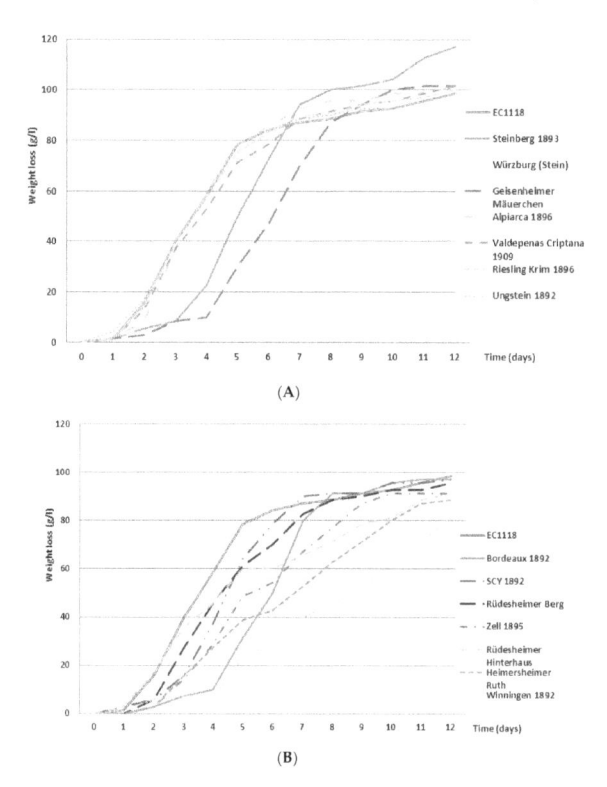

Figure 2. Fermentation curves based on CO_2 release/weight loss of *Saccharomyces* cultures derived from isolates of the old collection. Release of CO_2 was measured daily. (**A**) Strains are shown that released more CO_2 than the EC1118 control wine yeast strain. (**B**) Strains are shown that released less CO_2 than the EC1118 control wine yeast strain.

To analyze the fermented liquids in more detail, the residual sugars, organic acids, and final ethanol content were determined (Table 2). Most of the *Saccharomyces* strains reached complete fermentation with around 7% alcohol content. Two glucophilic strains, however, failed to utilize all of the fructose in the 12-day fermentation time. The *S. kudriavzevii* strains produced less alcohol than the *S. cerevisiae* strains. The two *S. cerevisiae* strains Rüdesheimer Hinterhaus 1893 and Heimersheimer Ruth, while using up glucose, did not utilize fructose completely during the 12-day fermentation. All strains showed a similar organic acid profile, with malate being the pronounced acid. Rüdesheimer Hinterhaus 1893, on the other hand, showed a surprising amount of shikimic acid (Table 2).

Table 2. HPLC analyses of residual sugars, organic acids, and total alcohol of *Saccharomyces* wine yeasts.

Strains	Glucose	Fructose	Total Sugar	Malate	Shikimic Acid	Lactate	Acetate	Citric acid	Ethanol	Ethanol
	[g/L]	[g/L]	[g/L]	[g/L]	[mg/L]	[g/L]	[g/L]	[g/L]	[g/L]	[%]
EC1118 wine yeast	<1	<1		4.6	38	0.2	0.1	0.2	60.7	7.7
GHM wine yeast	<1	<1		4.4	40	0.3	0.2	0.2	56.3	7.1
Saccharomyces cerevisiae										
Zell 1895	<1	<1		3.7	41	0.4	0.1	0.2	60.2	7.6
Valdepenas Criptana 1909	<1	<1		3.5	39	0.2	0.1	0.2	54.9	7.0
Ungstein 1892	<1	<1		2.9	28	0.2	0.2	0.2	56.6	7.2
Riesling Krim 1896	<1	<1		2.2	20	0.1	0.2	0.2	57.1	7.3
Rüdesheimer Hinterhaus 1893	<1	4.1	4.1	3.9	297	0.9	0.2	0.4	53.8	6.9
Alpiarca II 1896	<1	<1		4.1	38	0.2	0.1	0.2	56.8	7.2
Heimersheimer Ruth 1895	<1	6.9	6.9	3.4	34	0.3	0.2	0.2	56.4	7.2
Steinberg 1893	<1	<1		3.8	32	0.1	0.1	0.2	56.4	7.1
Rüdesheimer Berg	<1	<1		4.0	39	0.5	0.2	0.2	55.9	7.1
Scy 1892	<1	<1		3.6	31	0.1	0.2	0.2	58.7	7.4
Bordeaux 1892	<1	<1		2.9	29	0.2	0.1	0.2	56.9	7.2
Geisenheimer (Mäuerchen)	<1	<1		3.8	27	0.1	0.4	0.2	57.1	7.2
S. kudriavzevii										
Würzburg (Stein)	<1	<1		3.7	26	<0.1	0.2	0.2	54.2	6.9
Winningen 1892	2.1	19.2	20.8	4.1	37	0.5	0.2	0.2	49.7	6.3

3.3. Production of Aroma Compounds

We routinely examined 28 aroma compounds, specifically alcohols and esters (Table S1). A comparison within strains using a selection of eight major compounds is shown in Figure 3 and Table 3. While all species produced a range of compounds, it was interesting to see that a major current wine production strain, EC1118, was actually not the highest producer of certain aroma compounds in our assay. The three strains that produced most fruity esters were Rüdesheimer Hinterhaus 1893, Alpiarca 1896, and Valdepenas Criptana 1909. Additionally, Rüdesheimer Hinterhaus 1893 championed the production of isoamyl acetate (acetic acid 3-methyl butyl ester) and 2-phenylethyl acetate (acetic acid 2-phenylethylester), which is the acetate ester of 2-phenylethanol. This is apparently a consequence of Ehrlich pathway output as regarding the production of alcohols, particularly i-butanol, isoamylalcohol (3-methyl-butanol), and 2-phenlyethanol, the strain that came on top for each of the compounds was also the Rüdesheimer Hinterhaus 1893 yeast (see Table S1).

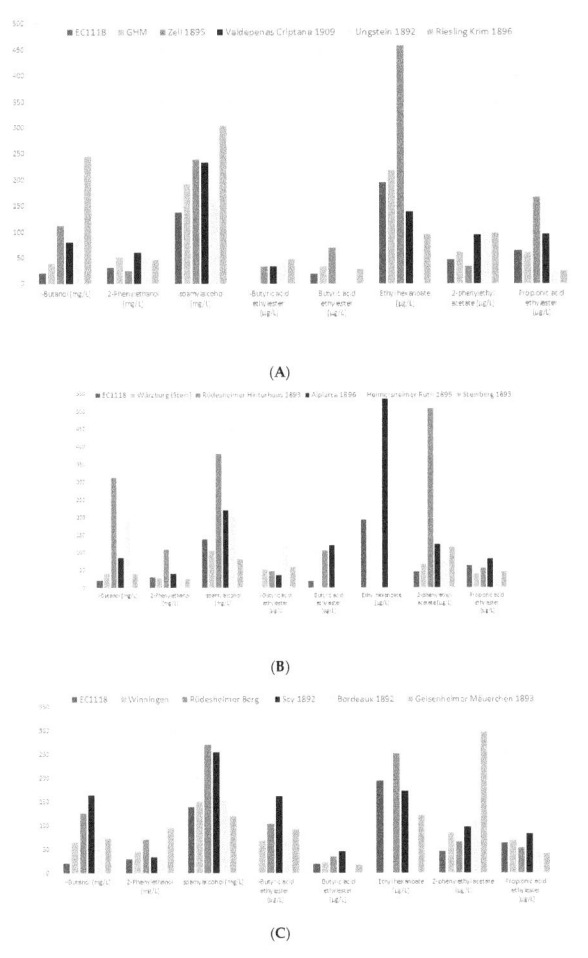

Figure 3. (**A–C**). Bar charts with selected alcohol (in mg/L) and ester (in μg/L) aroma compounds of *Saccharomyces* strains compared to the EC1118 wine yeast. Flavor compounds were measured using gas chromatography at the end of fermentation. The full list of aroma compounds is shown in Table S1.

Table 3. Aroma compound generation determined at the end of fermentation.*

	i-Butanol [mg/L]	2-Phenyl-Ethanol [mg/L]	Isoamyl Alcohol [mg/L]	i-Butyric Acid Ethyl-Ester [µg/L]	Butyric Acid Ethylester [µg/L]	Hexanoic Acid Ethyl-Ester [µg/L]	2-Phenylethyl Acetate [µg/L]	Propionic Acid Ethyl-Ester [µg/L]
EC1118	21	30	138	nq	20	195	46	64
GHM	40	51	192	nq	34	218	62	60
Saccharomyces cerevisiae								
Zell 1895	112	24	238	34	69	**457+**	34	**167**
Valdepenas Criptana 1909	81	60	233	33	nq	138	94	**95**
Ungstein 1892	90	52	163	41	nq	nq	106	23
Riesling Krim 1896	**246**	46	303	48	28	96	98	25
Rüdesheimer Hinterhaus 1893	**314**	**108**	**380**	47	**106**	nq	**507**	57
Alpiarca 1896	85	41	220	35	**120**	**536**	123	82
Heimersheimer Ruth 1895	187	44	202	**122**	nq	nq	74	46
Steinberg 1893	40	25	83	59	nq	nq	116	46
Rüdesheimer Berg	126	72	**271**	104	35	252	67	54
Scy 1892	164	34	255	**162**	46	174	98	82
Bordeaux 1892	83	31	155	77	nq	nq	110	46
Geisenheimer Mäuerchen 1893	74	**95**	120	92	18	121	**297**	41
S. kudriavzevii								
Würzburg (Stein)	42	29	106	54	nq	nq	68	40
Winningen	65	46	151	68	23	nq	85	68

* nq = not quantified; + = pronounced values are shown in bold and highlighted in orange.

What was interesting to note is that some strains produced either high levels of hexanoic acid ethylester (ethyl hexanoate) or high levels of acetic acid phenylethylester, but not of both substances. This was observed when comparing Rüdesheimer Hinterhaus 1893 with Zell 1895 and Alpiarca 1896. The first was low on hexanoic acid ethylester but produced high levels of 2-phenylethyl acetate, while the latter strains produced high levels of hexanoic acid ethylester and much lower levels of 2-phenylethyl acetate (Table 3, Figure 3; see Section 4).

4. Discussion

Alcoholic beverage production has been carried out by spontaneous fermentation throughout the ages, and today, often still the preferred method of fermentation for some. In beer and baking enterprises, it was already realized in the middle ages that there are special properties in the slurry that leavens bread and makes beer ferment and foam. It was the traditional occupation of a 'Hefner' in Germany to maintain and provide sufficient supplies of this leavening activity [16]. Yet, as it was not clear what the causal activity in the slurry was, the German 'Reinheitsgebot' (purity law) from 1516 only stated that beer should be brewed using barley malt, hops, and water. With the work of Pasteur, published in his Etudes sur le vin and Etudes sur la biere, and the work of others, it became evident that yeast, *Saccharomyces cerevisiae*, was responsible for the observed fermenting power.

The pure yeast strain isolation procedure worked out by Hansen was transferred to the wine industry by the German Julius Wortmann, who founded the 'Geisenheim Yeast Breeding Center' in 1894 [1,2]. The isolation of pure yeast strains in Geisenheim had been initiated in the early 1890s and these strains are still preserved today. Today's strain maintenance relies on deep freezing of culture collections at −80 °C, while supplies for the industry are generally provided as instant dry yeasts [17,18].

Yet, at the Geisenheim Yeast Breeding Center, stocks were originally maintained as slants with a paraffin overlay. Cultures were stored either in this manner or, even simpler, in plain water [5,6]. A recent report analyzed long-term storage (12 years) of >1000 stocks and provided evidence that water storage yielded survival rates of 98.9%, closely resembling that of frozen stocks (99.5%), while survival rates under a mineral oil layer were a bit less with 88.2% [10]. Two racks of Geisenheim yeast stocks were kept over the years, more as display items than out of necessity. Such a long-term storage has not been previously reported. In fact, it is not entirely clear anymore when exactly these stocks were generated (besides the fact that they are very old and between 30 and 60 years of age). Collectively, these studies show, however, that yeast strain collections can be routinely kept under a zero-emission regime, which could be used as an incentive to reduce the energy-demanding storage of cultures at −80 °C.

The 'old' Geisenheim yeast strains can be both a heritage and a source of new strains for yeast breeding programs. As a heritage, they could be used to strengthen the local character of wines produced with them and thus contribute to the terroir of these wines [19–21]. To be useful as a breeding stock, these old strains need to be characterized in more detail, preferably including their genomes [22,23]. In our study, the yeast with the most pronounced flavor production capability of alcohols and esters was the *S. cerevisiae* strain Rüdesheimer Hinterhaus 1893 ('Hinterhaus' refers to backyard). It was apparent that this strain is a remarkable producer of acetate esters, but not so of medium chain fatty esters. The former are produced by the acetyl transferases Atf1 and Atf2, while the latter are generated by the acyl-coenzymeA:ethanol O-acyltransferases Eeb1 and Eht1 [24,25]. This suggests strong activity of the ATF alcohol acetyl-coA transferases in the Rüdesheimer Hinterhaus strain isolated already in 1893, with reduced formation of medium chain fatty acids. It will be interesting to explore these properties through a full-scale wine fermentation trial and use in-depth genomics and transcriptomics to generate molecular markers for yeast breeding.

The strive for more volatile aromas has spurred the search for alternative yeasts and renewed the interest in spontaneous fermentations [26]. Yet, due to the unpredictable and inconsistent outcomes of spontaneous fermentations, the development of improved starter cultures or consortia may provide better alternatives [27,28].

5. Conclusions

In conclusion, our study of yeast isolates stored for a very long period opens several new research avenues in the use of these strains, either directly as starter cultures or as stocks for the Geisenheim Yeast Breeding Center; both will be interesting to exploit in the future. For general use, our data and previous work, e.g., on water cultures show that zero-emission strain-keeping of yeast cultures is feasible and should be more generally exploited.

Supplementary Materials: The following are available online at http://www.mdpi.com/2311-5637/6/1/9/s1, Table S1. Full list of aroma compounds.

Author Contributions: Conceptualization, J.W., D.R., and C.v.W.; methodology, K.M., B.B., S.B., H.S.; validation, J.W., K.M., B.B., S.B., H.S.; formal analysis, J.W., K.M., S.B., H.S.; investigation, J.W., K.M., B.B., S.B., H.S.; resources, J.W.; D.R., C.v.W. data curation, J.W., K.M., S.B., H.S.; writing—original draft preparation, J.W.; writing—review and editing, J.W., D.R., C.v.W., K.M., B.B., S.B., H.S.; visualization, K.M. and J.W. supervision, J.W. and D.R.; project administration, J.W. and D.R.; funding acquisition, J.W. All authors have read and agreed to the published version of the manuscript.

Funding: This research was funded in part by the European Union Marie Curie Initial Training Network Aromagenesis 764364 (http://www.aromagenesis.eu).

Acknowledgments: We thank Judith Muno-Bender for taking photographs of the old yeast samples; Stefanie Fritsch for calibrations of the analytical equipment and Bettina Mattner for excellent technical service and all Department members for input and support of this project.

Conflicts of Interest: The authors declare no conflict of interest.

References

1. Hansen, E.C. Recherches sur la physiologie et la morphologie des ferments alcooliques V. Methodes pour obtenir des cultures pures de *Saccharomyces* et de mikroorganismes analogues. *C R Trav. Lab. Carlsberg* **1888**, *2*, 143–167.
2. Wortmann, J. *Anwendung und Wirkung Reiner Hefen*; Verlagsbuchhandlung Paul Parey: Berlin, Germany, 1895.
3. Gélinas, P. Active dry yeast: Lessons from patents and science. *Compr. Rev. Food Sci. Saf.* **2019**, *18*, 1227–1255. [CrossRef]
4. Wellman, A.M.; Stewart, G.G. Storage of brewing yeasts by liquid nitrogen refrigeration. *Appl. Microbiol.* **1973**, *26*, 577–583. [CrossRef] [PubMed]
5. Castellani, A. Viability of some pathogenic fungi in distilled water. *J. Trop. Med. Hyg.* **1939**, *42*, 225–226.
6. Hartsell, S.E. The longevity of bacterial cultures under paraffin oil. *J. Bacteriol.* **1947**, *53*, 801. [PubMed]
7. Hartung de Capriles, C.; Mata, S.; Middelveen, M. Preservation of fungi in water (castellani): 20 years. *Mycopathologia* **1989**, *106*, 73–79. [CrossRef] [PubMed]
8. Rodrigues, E.G.; Lírio, V.S.; Lacaz, C.d.S. Preservation of fungi and actinomycetes of medical importance in distilled water. *Rev. Inst. Med. Trop. Sao Paulo* **1992**, *34*, 159–165. [CrossRef]
9. Richter, D.L. Revival of saprotrophic and mycorrhizal basidiomycete cultures after 20 years in cold storage in sterile water. *Can. J. Microbiol.* **2008**, *54*, 595–599. [CrossRef]
10. Karabicak, N.; Karatuna, O.; Akyar, I. Evaluation of the viabilities and stabilities of pathogenic mold and yeast species using three different preservation methods over a 12-year period along with a review of published reports. *Mycopathologia* **2016**, *181*, 415–424. [CrossRef]
11. Henry, B.S. Viability of yeast cultures preserved under mineral oil. *J. Bacteriol.* **1947**, *54*, 264. [CrossRef]
12. Hamelin, R.C.; Berube, P.; Gignac, M.; Bourassa, M. Identification of root rot fungi in nursery seedlings by nested multiplex pcr. *Appl. Environ. Microbiol.* **1996**, *62*, 4026–4031. [CrossRef] [PubMed]
13. Schneider, A.; Gerbi, V.; Redoglia, M. A Rapid HPLC method for separation and determination of major organic acids in grape musts and wines. *Am. J. Enol. Vitic* **1987**, *38*, 151–155.
14. Belda, I.; Ruiz, J.; Esteban-Fernández, A.; Navascués, E.; Marquina, D.; Santos, A. Microbial contribution to wine aroma and its intended use for wine quality improvement. *Molecules* **2017**, *22*, 189. [CrossRef] [PubMed]

15. Camara, J.S.; Alves, M.A.; Marques, J.C. Development of headspace solid-phase microextraction-gas chromatography–mass spectrometry methodology for analysis of terpenoids in Madeira wines. *Anal. Chim. Acta* **2006**, *555*, 191–200. [CrossRef]
16. Meusdoerffer, F.G. A Comprehensive History of Beer Brewing. In *Handbook of Brewing: Processes, Technology, Markets*; Eßlinger, H.M., Ed.; WILEY-VCH: Weinheim, Germany, 2009.
17. Querol, A.; Barrio, E.; Huerta, T.; Ramon, D. Molecular monitoring of wine fermentations conducted by active dry yeast strains. *Appl. Environ. Microbiol.* **1992**, *58*, 2948–2953. [CrossRef]
18. Beltran, G.; Torija, M.J.; Novo, M.; Ferrer, N.; Poblet, M.; Guillamon, J.M.; Rozes, N.; Mas, A. Analysis of yeast populations during alcoholic fermentation: A six year follow-up study. *Syst. Appl. Microbiol.* **2002**, *25*, 287–293. [CrossRef]
19. Bokulich, N.A.; Thorngate, J.H.; Richardson, P.M.; Mills, D.A. Microbial biogeography of wine grapes is conditioned by cultivar, vintage, and climate. *Proc. Natl. Acad. Sci. USA* **2014**, *111*, E139–E148. [CrossRef]
20. Gilbert, J.A.; van der Lelie, D.; Zarraonaindia, I. Microbial terroir for wine grapes. *Proc. Natl. Acad. Sci. USA* **2014**, *111*, 5–6. [CrossRef]
21. Capozzi, V.; Garofalo, C.; Chiriatti, M.A.; Grieco, F.; Spano, G. Microbial terroir and food innovation: The case of yeast biodiversity in wine. *Microbiol. Res.* **2015**, *181*, 75–83. [CrossRef]
22. Peter, J.; De Chiara, M.; Friedrich, A.; Yue, J.X.; Pflieger, D.; Bergstrom, A.; Sigwalt, A.; Barre, B.; Freel, K.; Llored, A.; et al. Genome evolution across 1,011 Saccharomyces cerevisiae isolates. *Nature* **2018**, *556*, 339–344. [CrossRef]
23. Langdon, Q.K.; Peris, D.; Baker, E.P.; Opulente, D.A.; Nguyen, H.V.; Bond, U.; Goncalves, P.; Sampaio, J.P.; Libkind, D.; Hittinger, C.T. Fermentation innovation through complex hybridization of wild and domesticated yeasts. *Nat. Ecol. Evol.* **2019**, *3*, 1576–1586. [CrossRef] [PubMed]
24. Saerens, S.M.; Delvaux, F.R.; Verstrepen, K.J.; Thevelein, J.M. Production and biological function of volatile esters in saccharomyces cerevisiae. *Microb. Biotechnol.* **2010**, *3*, 165–177. [CrossRef]
25. Saerens, S.M.; Verstrepen, K.J.; Van Laere, S.D.; Voet, A.R.; Van Dijck, P.; Delvaux, F.R.; Thevelein, J.M. The *Saccharomyces cerevisiae EHT1* and *EEB1* genes encode novel enzymes with medium-chain fatty acid ethyl ester synthesis and hydrolysis capacity. *J. Biol. Chem.* **2006**, *281*, 4446–4456. [CrossRef] [PubMed]
26. Bougreau, M.; Ascencio, K.; Bugarel, M.; Nightingale, K.; Loneragan, G. Yeast species isolated from Texas High Plains vineyards and dynamics during spontaneous fermentations of Tempranillo grapes. *PLoS ONE* **2019**, *14*, e0216246. [CrossRef] [PubMed]
27. Jolly, N.P.; Varela, C.; Pretorius, I.S. Not your ordinary yeast: Non-*Saccharomyces* yeasts in wine production uncovered. *FEMS Yeast Res.* **2014**, *14*, 215–237. [CrossRef] [PubMed]
28. Bagheri, B.; Bauer, F.F.; Setati, M.E. The impact of *Saccharomyces cerevisiae* on a wine yeast consortium in natural and inoculated fermentations. *Front. Microbiol.* **2017**, *8*, 1988. [CrossRef] [PubMed]

© 2020 by the authors. Licensee MDPI, Basel, Switzerland. This article is an open access article distributed under the terms and conditions of the Creative Commons Attribution (CC BY) license (http://creativecommons.org/licenses/by/4.0/).

fermentation

Article
Identification of Yeasts with Mass Spectrometry during Wine Production

Miroslava Kačániová [1,2,*], Simona Kunová [3], Jozef Sabo [1], Eva Ivanišová [4], Jana Žiarovská [5], Soňa Felsöciová [6] and Margarita Terentjeva [7]

[1] Department of Fruit Sciences, Viticulture and Enology, Faculty of Horticulture and Landscape Engineering, Slovak University of Agriculture, Tr. A. Hlinku 2, 94976 Nitra, Slovakia; sabododik@gmail.com
[2] Department of Bioenergy, Food Technology and Microbiology, Institute of Food Technology and Nutrition, University of Rzeszow, Zelwerowicza St. 4, 35601 Rzeszow, Poland
[3] Department of Food Hygiene and Safety, Faculty of Biotechnology and Food Sciences, Slovak University of Agriculture, Tr. A. Hlinku 2, 94976 Nitra, Slovakia; simona.kunova@uniag.sk or simona.kunova@gmail.com
[4] Department of Technology and Quality of Plant Products, Faculty of Biotechnology and Food Sciences, Slovak University of Agriculture, Tr. A. Hlinku 2, 94976 Nitra, Slovakia; eva.ivanisova@uniag.sk
[5] Department of Plant Genetics and Breeding, Faculty of Agrobiology and Food Resources, Slovak University of Agriculture, Tr. A. Hlinku 2, 94976 Nitra, Slovakia; jana.ziarovska@uniag.sk
[6] Department of Microbiology, Faculty of Biotechnology and Food Sciences, Slovak University of Agriculture, Tr. A. Hlinku 2, 94976 Nitra, Slovakia; sona.felsociova@uniag.sk
[7] Institute of Food and Environmental Hygiene, Faculty of Veterinary Medicine, Latvia University of Life Sciences and Technologies, K. Helmaņa iela 8, LV-3004 Jelgava, Latvia; margarita.terentjeva@llu.lv
* Correspondence: miroslava.kacaniova@gmail.com; Tel.: +421-905-499-166

Received: 16 November 2019; Accepted: 3 January 2020; Published: 7 January 2020

Abstract: The aim of the present study was to identify yeasts in grape, new wine "federweisser" and unfiltered wine samples. A total amount of 30 grapes, 30 new wine samples and 30 wine samples (15 white and 15 red) were collected from August until September, 2018, from a local Slovak winemaker, including Green Veltliner (3), Müller Thurgau (3), Palava (3), Rhein Riesling (3), Sauvignon Blanc (3), Alibernet (3), André (3), Blue Frankish (3), Cabernet Sauvignon (3), and Dornfelder (3) grapes; federweisser and unfiltered wine samples were also used in our study. Wort agar (WA), yeast extract peptone dextrose agar (YPDA), malt extract agar (MEA) and Sabouraud dextrose agar (SDA) were used for microbiological testing of yeasts. MALDI-TOF Mass Spectrometry (Microflex LT/SH) (Bruker Daltonics, Germany) was used for the identification of yeasts. A total of 1668 isolates were identified with mass spectrometry. The most isolated species from the grapes was *Hanseniaspora uvarum*, and from federweisser and the wine—*Saccharomyces cerevisiae*.

Keywords: yeasts; grape; federweisser; wine; microbiota identification; MALDI-TOF MS Biotyper

1. Introduction

Yeasts naturally occur in wines and vineyards and are especially common on the grapes. Population of yeast species on the grape is not constant and increases during the ripening process. *Kloeckera apiculata* is a lemon-like cell shape yeast, which colonizes the grape surface [1]. *Kloeckera apiculata* comprises more than 50% of the total healthy grape microbiota. Other yeasts like *Kloeckera* were isolated from the surface of the grapes, which included mainly genera *Metschnikowia*, *Candida*, *Cryptococcus*, *Pichia*, *Rhodotorula*, *Zygosaccharomyces* or *Kluyveromyces* [2]. The presence of yeasts of the genus *Aureobasidium* attracted attention as a transitional genus between yeast and microscopic fungi. All the yeasts associated with natural microbiota of grapes are wild yeast strains or non-saccharomyces. Despite the presence of those yeasts on the surface of grapes, the wine production consists of subsequent fermentation stages, which are typical for only particular yeast genera [3]. The *Saccharomyces* genus is the most important for the wine making process; however, this yeast is found on the grapes only in very small amounts. Previous studies that counted *Saccharomyces* on grapes found as little as 50 CFU/g. Mostly wild yeasts cultures could be found on the grapes and in freshly pressed must with colonization rates of 10^3 to 10^5 CFU/mL. During alcoholic fermentation, *Saccharomyces cerevisiae* is dominant, while yeasts in the *Pichia* and *Candida* genera are widespread in finished wine. The osmotolerant yeasts *Zygosaccharomyces* were reported in wines with higher content of residual sugar; yeasts of the *Brettanomyces* genus were common for wines in barrels [4,5].

The most important yeasts associated with wine production were Hanseniaspora uvarum (anamorph Kloeckera apiculata), Metschnikowia pulcherrima, Rhodotorula mucilaginosa, Rhodotorula glutinis, Aureobasidium pullulans, Cryptococcus magnus, Pichia manshurica, Pichia membranifaciens (anamorph Candida valida), Pichia fermentans, Pichia kluyveri, Pichia occidentalis (anamorph Candida sorbosa), Wickerhamomyces anomalus (anamorph Candida pelliculosa; Pichia anomala is synonymous), Cyberlindnera jadinii (Pichia jadinii is synonymous), Kregervanrija fluxuum (anamorph Candida vini), Candida stellata, Candida inconspicua, Meyerozyma guilliermondii, Zygosaccharomyces bailii, Brettanomyces bruxellensis (teleomorph Dekkera bruxellensis), Saccharomycodes ludwigii, Torulaspora delbrueckii and Saccharomyces cerevisiae. Kluyveromyces marxianus and Debaryomyces hansenii associated with grapes and are known as a contaminant in wine production. The microbiota of grapes creates better conditions for the growth of yeasts rather than bacteria. Low pH (pH 3–3.3), high content of sugars (mainly glucose) in grapes, and an anaerobic environment in must are necessary for ethanol fermentation of sugars, converting them into alcohol (ethanol) and CO_2 [5–8].

The aim of this study was to identify yeasts in grapes, federweisser and wine samples.

2. Materials and Methods

2.1. Collection of Grape, Federweisser and Wine Samples

An amount of 90 samples, including grape berries (n = 30), federweisser (n = 30) and wine (n = 30) of *Vitis vinifera* were collected aseptically in the viticultural area of Vrbové (approximately 48°37′12″ N and 017°43′25″ E) in 2018. The grape berry samples were transported on ice and stored at −20 °C until processing. The white grape varieties Green Veltliner, Müller Thurgau, Palava, Rhein Riesling and Sauvignon Blanc as well as red grape varieties Alibernet, André, Blue Frankish, Cabernet Sauvignon and Dornfelder were collected. Three sampling points in distal spatial points of different rows were used for sampling of grape berries. Grape samples were collected in August, and processed independently.

Samples of "federweisser" were collected at the end of August 2018 and in the middle of September 2018 from the same winery as the grapes. Samples were collected into 200 mL sterile plastic bottles and stored at 8 ± 1 °C in a refrigerator. Before testing, the samples (n = 30) were diluted with sterile physiological saline (0.85%). A total of 100 μL of each dilution (10^{-1} to 10^{-5}) was used for microbiological testing.

An amount of 200 mL of each unfiltered wine (before microfiltration) and immediately after were stored at 6–8 °C in a refrigerator. Collected wine samples were fermented with *Saccharomyces cerevisiae* in the producing process. The samples were later incubated in the laboratory at room temperature (25 ± 2 °C) for one week until the laboratory testing was initiated.

2.2. Cultivation Media

Wort agar (WA) (HiMedia, Mumbai, India), yeast extract peptone dextrose agar (YPDA) (Conda, Madrid, Spain), malt extract agar (MEA) (Biomark, Maharashtra, India) and Sabouraud dextrose agar (SDA) (Conda, Madrid, Spain) were used for identification of yeasts. All media were supplemented with chloramphenicol (100 mg/L) to inhibit bacterial growth. Chloramphenicol (Biolife, Monza, Italy) was added into cultivation media before sterilization by autoclaving at 115–121 °C for 15 min. The acid base indicator bromocresol green (BG, Biolofe, Monza, Italy) (20 mg/L) (pH range: 3.8–5.4) was added into the MEA and WA cultivation media before sterilization. Media for yeast cultivation were inoculated with 100 µL of the sample suspension. Inoculated agars were incubated at 25 °C for 3–5 days and the yeasts were identified by colony morphology (colour, surface, edge and elevation) and reinoculated onto trypton soya agar (TSA) (Oxoid, Basingstoke, UK). Yeast species were identified with a MALDI-TOF MS Biotyper.

2.3. Identification of Isolates with Mass Spectrometry

Qualitative analysis of yeasts isolates was performed with MALDI-TOF mass spectrometry (Bruker Daltonics, Bremen, Germany). Isolates were put in 300 µL of distilled water and 900 µL of ethanol, and the suspension centrifuged for 2 min at 14,000 rpm. The pellet was centrifuged repeatedly and allowed to dry. An amount of 30 µL of 70% formic acid was added to the pellet and 30 µL of acetonitrile. Tubes were centrifuged for 2 min at 14,000 rpm and 1 µL of the supernatant was used for MALDI identification. Once dry, every spot was overlaid with 1 µL of an HCCA matrix and left to dry at room temperature before analysis. Generated spectra were analyzed on a MALDI-TOF Microflex LT (Bruker Daltonics, Bremen, Germany) instrument using Flex Control 3.4 software and Biotyper Realtime Classification 3.1 with BC-specific software. Criteria for reliable identification were a score of ≥2.0 at species level [9].

2.4. Statistical Analysis

The statistical processing of the data obtained from each evaluation was done with Statgraphics Plus version 5.1 (AV Trading, Umex, Dresden, Germany). For each replication the mean was calculated, and the data set were log transformed. Descriptive statistics and logical-cognitive methods and one-way analysis ANOVA were used in the evaluation and statistical analysis.

3. Results and Discussion

Grapes are inhabited by versatile microbial groups and have a complex microbial ecology, including filamentous fungi, yeasts and bacteria. These microorganisms pose different physiological characteristics and may affect the wine quality. Some species of parasitic fungi or environmental bacteria might be only found in grapes, while other microorganisms like yeast, lactic acid and acetic acid bacteria occur during the winemaking process [10].

The yeast count in grape ranged from 2.34 (Greener Veltliner) to 2.67 (Dornfelder) log CFU/g on MEA, from 2.19 (Müller Thurgau) to 2.38 (Dornfelder) log CFU/g on WA, from 2.46 (Greener Veltliner) to 2.66 (Dornfelder) log CFU/g on YPDA, and from 1.55 (Greener Veltliner) to 1.88 (Dornfelder) log CFU/g on SDA. The colonization of grapes with yeasts is shown in Table 1.

Table 1. Yeasts counts in grape berries on different media.

Sample	MEA	WA	YPDA	SDA
	\multicolumn{4}{c}{Microbial Counts log CFU/g}			
Green Veltliner	2.37 ± 0.14	2.22 ± 0.06	2.46 ± 0.05	1.55 ± 0.14
Müller Thurgau	2.34 ± 0.09	2.19 ± 0.04	2.49 ± 0.04	1.66 ± 0.22
Palava	2.36 ± 0.13	2.20 ± 0.07	2.52 ± 0.01	1.69 ± 0.17
Rhein Riesling	2.43 ± 0.01	2.17 ± 0.06	2.51 ± 0.02	1.67 ± 0.16
Sauvignon Blanc	2.40 ± 0.03	2.19 ± 0.04	2.49 ± 0.02	1.65 ± 0.12
Alibernet	2.64 ± 0.10	2.26 ± 0.08	2.53 ± 0.03	1.73 ± 0.16
André	2.66 ± 0.07	2.33 ± 0.10	2.57 ± 0.06	1.79 ± 0.05
Blue Frankish	2.64 ± 0.03	2.36 ± 0.06	2.59 ± 0.06	1.82 ± 0.04
Cabernet Sauvignon	2.66 ± 0.03	2.34 ± 0.09	2.64 ± 0.01	1.84 ± 0.02
Dornfelder	2.67 ± 0.05	2.38 ± 0.04	2.66 ± 0.04	1.88 ± 0.06

WA—wort agar; YPDA—yeast extract peptone dextrose agar; MEA—malt extract agar; SDA—Sabouraud dextrose agar.

ANOVA analysis was performed to inspect the significant differences among the microbial count for individual wine varieties when different cultivation media were used (Table 2).

Table 2. One-way ANOVA for analyzed wine varieties—grapes.

Cultivation Media	Source	Sum of Squares	Degrees of Freedom	Mean Square	F Statistic	p-Value
MEA	treatment	0.5779	9	0.0642	9.55	1.65×10^{-5}
	error	0.1347	20	0.0067		
	total	0.7125	29			
WA	treatment	0.1755	9	0.0195	4.26	0.0025
	error	0.0868	20	0.0043		
	total	0.2623	29			
YPDA	treatment	0.1179	9	0.0131	8.74	3.77×10^{-5}
	error	0.0307	20	0.0015		
	total	0.1487	29			
SDA	treatment	0.2863	9	0.0318	1.13	0.1287
	error	0.3513	20	0.0176		
	total	0.6376	29			

WA—wort agar; YPDA—yeast extract peptone dextrose agar; MEA—malt extract agar; SDA—Sabouraud dextrose agar.

Statistically significant differences among microbial counts for individual cultivation media were found in three of the four cultivation media used (Table 3).

Table 3. Significant differences among analyzed grape varieties for individual cultivation media.

Treatments Pair	Tukey HSD p-Value	Tukey HSD Inferfence
MEA		
A vs. F	0.0180304	* $p < 0.05$
A vs. G	0.0085086	** $p < 0.01$
A vs. H	0.0180304	* $p < 0.05$
A vs. I	0.0076352	** $p < 0.01$
A vs. J	0.0068497	** $p < 0.01$
B vs. F	0.0085086	** $p < 0.01$
B vs. G	0.0039776	** $p < 0.01$
B vs. H	0.0085086	** $p < 0.01$
B vs. I	0.0035659	** $p < 0.01$
B vs. J	0.0032002	** $p < 0.01$
C vs. F	0.0130929	* $p < 0.05$
C vs. G	0.0061450	** $p < 0.01$
C vs. H	0.0130929	* $p < 0.05$
C vs. I	0.0055127	** $p < 0.01$
C vs. J	0.0049454	** $p < 0.01$
E vs. G	0.0247440	* $p < 0.05$
E vs. I	0.0222763	* $p < 0.05$
E vs. J	0.0200452	* $p < 0.05$
WA		
D vs. H	0.0395384	* $p < 0.05$
D vs. J	0.0206542	* $p < 0.05$
YPDA		
A vs. H	0.0262122	* $p < 0.05$
A vs. I	0.0010053	** $p < 0.01$
A vs. J	0.0010053	** $p < 0.01$
B vs. I	0.0068426	** $p < 0.01$

Table 3. Cont.

Treatments Pair	Tukey HSD p-Value	Tukey HSD Inferference
YPDA		
B vs. J	0.0011070	** $p < 0.01$
C vs. J	0.0085840	** $p < 0.01$
D vs. I	0.0210350	* $p < 0.05$
D vs. J	0.0034566	** $p < 0.01$
E vs. I	0.0043409	** $p < 0.01$
E vs. J	0.0010053	** $p < 0.01$
F vs. J	0.0168484	* $p < 0.05$

A—Green Veltliner; B—Müller Thurgau; C—Palava; D—Rhein Riesling; E—Sauvignon Blanc; F—Alibernet; G—André; H—Blue Frankish; I—Cabernet Sauvignon; J—Dornfelder; WA—wort agar; YPDA—yeast extract peptone dextrose agar; MEA—malt extract agar; SDA—Sabouraud dextrose agar.

Different studies have evaluated the surface microbiota of grape berries due to a possible impact on the hygienic state of the grapes and the direct influence on the winemaking process and wine quality [11–18].

The yeasts count in "federweisser" ranged from 3.51 in Greener Veltliner and Palava to 3.80 log CFU/mL in Dornfelder on MEA. On WA, the yeasts count from 3.30 in Palava to 3.53 log CFU/mL in Dornfelder were observed. On YPDA, the yeasts count varied from 3.24 in Rhein Riesling to 3.45 log CFU/mL in Dornfelder, and from 3.13 (Sauvignon Blanc) to 3.33 (Dornfelder) log CFU/mL on SDA. Yeasts counts in federweisser are summarized in Table 4.

Table 4. Yeast counts in "federweisser" on different media.

Sample	MEA	WA	YPDA	SDA
	log CFU/g			
Green Veltliner	3.51 ± 0.15	3.41 ± 0.06	3.37 ± 0.14	3.21 ± 0.01
Müller Thurgau	3.54 ± 0.10	3.38 ± 0.06	3.30 ± 0.04	3.19 ± 0.03
Palava	3.51 ± 0.05	3.30 ± 0.06	3.27 ± 0.04	3.16 ± 0.05
Rhein Riesling	3.58 ± 0.06	3.34 ± 0.01	3.24 ± 0.01	3.14 ± 0.02
Sauvignon Blanc	3.56 ± 0.10	3.36 ± 0.05	3.27 ± 0.05	3.13 ± 0.02
Alibernet	3.67 ± 0.08	3.40 ± 0.06	3.29 ± 0.06	3.17 ± 0.03
André	3.70 ± 0.07	3.43 ± 0.02	3.36 ± 0.05	3.18 ± 0.06
Blue Frankish	3.74 ± 0.02	3.46 ± 0.05	3.39 ± 0.05	3.24 ± 0.09
Cabernet Sauvignon	3.76 ± 0.05	3.48 ± 0.06	3.41 ± 0.01	3.30 ± 0.04
Dornfelder	3.80 ± 0.07	3.53 ± 0.03	3.45 ± 0.06	3.33 ± 0.01

WA—wort agar; YPDA—yeast extract peptone dextrose agar; MEA—malt extract agar; SDA—Sabouraud dextrose agar.

In study in Slovakia [19], the highest yeasts counts were on MEA for Pinot Noir—6.43 log CFU/mL and the lowest for Moravian Muscat—4.62 log CFU/mL. The highest yeasts count on WA were in Pinot Noir—6.39 log CFU/mL, but the lowest in Irsai Oliver—5.38 log CFU/mL. The highest count of yeasts on wild yeast medium (WYM) was in Blue Frankish 6.33 log CFU/mL and the lowest in Dornfelder 4.20 log CFU/mL [19].

As the results show, a higher number of yeasts were detected in "federweisser" than in grape. The young wine is a product of fermentation where *S. cerevisiae* was mostly found. Other species like *Hanseniaspora uvarum*, *Metschnikowia pulcherrima* or the genera *Pichia* or *Candida* may be present during the individual fermentation stages when the alcohol content do not exceed 4–6% [5,20]. The main microbiota of the grape is the yeast *Hanseniaspora uvarum* followed by *Metschnikowia pulcherrima* [4]. These species also initiate the pre-alcoholic fermentation but are being replaced by the dominant *S. cerevisiae* 3–4 days after fermentation. *Saccharomyces cerevisiae* starts to multiply within 20 days after inoculation into the must [21].

ANOVA analysis was performed to inspect the significant differences among the microbial count for individual wine varieties when different cultivation media were used (Table 5).

Table 5. One-way ANOVA results for the analyzed wine varieties—federweisser.

Cultivation Media	Source	Sum of Squares	Degrees of Freedom	Mean Square	F Statistic	p-Value
MEA	treatment	0.3717	9	0.0413	6.36	0.0002
	error	0.1234	20	0.0062		
	total	0.4951	29			
WA	treatment	0.1305	9	0.0145	5.23	0.0007
	error	0.0525	20	0.0026		
	total	0.1831	29			
YPDA	treatment	0.1426	9	0.0158	3.36	0.0056
	error	0.0818	20	0.0041		
	total	0.2244	29			
SDA	treatment	0.1194	9	0.0133	6.91	0.0002
	error	0.0397	20	0.0020		
	total	0.1591	29			

WA—wort agar; YPDA—yeast extract peptone dextrose agar; MEA—malt extract agar; SDA—Sabouraud dextrose agar.

Statistically significant differences among microbial counts for individual cultivation media were found in three of the four cultivation media used (Table 6).

Table 6. Significant differences among analyzed federweisser samples for individual cultivation media.

Reatments Pair	Tukey HSD p-Value	Tukey HSD Inferfence
MEA		
A vs. H	0.0368638	* $p < 0.05$
A vs. I	0.0238533	* $p < 0.05$
A vs. J	0.0055561	** $p < 0.01$
B vs. H	0.0456379	* $p < 0.05$
B vs. I	0.0296858	* $p < 0.05$
B vs. J	0.0069685	** $p < 0.01$
C vs. H	0.0410343	* $p < 0.05$
C vs. I	0.0266151	* $p < 0.05$
C vs. J	0.0062207	** $p < 0.01$
E vs. H	0.0456379	* $p < 0.05$
E vs. I	0.0296858	* $p < 0.05$
E vs. J	0.0069685	** $p < 0.01$
WA		
C vs. H	0.0329699	* $p < 0.05$
C vs. I	0.0100494	* $p < 0.05$
C vs. J	0.0010463	** $p < 0.01$
D vs. J	0.0084520	** $p < 0.01$
E vs. J	0.0071077	** $p < 0.01$
YPDA		
D vs. J	0.0105986	* $p < 0.05$
E vs. J	0.0359336	* $p < 0.05$
SDA		
B vs. J	0.0319805	* $p < 0.05$
C vs. I	0.0217427	* $p < 0.05$
C vs. J	0.0044497	** $p < 0.01$

Table 6. *Cont.*

Reatments Pair	Tukey HSD *p*-Value	Tukey HSD Inferrence
SDA		
D vs. I	0.0081105	** $p < 0.01$
D vs. J	0.0016306	** $p < 0.01$
E vs. I	0.0036406	** $p < 0.01$
E vs. J	0.0010053	** $p < 0.01$
F vs. I	0.0466723	* $p < 0.05$
F vs. J	0.0098975	** $p < 0.01$
G vs. J	0.0178947	* $p < 0.05$

A—Green Veltliner; B—Müller Thurgau; C—Palava; D—Rhein Riesling; E—Sauvignon Blanc; F—Alibernet; G—André; H—Blue Frankish; I—Cabernet Sauvignon; J—Dornfelder; WA—wort agar; YPDA—yeast extract peptone dextrose agar; MEA—malt extract agar; SDA—Sabouraud dextrose agar.

The yeast counts in the unfiltered wines are summarized in Table 7. The yeast counts in wine ranged from 1.51 (Greener Veltliner) to 3.23 (Dornfelder) log CFU/mL on MEA, from 1.43 (Greener Veltliner) to 2.89 (Dornfelder) log CFU/mL on WA, from 1.18 (Greener Veltliner) to 2.65 (Dornfelder) log CFU/mL on YPDA and from 1.09 (Rhein Riesling) to 2.21 (Dornfelder) log CFU/mL on SDA.

Table 7. Yeast counts in wine on different media.

Sample	MEA	WA	YPDA	SDA
	\multicolumn{4}{c}{log CFU/g}			
Green Veltliner	1.51 ± 0.27	1.43 ± 0.15	1.18 ± 0.06	1.13 ± 0.03
Müller Thurgau	1.55 ± 0.32	1.52 ± 0.01	1.21 ± 0.06	1.18 ± 0.06
Palava	1.62 ± 0.34	1.49 ± 0.03	1.27 ± 0.11	1.12 ± 0.02
Rhein Riesling	1.73 ± 0.17	1.46 ± 0.05	1.27 ± 0.14	1.09 ± 0.06
Sauvignon Blanc	1.76 ± 0.11	1.48 ± 0.06	1.32 ± 0.23	1.14 ± 0.02
Alibernet	2.57 ± 0.50	2.37 ± 0.14	2.21 ± 0.05	1.51 ± 0.64
André	2.67 ± 0.42	2.41 ± 0.10	2.31 ± 0.13	1.83 ± 0.62
Blue Frankish	2.59 ± 0.28	2.44 ± 0.16	2.38 ± 0.11	2.13 ± 0.12
Cabernet Sauvignon	2.89 ± 0.37	2.51 ± 0.14	2.47 ± 0.04	2.17 ± 0.07
Dornfelder	3.23 ± 0.02	2.89 ± 0.27	2.65 ± 0.22	2.21 ± 0.06

WA—wort agar; YPDA—yeast extract peptone dextrose agar; MEA—malt extract agar; SDA—Sabouraud dextrose agar.

ANOVA analysis was performed to inspect the significant differences among the microbial count for individual wine varieties when different cultivation media were used (Table 8).

Table 8. One-way ANOVA results for analyzed wine varieties—unfiltered wine.

Cultivation Media	Source	Sum of Squares	Degrees of Freedom	Mean Square	F Statistic	*p*-Value
MEA	treatment	11.0908	9	1.23	12.60	1.95×10^{-6}
	error	1.45	20	0.0982		
	total	13.0552	29			
WA	treatment	8.05	9	0.9745	54.8807	3.84×10^{-12}
	error	0.3551	20	0.0178		
	total	9.1256	29			
YPDA	treatment	10.74	9	1.1542	65.9142	6.70×10^{-13}
	error	0.3502	20	0.0175		
	total	10.76	29			
SDA	treatment	6.61	9	0.7118	8.13	3.51×10^{-5}
	error	1.31	20	0.0827		
	total	8.0592	29			

WA—wort agar; YPDA—yeast extract peptone dextrose agar; MEA—malt extract agar; SDA—Sabouraud dextrose agar.

Statistically significant differences among microbial count for individual cultivation media were found in three of the four cultivation media used (Table 9).

Table 9. Significant differences among unfiltered wine samples for individual cultivation media.

Treatments Pair	Tukey HSD p-Value	Tukey HSD Inferfence
MEA		
A vs. F	0.0141715	* $p < 0.05$
A vs. G	0.0064264	** $p < 0.01$
A vs. H	0.0119733	* $p < 0.05$
A vs. I	0.0010053	** $p < 0.01$
A vs. J	0.0010053	** $p < 0.01$
B vs. F	0.0192566	* $p < 0.05$
B vs. G	0.0087792	** $p < 0.01$
B vs. H	0.0162964	* $p < 0.05$
B vs. I	0.0012652	** $p < 0.01$
B vs. J	0.0010053	** $p < 0.01$
C vs. F	0.0342577	* $p < 0.05$
C vs. G	0.0158481	* $p < 0.05$
C vs. H	0.0290959	* $p < 0.05$
C vs. I	0.0023022	** $p < 0.01$
C vs. J	0.0010053	** $p < 0.01$
D vs. G	0.0381648	* $p < 0.05$
D vs. I	0.0057354	** $p < 0.01$
D vs. J	0.0010053	** $p < 0.01$
E vs. I	0.0078360	** $p < 0.01$
E vs. J	0.0010053	** $p < 0.01$
WA		
A vs. F	0.0010053	** $p < 0.01$
A vs. G	0.0010053	** $p < 0.01$
A vs. H	0.0010053	** $p < 0.01$
A vs. I	0.0010053	** $p < 0.01$
A vs. J	0.0010053	** $p < 0.01$
B vs. F	0.0010053	** $p < 0.01$
B vs. G	0.0010053	** $p < 0.01$
B vs. H	0.0010053	** $p < 0.01$
B vs. I	0.0010053	** $p < 0.01$
B vs. J	0.0010053	** $p < 0.01$
C vs. F	0.0010053	** $p < 0.01$
C vs. G	0.0010053	** $p < 0.01$
C vs. H	0.0010053	** $p < 0.01$
C vs. I	0.0010053	** $p < 0.01$
C vs. J	0.0010053	** $p < 0.01$
D vs. F	0.0010053	** $p < 0.01$
D vs. G	0.0010053	** $p < 0.01$
D vs. H	0.0010053	** $p < 0.01$
D vs. I	0.0010053	** $p < 0.01$
D vs. J	0.0010053	** $p < 0.01$
E vs. F	0.0010053	** $p < 0.01$
E vs. G	0.0010053	** $p < 0.01$
E vs. H	0.0010053	** $p < 0.01$
E vs. I	0.0010053	** $p < 0.01$
E vs. J	0.0010053	** $p < 0.01$
F vs. J	0.0037886	** $p < 0.01$
G vs. J	0.0079079	** $p < 0.01$
H vs. J	0.0143649	* $p < 0.05$

Table 9. Cont.

Treatments Pair	Tukey HSD p-Value	Tukey HSD Inferfence
YPDA		
A vs. F	0.0010053	** $p < 0.01$
A vs. G	0.0010053	** $p < 0.01$
A vs. H	0.0010053	** $p < 0.01$
A vs. I	0.0010053	** $p < 0.01$
A vs. J	0.0010053	** $p < 0.01$
B vs. F	0.0010053	** $p < 0.01$
B vs. G	0.0010053	** $p < 0.01$
B vs. H	0.0010053	** $p < 0.01$
B vs. I	0.0010053	** $p < 0.01$
B vs. J	0.0010053	** $p < 0.01$
C vs. F	0.0010053	** $p < 0.01$
C vs. G	0.0010053	** $p < 0.01$
C vs. H	0.0010053	** $p < 0.01$
C vs. I	0.0010053	** $p < 0.01$
C vs. J	0.0010053	** $p < 0.01$
D vs. F	0.0010053	** $p < 0.01$
D vs. G	0.0010053	** $p < 0.01$
D vs. H	0.0010053	** $p < 0.01$
D vs. I	0.0010053	** $p < 0.01$
D vs. J	0.0010053	** $p < 0.01$
E vs. F	0.0010053	** $p < 0.01$
E vs. G	0.0010053	** $p < 0.01$
E vs. H	0.0010053	** $p < 0.01$
E vs. I	0.0010053	** $p < 0.01$
E vs. J	0.0010053	** $p < 0.01$
F vs. J	0.0154088	* $p < 0.05$
SDA		
A vs. H	0.0103393	* $p < 0.05$
A vs. I	0.0071377	** $p < 0.01$
A vs. J	0.0044827	** $p < 0.01$
B vs. H	0.0168657	* $p < 0.05$
B vs. I	0.0116918	* $p < 0.05$
B vs. J	0.0073639	** $p < 0.01$
C vs. H	0.0100244	* $p < 0.05$
C vs. I	0.0069202	** $p < 0.01$
C vs. J	0.0043452	** $p < 0.01$
D vs. H	0.0075929	** $p < 0.01$
D vs. I	0.0052344	** $p < 0.01$
D vs. J	0.0032871	** $p < 0.01$
E vs. H	0.0116918	* $p < 0.05$
E vs. I	0.0080777	** $p < 0.01$
E vs. J	0.0050746	** $p < 0.01$

A—Green Veltliner; B—Müller Thurgau; C—Palava; D—Rhein Riesling; E—Sauvignon Blanc; F—Alibernet; G—André; H—Blue Frankish; I—Cabernet Sauvignon; J—Dornfelder; WA—wort agar; YPDA—yeast extract peptone dextrose agar; MEA—malt extract agar; SDA—Sabouraud dextrose agar.

Altogether, 1668 isolates were identified with mass spectrometry with a score of ≥2.0 (Table 10). The most isolated species from grape was *Hanseniaspora uvarum* (70 isolates), and from "federweisser" and wine *S. cerevisiae* (85 and 120 isolates, respectively). Yeasts species of grape, "frederweisser" and wine are shown in Figures 1–3.

Table 10. Yeasts species in grape, "federweisser" and wine.

Yeast Species	Grape	"Federweisser" No. of Isolates	Wine
Aureobasidium pullulans	25	0	0
Candida inconspicua	5	0	5
Candida parapsilosis	5	0	10
Candida saitoana	5	0	5
Candida sake	5	0	5
Cyberlindnera jadinii	0	0	8
Debaryomyces hansenii	0	0	15
Dekkera bruxellensis	0	0	25
Filobasidium magnum	30	0	0
Hanseniaspora uvarum	70	25	0
Issatchenkia orientalis	38	0	0
Kazachstania exigua	33	0	0
Kluyveromyces marxianus	35	0	32
Kregervanrija fluxuum	0	0	25
Metschnikowia pulcherrima	28	50	55
Meyerozyma guilliermondii	0	0	52
Naganishia diffluens	25	0	0
Pichia fermentans	10	0	58
Pichia kluyveri	12	49	50
Pichia mandshurica	10	0	25
Pichia membranifaciens	25	0	15
Pichia norvegensis	10	0	5
Pichia occidentalis	10	35	45
Rhodotorula glutinis	40	0	20
Rhodotorula mucilaginosa	25	0	45
Saccharomyces cerevisiae	0	85	120
Starmerella magnolia	30	0	15
Torulaspora delbrueckii	12	0	39
Wickerhamomyces anomalus	15	0	75
Yarrowia lipolytica	20	0	10
Zygosaccharomyces bailii	15	0	52
Zygotorulaspora florentina	25	0	50
Total	563	244	861

Brettanomyces bruxellensis, Candida stellata, Saccharomyces cerevisiae and *Zygosaccharomyces bailii* were the yeasts identified in wine [22–25]. In our study, *Pichia mandshurica*—the main contaminant of wines—was present in 66% samples of white wines (10 out of 15) and in seven samples of red wines (46%). *Pichia membranifaciens* was isolated from five samples of white (33%) and five samples of red wines (33%). *Saccharomyces cerevisiae* was isolated from all white and red wines (100%). *Zygosaccharomyces bailii* was found in 14 samples of white (93%) and two samples of red (13%) wines. Our study shows that *Z. bailii* and *P. mandshurica* were isolated more frequently from white than from red wines, while *S. cerevisiae* was identified in white and red wines. The occurrence of *Pichia manshurica* and *S. cerevisiae* was different between the wine samples. According to Thomas [26], the presence of *Zygosaccharomyces* in wine is unacceptable in terms of wine quality. The author has stated that the minimum number of yeast present in wine spoils the product under appropriate conditions [26]. *Saccharomyces cerevisiae, Debaryomyces hansenii, Wickerhamomyces anomalus (Pichia anomala), Pichia membranifaciens, Rhodotorula glutinis, Rhodotorula mucilaginosa, Torulaspora delbrueckii, Kluyveromyces marxianus, Issatchenkia orientalis, Zygosaccharomyces bailii parapsilosis, Pichia fermentans* and *Hanseniaspora uvarum* are frequent contaminants of wines as well [27,28]. However, Renous [29] did not describe associations between wine and *Pichia manshurica, Kregervanrija fluxuum (Candida vini), Candida inconspicua* and *Zygotorulaspora florentina*. Saez [30] found that *S. cerevisiae* (13.93%),

Wickerhamomyces anomalus (8.72%), *Pichia fermentans* (6.74%) and *Metschnikowia pulcherrima* (6.39%) were the most abundant in wine. *Pichia* (*Pichia manshurica*, *P. membranifaciens*) and *Brettanomyces* are producing volatile phenols, thereby affecting the quality of the wine [30].

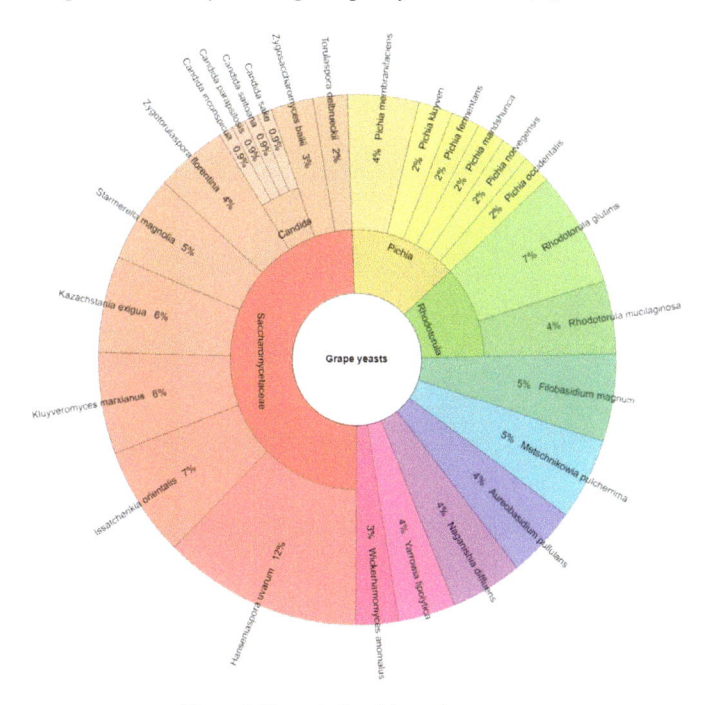

Figure 1. Yeasts isolated from the grapes.

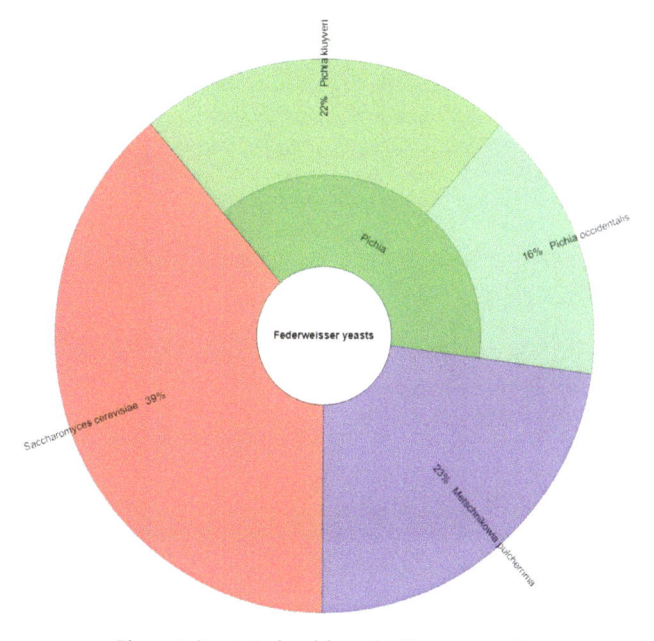

Figure 2. Yeasts isolated from the "federweisser".

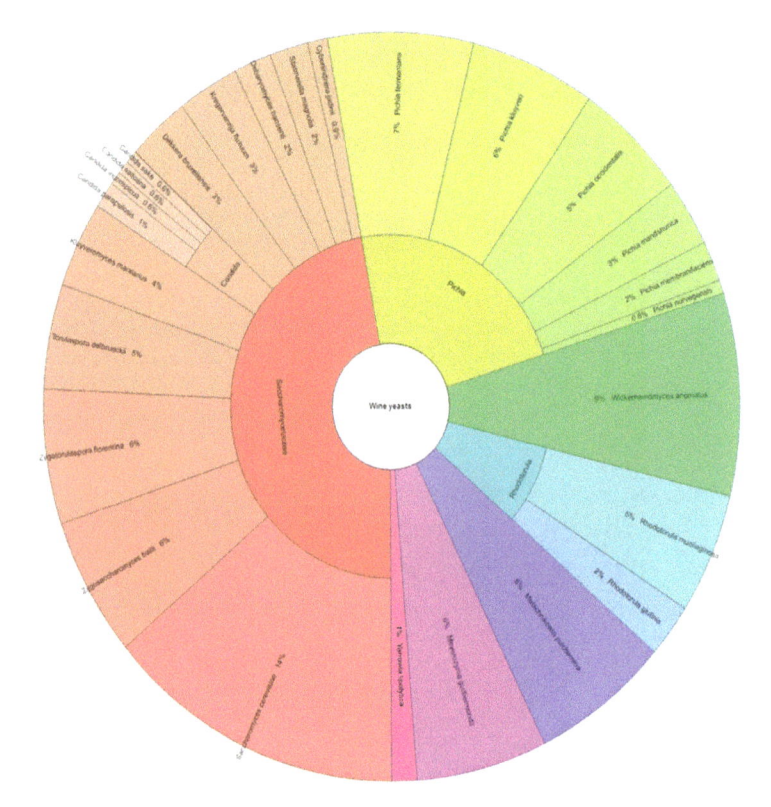

Figure 3. Yeasts isolated from the wine.

Sporadically, *Candida inconspicua* (5 isolates, 0.58%), *Candida saitoana* (5 isolates, 0.58%), *Candida sake* (5 isolates, 0.58%), *Pichia norvegensis* (5 isolates, 0.58%) and other species were isolated. Jolly et al. [31] noticed the importance of *Candida*, *Cryptococcus*, *Kloeckera* and *Rhodotorula* species in the wine making process. *Candida* was considered as the dominant genus, including their teleomorphic stages—*Candida pulcherrima* (*Metschnikowia pulcherrima*), *Candida vini* (*Kregervanrija fluxuum*) and *Candida valida* (*Pichia membranifaciens*) [31].

4. Conclusions

A total of 90 samples (30 from grapes, 30 of "federweisser" and 30 of wine) was studied for characterization of the yeast species. The mass spectrometry method was used for identification of 1668 grape, "federweisser" and wine isolates. From grape, 26 species of 17 genera within 9 families, and in "federweisser" 4 species of 3 genera and families were found. In wine, 26 species of 17 genera within 6 families were identified. *Rhodoturulla* species were not included in any family and they were classified as incertae sedis (not belonging anywhere).

Author Contributions: M.K., M.T., J.Ž. were responsible for the design of the study; M.K., S.K., J.S., S.F., conducted the study and collected the samples; M.K., S.K., J.S., J.Ž. performed the laboratory analysis; M.K., S.K., E.I., M.T. were responsible for writing and editing the manuscript; all authors have carefully revised and approved the final version of the manuscript. All authors have read and agreed to the published version of the manuscript.

Funding: This work has been supported by the grants of the Slovak Research and Development Agency No. VEGA 1/0411/17.

Acknowledgments: The Paper was supported by the project: The research leading to these results has received funding from the European Community under project no. 26220220180: Building Research Centre "AgroBioTech".

Conflicts of Interest: The authors declare no conflict of interest.

References

1. Padilla, B.; Gil, J.V.; Manzanares, P. Past and future of non-saccharomyces yeasts: From spoilage microorganisms to biotechnological tools for improving wine aroma complexity. *Front. Microbiol.* **2016**, *7*, 411. [CrossRef]
2. Kantor, A.; Kačániová, M.; Kluz, M. Natural microflora of wine grape berries. *J. Microbiol. Biotech. Food Sci.* **2015**, *4*, 32–36. [CrossRef]
3. Bozoudi, D.; Tsaltas, D. The multiple and versatile roles of *Aureobasidium pullulans* in the vitivinicultural sector. *Fermentation* **2018**, *4*, 85. [CrossRef]
4. Fugelsang, K.C.; Edwards, C.G. *Wine Microbiology: Practical Applications and Procedures*, 2nd ed.; Springer: New York, NY, USA, 2007; p. 393.
5. König, H.; Fröhlich, J.; Unden, G. *Biology of Microorganisms on Grapes, in Must and in Wine*; Springer: Berlin/Heidelberg, Germany, 2009; p. 522.
6. Fleet, G.H. Wine. In *Food Microbiology Fundamentals & Frontiers*; Doyle, M.P., Beuchat, L.R., Montville, T.J., Eds.; ASM Press: Washington, DC, USA, 2001; pp. 747–772.
7. Romano, P.; Fiore, C.; Paraggio, M.; Caruso, M.; Capece, A. Function of yeast species and strains in wine flavor. *Int. J. Food Microbiol.* **2003**, *86*, 169–180. [CrossRef]
8. Kurtzman, C.P.; Fell, J.W.; Boekhout, T. *The Yeasts: A Taxonomic Study*, 5th ed.; Elsevier Science: Amsterdam, The Netherlands, 2011; p. 2384.
9. Kačániová, M.; Mellen, M.; Vukovic, N.L.; Kluz, M.; Puchalski, C.; Haščík, P.; Kunová, S. Combined effect of vacuum packaging, fennel and svory essential oil treatment on the quality of chicken thighs. *Microorganisms* **2019**, *7*, 134. [CrossRef] [PubMed]
10. Barata, A.; Malfeito-Ferreira, M.; Loureiro, V. The microbial ecology of wine grape berries. *Int. J. Food Microbiol.* **2012**, *153*, 243–259. [CrossRef]
11. Martins, G.; Miot-Sertier, C.; Lauga, B.; Claisse, O.; Lonvaud-Funel, A.; Soulas, G.; Masneuf-Pomarède, I. Grape berry bacterial microbiota: Impact of the ripening process and the farming system. *Int. J. Food Microbiol.* **2012**, *158*, 93–100. [CrossRef]
12. Martins, G.; Vallance, J.; Mercier, A.; Albertin, W.; Stamatopoulos, P.; Rey, P.; Lonvaud, A.; Masneuf-Pomarède, I. Influence of the farming system on the epiphytic yeasts and yeast-like fungi colonizing grape berries during the ripening process. *Int. J. Food Microbiol.* **2014**, *177*, 21–28. [CrossRef]
13. Setati, M.E.; Jacobson, D.; Andong, U.C.; Bauer, F. The vineyard yeast microbiome, a mixed model microbial map. *PLoS ONE* **2012**, *7*, e52609. [CrossRef]
14. Bokulich, N.A.; Thorngate, J.H.; Richardson, P.M.; Mills, D.A. Microbial biogeography of wine grapes is conditioned by cultivar, vintage, and climate. *Proc. Natl. Acad. Sci. USA* **2014**, *111*, E139–E148. [CrossRef]
15. Portillo, M.; Del, C.; Franquès, J.; Araque, I.; Reguant, C.; Bordons, A. Bacterial diversity of Grenache and Carignan grape surface from different vineyards at Priorat wine region (Catalonia, Spain). *Int. J. Food Microbiol.* **2016**, *219*, 56–63. [CrossRef] [PubMed]
16. Salvetti, E.; Campanaro, S.; Campedelli, I.; Fracchetti, F.; Gobbi, A.; Tornielli, G.B.; Torriani, S.; Felis, G.E. Whole-metagenome-sequencing-based community profiles of *Vitis vinifera* L. cv. Corvina berries withered in two post-harvest conditions. *Front. Microbiol.* **2016**, *7*, 937. [CrossRef]
17. Grangeteau, C.; Roullier-Gall, C.; Rousseaux, S.; Gougeon, R.D.; Schmitt-Kopplin, P.; Alexandre, H.; Guilloux-Benatier, M. Wine microbiology is driven by vineyard and winery anthropogenic factors. *Microb. Biotechnol.* **2017**, *10*, 354–370. [CrossRef] [PubMed]
18. Oliveira, M.; Arenas, M.; Lage, O.; Cunha, M.; Amorim, M.I. Epiphytic fungal community in *Vitis vinifera* of the Portuguese wine regions. *Lett. Appl. Microbiol.* **2018**, *66*, 93–102. [CrossRef] [PubMed]
19. Kántor, A.; Petrová, J.; Hutková, J.; Kačániová, M. Yeast diversity in new, still fermenting wine "federweisser". *Potravinarstvo* **2016**, *10*, 120–125. [CrossRef]
20. Malík, F.; Furdíková, K.; Roman, T.; Malík, F. *Vinársky Lexicon*, 1st ed.; Fedor Malík a syn: Modra, Slovakia, 2012; p. 144.

21. Ribéreau-Gayon, P.; Dubourdieu, D.; Donèche, B.; Lonvaud, A. *Handbook of Enology, the Microbiology of Wine and Vinifications*; John Wiley & Sons Ltd.: West Sussex, UK, 2006; Volume 1, p. 512.
22. Millet, V.; Lonvaud-Funel, A. The viable but non-culturable state of wine microorganisms during storage. *Lett. Appl. Microbiol.* **2000**, *30*, 136–141. [CrossRef]
23. Divol, B.; Lonvaud-Funel, A. Evidence for viable but non-culturable yeasts in botrytis-affected wine. *J. Appl. Microbiol.* **2005**, *99*, 85–93. [CrossRef]
24. Du Toit, W.J.; Pretorius, I.S.; Lonvaud-Funel, A. The effect of suphur dioxide and oxygen on the viability and culturability of a strain of Acetobacter pasteurianus and a strain of *Brettanomyces bruxellensis* isolated from wine. *J. Appl. Microbiol.* **2005**, *98*, 862–871. [CrossRef]
25. Oliver, J.D. The viable but non-culturable state in bacteria. *J. Microbiol.* **2005**, *43*, 93–100.
26. Thomas, D.S. Yeasts as spoilage organisms in beverages. In *The Yeasts*, 2nd ed.; Rose, A.H., Harrison, J.S., Eds.; Academic Press: London, UK, 1993; Volume 5, pp. 517–561.
27. Deák, T.; Beuchat, L.R. Yeasts associated with fruit juice concentrates. *J. Food Protect.* **1993**, *56*, 777–782. [CrossRef]
28. Loureiro, V.; Malfeito-Ferreira, M. Spoilage yeasts in the wine industry (review). *Int. J. Food Microbiol.* **2003**, *86*, 23–50. [CrossRef]
29. Renouf, V.; Lonvaud-Funel, A. Development of an enrichment medium to detect *Dekkera/Brettanomyces bruxellensis*, a spoilage wine yeast, on the surface of grape berries. *Microbiol. Res.* **2007**, *162*, 154–167. [CrossRef] [PubMed]
30. Saez, J.S.; Lopes, C.H.A.; Kirs, V.E.; Sangorrín, M. Production of volatile phenols by *Pichia manshurica* and *Pichia membranifaciens* isolated from spoiled wines and cellar environment in Patagonia. *Food Microbiol.* **2011**, *28*, 503–509. [CrossRef] [PubMed]
31. Jolly, N.P.; Augustyn, O.P.H.; Pretorius, I.S. The occurrence of non-Saccharomyces yeast strains over three vintages in four vineyards and grape musts 130 from four production regions of the Western Cape, South Africa. *S. Afr. J. Enol. Vitic.* **2003**, *24*, 35–42.

© 2020 by the authors. Licensee MDPI, Basel, Switzerland. This article is an open access article distributed under the terms and conditions of the Creative Commons Attribution (CC BY) license (http://creativecommons.org/licenses/by/4.0/).

Review

Modulating Wine Pleasantness Throughout Wine-Yeast Co-Inoculation or Sequential Inoculation

Alice Vilela

Department of Biology and Environment, School of Life Sciences and Environment, University of Trás-os-Montes and Alto Douro, CQ-VR, Chemistry Research Centre, 5001-801 Vila Real, Portugal; avimoura@utad.pt; Tel.: +351-259-350-973; Fax: +351-259-350-480

Received: 17 November 2019; Accepted: 5 February 2020; Published: 9 February 2020

Abstract: Wine sensory experience includes flavor, aroma, color, and (for some) even acoustic traits, which impact consumer acceptance. The quality of the wine can be negatively impacted by the presence of off-flavors and aromas, or dubious colors, or sediments present in the bottle or glass, after pouring (coloring matter that precipitates or calcium bitartrate crystals). Flavor profiles of wines are the result of a vast number of variations in vineyard and winery production, including grape selection, winemaker's knowledge and technique, and tools used to produce wines with a specific flavor. Wine color, besides being provided by the grape varieties, can also be manipulated during the winemaking. One of the most important "tools" for modulating flavor and color in wines is the choice of the yeasts. During alcoholic fermentation, the wine yeasts extract and metabolize compounds from the grape must by modifying grape-derived molecules, producing flavor-active compounds, and promoting the formation of stable pigments by the production and release of fermentative metabolites that affect the formation of vitisin A and B type pyranoanthocyanins. This review covers the role of *Saccharomyces* and non-*Saccharomyces* yeasts, as well as lactic acid bacteria, on the perceived flavor and color of wines and the choice that winemakers can make by choosing to perform co-inoculation or sequential inoculation, a choice that will help them to achieve the best performance in enhancing these wine sensory qualities, avoiding spoilage and the production of defective flavor or color compounds.

Keywords: wine yeasts; lactic acid bacteria; co-inoculation; sequence inoculation; flavor compounds; color pigments

1. Introduction

1.1. The Human Senses in Wine Evaluation

Five senses are involved in perceiving wine sensory quality: sight, taste, hearing, touch, and smell. Color perception results from the stimulus of the retina by light (wavelengths 380 to 760 nm). In wine, color and appearance are the first attributes by which quality is assessed. According to Spence [1], color is the most important product-intrinsic indicator used by consumers when searching, purchasing, and subsequently consuming food or a libation. Color, clarity, and hue affect the perception of other attributes such as flavor due to the association with color. For example, a yellow/green beverage is expected to have a lemon flavor and an acidic taste.

Taste is a chemical sense and happens when taste stimuli contact with the taste receptors located on the tongue, called taste buds. Humans can distinguish six basic tastes: sweet, sour, salty, bitter, umami, and fatty [2,3]. Between 20 and 30 levels of intensity can be distinguished for each taste, and each taste quality represents different nutritional or physiological requirements, or a potential dietary risk [4].

Sound (waves which strike the eardrum, causing it to vibrate [5]) is also important when judging a wine. For instances, when we hear a champagne cork popping, it is a sign that the wine has an enjoyable gas.

Texture in wine can be defined as the total sum of kinesthetic sensations derived from oral manipulation. It encompasses mouthfeel, masticatory properties, residual properties, and even visual and auditory properties [6].

Aroma and flavor are chemical senses stimulated by the chemical properties of odor molecules which must reach the olfactory bulb to interact with olfactory cells in the olfactory mucosa [7]; therefore, to smell, molecules must be airborne (i.e., volatile). The sensory term which we call "flavor" is a mingled experience based on human judgment, built on personal differences in perception thresholds.

In conclusion, and as reported by Swiegers et al. [8], all of the senses play a key role in wine/flavor development—color, aroma, mouthfeel, sound, and, ultimately, taste. Altogether, these sensory perceptions are very complex. Wine contains many flavor and aroma-active compounds. Terpenes, methoxypyrazines, esters, ethanol and other alcohols and aldehydes impart distinct flavors and aromas (floral, pepper, fruit, woody and vinylic flavors, among others) to wine [9–11]. The taste of wine can be described as sweet, sour, salty, umami, bitter, and, to a lesser extent, fat [12]. These properties are the result of the presence of sugars, polyols, salts, polyphenols, flavonoid compounds, amino acids, and fatty acids. Compounds such as glycerol, polysaccharides, and mannoproteins contribute to the viscosity and mouthfeel of wines [13]; grape anthocyanins contribute to the color [14], and ethanol, by sheer mass, also carries other alcohols along, promoting a mouth-warming effect [15].

1.2. Main Wine Aroma and Flavor Compounds from the Fermentative Origin

Yeast and bacteria are vital to the development of wine flavor. Many biosynthetic pathways, in wine yeast and malolactic bacteria, are responsible for the formation of wine aroma and flavor. However, we cannot discard the other factors that can also influence the wine chemical composition, such as viticultural practices, grape-must composition, pH, fermentation temperature, and technological aspects associated with the vinification process [8]. So, depending on their origin, wine aroma and flavor compounds can be named varietal aromas (originating from the grapes), fermentative aromas (originating during alcoholic and malolactic fermentations), and aging aromas (developed during the reductive or oxidative wine-aging that depends on storage conditions) [16].

Most of the wine aroma and flavor compounds are produced or released during wine fermentation due to microbial activities of *Saccharomyces* and non-*Saccharomyces* yeast genera (*Brettanomyces, Candida, Debaryomyces, Hanseniaspora, Hansenula, Kloeckera, Kluyveromyces, Lachancea, Metschnikowia, Pichia, Saccharomycodes, Schizosaccharomyces, Torulaspora,* and *Zygosaccharomyces*). Both in spontaneous and inoculated wine fermentations, non-*Saccharomyces* are important in early stages of the fermentation, before *Saccharomyces* becomes dominant in the culture, and contribute meaningfully to the global aroma profile of wines by producing flavor-active compounds [17,18].

A group of aroma compounds has been directly linked to specific varietal flavors and aromas in wines [19,20]. Most of these compounds are present at low concentrations in both grapes and fermented wine. These aroma compounds are found in grapes in the form of non-odorant precursors that, due to the metabolic activity of *Saccharomyces* and non-*Saccharomyces* yeast during fermentation, are transformed to aromas and flavor that are of great relevance in the sensory perception of wines [20] (Table 1).

Table 1. Main odorants contributing to varietal aromas of some monovarietal wines.

Compounds	Main Cultivars	Odour Descriptor	Ref.
Geraniol	Muscat, Gewurztraminer	Citrus, floral, geranium	[21]
Linalool	Muscat, Gewurztraminer	Floral, lavender	[21]
Nerol	Muscat	Floral	[21]
Tetrahydro-4-methyl-2-(2-methyl-1-propenyl)-2,5-cis-2h-pyran (cis-rose oxide)	Gewurztraminer	Geranium oil	[22]
3,6-Dimethyl-3a,4,5,7a-Tetrahydro-3h-1-Benzofuran-2-One	Gewurztraminer	Coconut, woody, sweet	[23]
3-Isobutyl-2-Methoxypyrazines	Sauvignon blanc	Bell pepper	[24,25]
4-Methyl-4-Mercaptopentan-2-One	Sauvignon blanc	Black currant	[24,25]
3-Mercapto-1-Hexanol (R Isomer)	Sauvignon blanc	Grapefruit, citrus peel	[24,25]
1,1,6-Trimethyl-1,2-Dihydronaphthalene	Riesling	Kerosene	[26]
3-Mercapto-1-Hexanol (S Isomer)	Semillon	Passion fruit	[27]
Rotundone	Shiraz	Black pepper	[28]

During alcoholic fermentation, some yeast, mainly non-*Saccharomyces* yeasts, can release β-glucosidases that hydrolyze the glycosidic bonds of the odorless non-volatile glycoside linked forms of monoterpenes (geraniol, linalool, nerol, among others), releasing the odor compounds to the wine [29]. Volatile thiols that give Sauvignon blanc wines their characteristic aroma (bell pepper, black currant, grapefruit, and citrus peel) are not present in grape juice but occur in grape must as odorless, non-volatile, cysteine-bound conjugates. During fermentation, the wine yeasts are responsible for the cleaving of the thiol from the precursor [30].

However, the major groups of aromas and flavor compounds from the fermentative origin are ethanol, higher alcohols or fusel alcohols, and esters. The biosynthetic pathways responsible for the formation of higher alcohols, the Ehrlich pathway, or the enzymes responsible for the formation of esters, have been studied in wine yeasts [31].

Higher alcohols are derived from amino acid catabolism via a pathway that was first described by Ehrlich [32] and later revised by Neubauer and Fromherz in 1911 [33]. Amino acids that are assimilated by the Ehrlich pathway (valine, leucine, isoleucine, methionine, and phenylalanine), present in grape must are metabolized by yeasts, sequentially, throughout the fermentation. Figure 1 shows the metabolism of phenylalanine with the production of 2-phenylethanol and, after oxidation of phenylacetaldehyde, the formation of phenylacetate. Both compounds possess a pleasant rose-like aroma/flavor.

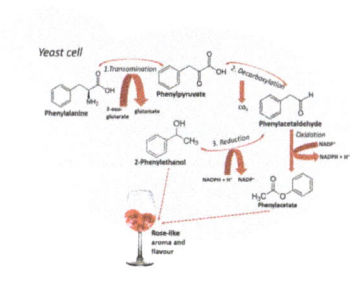

Figure 1. Schematic representation of the Ehrlich pathway for the catabolism of the aromatic amino acid, phenylalanine leading to the formation of 2-phenylethanol [34]. This biosynthetic pathway consists of three steps (reactions 1, 2 and 3): first, amino acids are deaminated to the corresponding α-ketoacids, in reactions catalyzed by transaminases. In a second step, α-ketoacids are decarboxylated and converted to their corresponding aldehydes (five decarboxylases are involved in this process), in a third step, alcohol dehydrogenases (Adh1p to Adh6p and Sfa1p) catalyze the reduction of aldehydes to their corresponding higher alcohols [35].

Studies have shown that profiles and concentrations of higher alcohols produced vary by yeast species, even when the fermentation conditions are similar, which indicates that the mechanisms that regulate the Ehrlich pathway are diverse in non-*Saccharomyces* yeasts compared to *Saccharomyces* [16,36]. So, Ehrlich pathway mechanisms should be explored in detail in non-*Saccharomyces* yeasts as it contributes to the formation of important and flavorful wine aromas [36].

The most important esters are synthetized by yeasts during alcoholic fermentation as a detoxification mechanism since they are less toxic than their correspondent alcohol or acidic precursors. Moreover, their synthesis serves as a mechanism for the regeneration of free CoA from its conjugates [16,37].

Esters (Figure 2) that contribute to wine aroma, derived from fermentation, belong to two categories: the acetate esters of higher alcohols and the ethyl esters of medium-chain fatty acids (MCFA). Acetate esters are formed inside the yeast cell, and in *S. cerevisiae* the reaction is metabolized by two alcohol acetyltransferases, AATase I and AATase II (encoded by genes *ATF1* and *ATF2* [35,38]). Eat1p is responsible for the production of acetate and propanoate esters [39,40]. Most medium-chain fatty acid ethyl ester biosynthesis during fermentation is catalyzed by two enzymes, Eht1p and Eeb1p [38,41].

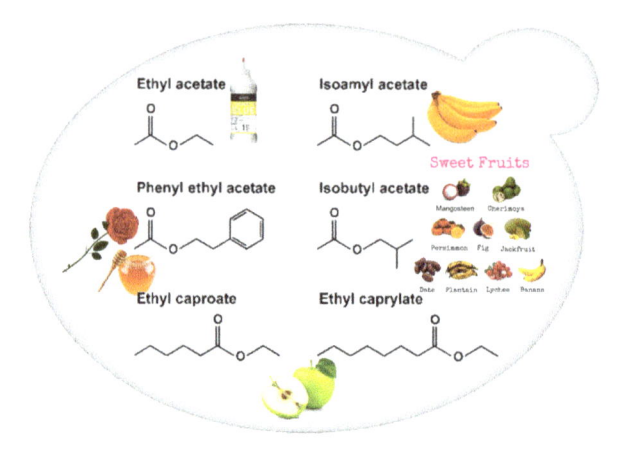

Figure 2. Schematic representation of the most important wine esters: ethyl acetate (glue-like aroma), isoamyl acetate (banana aroma), 2-phenylethyl acetate (roses and honey aromas), isobutyl acetate (sweet-fruits aromas), and ethyl caproate and ethyl caprylate with a sour-apple aroma [38].

Volatile fatty acids also contribute to the flavor and aroma of the wine. During yeast fermentation, long-chain fatty acids (LCFAs) are also formed via the fatty-acid synthesis pathway from acetyl-CoA in concentrations varying from ng/L to g/L [42]. Medium-chain fatty acids (MCFAs (C_6 to C_{12})) are produced primarily by yeasts as intermediates in the biosynthesis of LCFAs that are prematurely released from the fatty acid synthase complex. These acids (Table 2) directly contribute to the flavor of wine or serve as substrates that participate in the formation of ethyl acetates [43]. As most have unpleasant aromas (see Table 2), their formation should be minimized.

Table 2. Main medium-chain fatty acids (MCFAs (C_6 to C_{12})), produced by yeasts during alcoholic fermentation.

Fatty Acid	Associated Aroma	Odor Threshold (µg/L) [1]
Butanoic Acid	Rancid butter or baby vomit aroma	173
Hexanoic Acid	Sour, fatty, sweat, cheese	420
Octanoic Acid	Fatty, waxy, rancid oily, vegetable, cheesy	500
Decanoic Acid	Unpleasant rancid, sour, fatty, citrus	1000
2-Methylpropanoic Acid	Acidic sour, cheese, dairy, buttery, rancid	2300
2- and 3-Methylbutyric Acid	Sour, stinky feet, sweaty, cheese, tropical	33

[1] Measured in model wine, water/ethanol (90 + 10, w/w) [44].

Sulfur-containing compounds can also be formed by yeasts during alcoholic fermentation. They are usually perceived as off-flavors. The sulfur-containing compounds can be derived from the grape and the metabolic activities of yeast and bacteria. They can also occur due to the chemical reactions during the wine aging and storage and also due to environmental contamination [45]. They can be formed by enzymatic mechanism as the products of metabolic and fermentative pathways whose substrates are both amino acids and some sulfur-containing pesticides. When wine microorganisms metabolize these thiols, the sulfur compounds formed are considered off-flavors [46] which convey negative notes such as cabbage, garlic, onion, rotten eggs, rubber, and sulfur to wines [47]. However, there are some volatile thiols that may confer enjoyable aromatic notes at trace levels, such as 4-mercapto-4-methylpentan-2-one (4MMP), 3-mercaptohexan-1-ol (3MH), already mentioned in Table 1, and 3-mercaptohexyl acetate (3MHA), important for the characterization of the typical Sauvignon Blanc wine aroma [24,25,48].

Finally, another important family of aromatic compounds present in wines are the carbonyl compounds. In this group we may include acetaldehyde, acrolein, ethyl carbamate, formaldehyde, and furfural [49]. Several factors may contribute to the presence of carbonyl compounds in wines, including the fermentation of over-ripe grapes and increasing the maceration time, probably due to increased concentration of the precursors like amino acids and glucose in the must [50]. Due to their carbonyl group, carbonyl compounds present a high reactivity with the nucleophile's cellular constituents [51] and may cause cell damage. So, these compounds are toxic, and their formation should be avoided.

2. Yeast Modulation of Wine Aroma and Flavor Compounds

2.1. Non-Saccharomyces and Saccharomyces Co-inoculation vs. Sequential Inoculation

The wine industry attempts to diversify producing wines with distinctive characteristics and creating high-quality new products. A true test for winemakers is to blend several grapes, grown on different soil and climate conditions (terroir), with a developing science of yeast and bacterial metabolism, to produce the most enjoyable wine [52]. Many winemakers today use commercial yeast and bacteria starter cultures for alcoholic and malolactic fermentation, respectively. The selection of a "fit-for-purpose" starter strain has a pivotal role in optimizing flavor and aroma.

There is no consensus on the impact of indigenous yeasts on wine sensory properties; while some researchers show a positive effect, others show negative effects on the wine chemical composition and sensory properties [53,54]. For example, Varela et al. [55] showed an increase in the concentration of some higher alcohols and esters in wines produced with autochthonous yeasts compared to the wines produced with commercial yeasts. Moreover, some non-*Saccharomyces* yeast may increase the concentration of biogenic amines in wines [56].

In red winemaking, significant increases in the concentrations of desirable compounds such as ethyl lactate (sweet, fruity, acidic, ethereal with a brown nuance aroma), 2,3-butanediol (buttery aroma), 2-phenylethanol, and 2-phenylethyl acetate (both with a rose-like scent) can be obtained when non-*Saccharomyces* yeasts are introduced into the fermentation process [57,58].

Another selection criterion for non-*Saccharomyces* yeasts, when aiming to improve the wine aroma, is the presence of β-glucosidase activity that favors the hydrolysis of the non-volatile aromatic precursors from the grape [59,60]. Non-*Saccharomyces* species display superior β-glucosidase activity to that of *Saccharomyces* species, which has been defined as intracellular and strain-dependent [23].

Budic-Leto et al. [61] found that Prosek, a traditional Croatian dessert wine, produced with native and inoculated yeasts differed in its volatile compounds. Using descriptive sensory analysis, it was shown that the sensory properties of the wines were significantly different depending on the type of fermentation, namely determined for the attributes strawberry jam aroma, and fullness. So, recently, non-*Saccharomyces* yeast species have been suggested for winemaking as they could contribute to the improvement of wine quality mostly in terms of aromatic characteristics [62,63]. Thus, starter cultures composed of non-*Saccharomyces* yeasts together with *S. cerevisiae* have been used for co-, or sequential wine fermentations [64].

Co-inoculation involving *S. cerevisiae* and non-*Saccharomyces* yeasts species typically results in the disappearance (or the presence in relative low amounts) and loss of viability of non-*Saccharomyces* [65,66]. Though *S. cerevisiae* dominance can be explained by the depletion of sugar and nutrients from the grape must followed by ethanol production and lack of oxygen, some direct mechanisms for yeast species antagonism have also been described: (i) killer factors (so-called killer toxins or killer proteins), which are secreted peptides, encoded by extrachromosomal elements of *S. cerevisiae* that affect other yeast species [67]; (ii) similar compounds have also been described for *Torulaspora delbrueckii* species [68] and for the genera *Pichia, Kluyveromyces, Lachancea, Candida, Cryptococcus, Debaryomyces, Hanseniaspora, Hansenula, Kluyveromyces, Metschnikowia, Torulopsis, Ustilago, Williopsis,* and *Zygosaccharomyces,* indicating that the killer phenomenon is indeed widespread among yeasts. [69]; (iii) *S. cerevisiae* are also able to secret antimicrobial peptides (AMPs) during alcoholic fermentation that are active against wine-related yeasts (e.g., *Dekkera bruxellensis*) and bacteria (e.g., *Oenococcus oeni*). These AMPs correspond to fragments of the *S. cerevisiae* glyceraldehyde 3-phosphate dehydrogenase (GAPDH) protein [70].

Several authors agree that the sequential culture is better than the mixed culture, especially because it allows for a greater expression of the metabolism of non-*Saccharomyces* yeasts at the beginning of fermentation [71,72]. However, as recently reported by Loira et al. [73], the winemaker selection criteria for performing co-inoculation or sequential inoculation with the appropriate non-*Saccharomyces* is dependent on the characteristics of the wine to be produced, including the desired sensory properties. The ratio of inoculation (non-*Saccharomyces* vs. *Saccharomyces*) is also a subject that must be considered. Moreover, the contribution of the inoculation of non-*Saccharomyces* strains to wine fermentation can be direct or indirect, through biological interactions with *S. cerevisiae*. Recently, Renault et al. [74,75] described a synergic interaction between *S. cerevisiae* and *T. delbrueckii* resulting in increased levels of 3-sulfanylhexan-1-ol a compound that presents a sulphurous aroma and an initially fruity flavor. However, with over-aging, the aroma/flavor evolves to savory and chicken meaty with roasted coffee shades and a hint of fruitiness [76].

Not long ago, García et al. [63] performed small-scale fermentations where they studied the oenological characterization of five non-*Saccharomyces* native yeast species under several co-culture conditions in combination with a selected strain of *S. cerevisiae*, aiming to improve the sensory characteristics of the Malvar wines. Sequential inoculations were elaborated with *S. cerevisiae* CLI 889 in combination with several non-*Saccharomyces*: (i) *T. delbrueckii* CLI 918, which produced wines with a lower ethanol content and higher fruity and floral aroma; (ii) *C. stellata* CLI 920, which augmented the aroma complexity and glycerol content; (iii) *L. thermotolerans* 9-6C, which produced an increase in acidity and floral and ripe-fruit aroma and a lower volatile acidity; (iv) *Schizosaccharomyces pombe*, which produced wines with fruity aromas; and (v) *M. pulcherrima*, which produced wines with lower volatile acidity and an increase of glycerol and ripe-fruit aroma.

Continuing their work on Malvar wines, García et al. [77] performed fermentations at the pilot scale, using sequential-inoculation strategies which resulted in wines that tasters were able to distinguish from the controls. Moreover, the wines were most appreciated, namely, those produced in sequential cultures with *T. delbrueckii* CLI 918/*S. cerevisiae* CLI 889 and *C. stellata* CLI 920/*S. cerevisiae* CLI 889 and, also, with mixed and sequential cultures of *L. thermotolerans* 9-6C/*S. cerevisiae* CLI 889 strains. Studies have shown that sequential cultures can produce more different wines, when compared with the controls, providing sensory properties associated with the non-*Saccharomyces* strains. Some strains of *T. delbrueckii* in sequential fermentation with *S. cerevisiae* can produce significant amounts of 3-ethoxy-1-propanol [78], with a fruity-like aroma with a low perception threshold, 0.1 mg/L [79].

However, sequential inoculation is not only favorable for positive aromas sequential. It is a fermentation technique that can be used to prevent or diminish the production of some undesirable compounds, augmenting the production of others. Viana et al. [80] reported a decrease in the higher alcohol's concentration (considered as possessing a fuel-like aroma) from 452.5 mg/L (control) to 306.2 mg/L when carrying out mixed fermentations with *H. osmophila* and *S. cerevisiae*. Moreover, higher concentrations of 2-phenylethyl acetate (rose-like aroma) were obtained. A higher intensity of fruitiness was also detected in these wines when compared to the control wine, obtained throughout the fermentation of a pure *S. cerevisiae* culture.

2.2. Saccharomyces and Lactic Acid Bacteria co inoculation vs. Sequential Inoculation

The vinification involves different microbiological processes, mainly alcoholic fermentation (AF) conducted by *Saccharomyces cerevisiae* and malolactic fermentation (MLF) conducted by lactic acid bacteria (*Oenococcus oeni*). These two distinctive fermentation processes represent an essential step in the improvement of the quality of red wines [81].

MLF naturally occurs after AF, however, the timing of the start of MLF depends on several parameters like temperature, pH, alcoholic degree and the concentration of sulfur dioxide (SO_2), as well as on certain yeast metabolites available, such as medium-chain fatty acids and peptidic fractions [82,83]. The success of spontaneous MLF is not always guaranteed, and the addition of starter culture can improve its viability. To overcome this issue, the winemakers may carry out traditional LAB inoculation after alcoholic fermentation (sequential inoculation), or simultaneous inoculation in the must with yeast (co-inoculation). The co-inoculation has, in the last several years, been adopted by some winemakers, particularly in warm climates with higher temperatures, where high concentrations of ethanol can inhibit LAB growth [84].

There are not many studies focusing on the impact of the co-inoculation technique on the aromatic and biochemical profile of wines. Abrahamse and Bartowsky [85] and Knoll et al. [86] have shown that the timing of inoculation with LAB in both white and red wines could influence the profile of aromatic yeast-derived compounds such as higher alcohols, terpene, esters, and fatty acids. However, these works were performed at the lab scale, and no sensory evaluation was performed on the wines.

Antalick and collaborators [87] demonstrated the impact of the timing of inoculation with LAB on the metabolic profile of wines manufactured at production scale. They have clearly shown that this technique has an impact on the aromatic profile of the wines, mainly in the presence of lactic and fruity notes. Co-inoculation can modulate the intensity of these descriptors, due to the production/degradation of metabolites or by the development of an aromatic mask over the short and long term. Moreover, they discuss that the metabolic and aromatic changes that occur with co-inoculation depend strongly on the yeast and LAB strains, as well as on the composition of the must. Co-inoculation of musts at the beginning of vinification can also lead to a faster vinification process without an excessive increase in volatile acidity [88].

Due to the importance of the interaction between yeasts (*Saccharomyces* and non-*Saccharomyces*) and bacteria during wine processing, Berbegal et al. [89] applied a next-generation sequencing (NGS) analysis to several fermentation modalities: uninoculated must, pied-de-cuve, *S. cerevisiae*, *S. cerevisiae*, and *Torulaspora delbrueckii* co-inoculated and sequentially inoculated, along with *S. cerevisiae* and *Metschnikowia pulcherrima* co-inoculated and sequentially inoculated, continued by spontaneous malolactic consortium to perform MLF. Each experimental trial led to the different taxonomic composition of the bacterial communities of the malolactic consortia, in terms of prokaryotic phyla and genera. Among other interesting findings, they found that MLF was delayed when *M. pulcherrima* was inoculated and was even inhibited when the inoculated yeast strain was *T. delbrueckii* [89]. Thus, an antagonistic effect of *M. pulcherrima* and especially *T. delbrueckii* on lactic bacteria population has been proven, which may be due to the ability of *T. delbrueckii* to produce "toxins" or killer factors that prevent bacteria growth [68] and to the ability of *M. pulcherrima* to produce high amounts of pulcherrimin (an iron chelator) that inhibits the growth of bacteria [90].

3. Yeast Modulation of Wine Color and Pigment Formation

Anthocyanins and their derivatives, originating in the grapes, are the main pigments responsible for the red wine color, and their structural modifications result in a characteristic variation of color in red wines, from pale ruby (young red wine) to deep purple-red color (aged red wine) [91,92]. Such variations can also result in changes in wine mouthfeel and flavor.

Saccharomyces yeasts can directly or indirectly contribute to wine color, altering color parameters such as intensity and tonality by: (i) increasing the formation of stable pigment precursors (vinyl phenols, acetaldehyde, and pyruvic acid); and (ii) modifying the pH due to organic acid metabolism (production or consumption) [73].

Pyruvic acid and acetaldehyde promote the formation of vitisins of types A and B, respectively [93,94], during must fermentation, (Figure 3A). Vitisins contribute more to wine color parameters than unmodified anthocyanins and exhibit a hypsochromic shift, i.e., a change of spectral band position (in the absorption, reflectance, transmittance, or emission spectrum) to a shorter wavelength corresponding to a higher frequency. Vitisins can change towards an orange-red hue due to the long conjugation afforded by pyran ring [95]. Moreover, their color expression remains stable against discoloration due to the presence of sulfur dioxide (bleaching capacity) or changes in pH [95] (Figure 4).

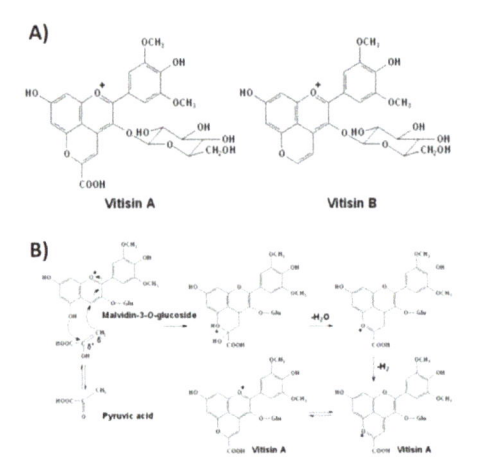

Figure 3. (**A**) Structures of vitisin A and vitisin B generated from malvidin-3-*O*-glucoside. (**B**) The formation mechanism of vitisin A produced by condensation of an anthocyanin (malvidin-3-*O*-glucoside) with pyruvic acid [92,96,97].

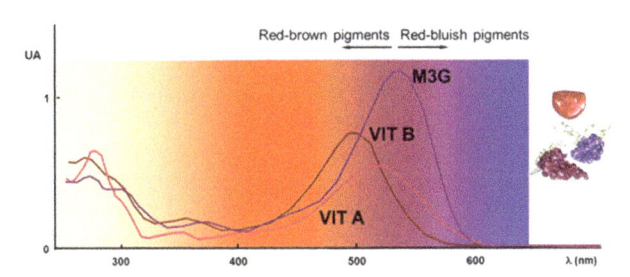

Figure 4. UV-visible spectra of malvidin-3-O-glucoside and vitisins A and B. Adapted from [98].

During the fermentation process, practices like pellicular maceration that increase the extraction of anthocyanins from the skins of grape berries will promote the formation of vitisins [14]; also, acetaldehyde and pyruvic acid, as mentioned before, are important intermediate compounds in yeast metabolism and influence wine vitisin content [93,99,100].

During sugars' fermentation by yeasts, pyruvate is metabolized into acetaldehyde, with the latter being the terminal electron acceptor in the formation of ethanol. Acetaldehyde and pyruvate, formed in yeast cytoplasm, are rapidly metabolized (the first is reduced to ethanol, and the second is either decarboxylated to acetaldehyde or used in the formation of acetyl-CoA). However, some of the acetaldehyde and pyruvate molecules, through cell lysis, pass to the wine medium and are sufficiently reactive to attack other molecules, enabling the transformation of anthocyanins into compounds such as pyranoanthocyanins and their secondary generated pigments (anthocyanin oligomers and polymeric anthocyanin) [92,97]. Pyroanthocyanins are the most important group of anthocyanin derivatives present in fermented beverages, including wine, and the A-type vitisins or carboxy-pyroanthocyanins are produced, as mentioned before, by condensation of anthocyanin with pyruvic acid [97] (Figure 3B).

In terms of vitisin kinetic formation, during *S. cerevisiae* fermentation, type-A vitisins are produced in the first six days of fermentation (when pyruvic acid is available). At the end of fermentation, when nutrients are limited, the amount of acetaldehyde is high enough to lead to the formation of type-B vitisins [93,99]. So, to generate a more pleasant red wine color, before fermentation the winemaker must select the wine yeast strains that will be able to increase anthocyanin extraction and/or can produce more pyruvic acid and acetaldehyde. Postponing or starting MLF early can prevent the consumption of pyruvic acid and acetaldehyde by lactic acid bacteria [101], increasing the possibility for vitisin synthesis.

Oxidative fermentation (fermentation in barrels or with micro-oxygenation) and wine aging in wood give rise to pyruvic acid and acetaldehyde, increasing the vitisin levels and, consequently, color intensity and stability [97]. Yellowish α-pyranone-anthocyanins called oxovitisins were also described by He et al. [102] in aged red wines derived from the direct oxidation of A-type vitisins. Moreover, A-type vitisins are the pyranoanthocyanins detected in higher concentrations in port wines. Port wine is made by stopping the fermentation process with the addition of wine spirit "aguardente", leaving the wine with a high concentration of sugars. So, when fermentation is stopped, the pyruvic acid concentration is relatively high and increases after wine fortification [97].

Vinylphenolic pyroanthocyanin adducts result from the condensation between vinyl phenols and anthocyanins. These color compounds also show high color stability [94]. Yeast strains are also able to affect the concentration and the composition of wine tannins as well as the degree of tannin polymerization [103], thus, indirectly throughout yeast actions affecting stabilization of anthocyanins, and consequently, stabilization of color can occur due to reaction between anthocyanins and tannins forming pigmented tannins and through copigmentation of anthocyanins [104].

Additionally, what about non-*Saccharomyces* wine yeasts? Well, we have already mentioned that an improvement in fermentation quality, efficiency, and wine pleasantness is obtained when sequential or co-inoculation of non-*Saccharomyces* and *Saccharomyces* yeast is performed.

Torulaspora delbrueckii is one non-*Saccharomyces* species available commercially (Viniflora® Harmony.nsac and Viniflora® Melody.nsac, Zymaflore® Alpha, BIODIVA®, and Viniferm NS-TD® are some commercial examples), therefore, this yeast could be a good candidate for wine color improvement as it is reported to have a positive influence on the taste and aroma of wines. For instance, Pinotage grape must inoculate with *T. delbrueckii* originated red wines improved in color intensity (anthocyanins) and mouthfeel (flavanols) when compared to the control (musts inoculated with *S. cerevisiae*) [105]. However, *T. delbrueckii* has poor production of acetaldehyde [106], thus being a poor candidate for wine color improvement in the context of B-type vitisins.

Other non-*Saccharomyces* yeasts, not yet available as commercial products, but studied in the academic community, could be good candidates for wine color improvement. Medina et al. [107] reported that in the case of co-fermentation of *Metschnikowia* and *Hanseniaspora* with *S. cerevisiae*, only an increase in the content of B-type vitisins occurred, probably due to the enhanced acetaldehyde formation.

Further experiments performed by sequential inoculation of *Schizosaccharomyces pombe* and *Lachancea thermotolerans* exposed an increase of type-A vitisins when compared with the control *S. cerevisiae* [108]. Also, several authors detected interesting features in non-*Saccharomyces* yeasts. *P. guilliermondii* strains presenting a high hydroxycinnamate decarboxylase activity may improve the formation of vinylphenolic pyranoanthocyanins; non-*Saccharomyces* yeasts, such as *Candida valida*, *Metschnikowia pulcherrima*, *Kloeckera apiculata* and *Starmerella bombicola*, which synthesize and release pectolytic enzymes, can improve wine color due to the extraction of a greater amount of polyphenolic compounds during fermentation and by facilitating clarification and filtration processes [73,98,109,110].

4. The Role of *Saccharomyces* and non-*Saccharomyces* Mannoproteins in Aroma and Color of Wines

Mannoproteins are highly glycosylated glycoproteins located on the external layer of the yeast cell wall, representing 35% to 40% of the *S. cerevisiae* cell wall (Figure 5) [111]. Mannoproteins are composed of 10% to 20% protein and 80% to 90% D-mannose associated with residues of D-glucose and N-acetylglucosamine [112].

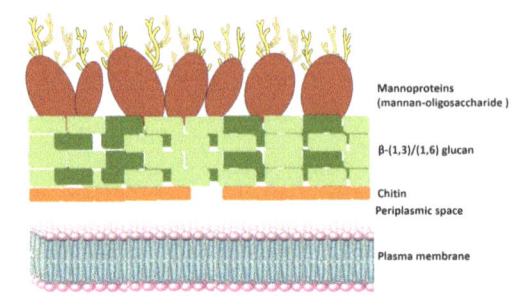

Figure 5. Schematic representation of the yeast cell wall. The yeast cell wall is composed of mannan–oligosaccharide (mannoproteins), complex polymers of β-(1,3)/(1,6) glucan, and chitin. As shown in Figure 5, mannoproteins are located on the surface of the cell wall.

In wine, we can find two groups of mannoproteins: one made up of those secreted into wine by yeast during alcoholic fermentation (100–150 mg/L) with molecular weights from 5000 to more than 800,000 Da [113], the other one composed by those that are released into the wine due to the autolysis of yeasts during aging on lees, probably through the cleavage of linkages between mannoproteins, glucans, and chitin [113,114].

The presence of these mannoproteins in wines has many positive consequences, from the reduction of the protein haze in white wines [115] to decreasing astringency of red wines, by increased inhibition of tannin aggregation [116,117]. Among other positive factors, mannoproteins also interact with wine volatile compounds [118]. So, these yeast-derived glycoprotein complexes can have positive effects on the technological and sensorial properties of wines [119].

In terms of improving wine palatability and mouth feel, yeast mannoproteins promote the increase of wine sweetness [120] and improve the aroma persistence and complexity [114,121]. However, the number of mannoproteins released by yeast into wine can vary concerning the strain and the chemical–physical and compositional conditions of the wine system [121,122].

Their presence can also affect the release of volatile compounds, affecting the final perception of the wine [121]. The physicochemical interactions between aroma compounds and mannoproteins depends on the nature of the volatile, since a greater degree of interaction is often observed with hydrophobic compounds [123], as well as the conformational structure of the mannoproteins [114]; moreover, Chalier et al. [114] demonstrated that both the glycosidic and peptidic parts of the mannoproteins may interact well with the aroma compounds.

It has been shown that the use of mannoproteins (in low amounts) or the contact of the wine with fine lees increases the levels of esters (ethyl hexanoate, methyl, and ethyl hexadecanoate, which present fruity aromas), due to the esterification of fatty acids released during yeast fermentation or yeast autolysis [121,124], while higher amounts increase, in excess, fatty acid content, producing yeasty, herbaceous, and cheese-like smells [125]. However, it has been suggested that mannoproteins can be used to remove or reduce the incidence of wine off-flavors—4-ethylguaiacol and 4-ethylphenol. The sorption of these compounds to the yeast walls could be due to their interactions with the functional groups of the mannoproteins and the free amino acids on the surface of the cell walls [126].

The release of mannoproteins into the wine is not a physiological characteristic of just *S. cerevisiae* yeast strains. In 2014, Domizio and collaborators [127], studied eight non-*Saccharomyces* wine strains (*H. osmophila, L. thermotolerans, M. pulcherrima, P. fermentans, S. ludwigii, S. bacillaris, T. delbrueckii*, and *Z. florentinus*) in mixed inocula fermentations of a synthetic polysaccharide-free grape juice for their ability to release mannoproteins. The eight non-*Saccharomyces* yeasts confirmed a higher capacity to release polysaccharides when compared to *S. cerevisiae*. Moreover, Pérez-Través et al. [128] also studied the ability of natural hybrids of *S. cerevisiae* × *S. krudriavzevii* to release mannoproteins. Interestingly, they found that this strain, in the fermentation conditions studied, was able to produce a higher quantity of mannoproteins when compared with the sample in which only *S. cerevisiae* was used. Furthermore, the authors also found that the genome interaction in hybrids creates a biological ecosystem that boosts the release of mannoproteins.

As mentioned before, the presence of mannoproteins in wines, namely red wines, promotes tannin stability and reduction of astringency [116,117]. The interaction between mannoproteins and wine phenolic compounds is a matter of interest, as studies show that they may have an impact on color stability [129,130]. However, results are contradictory, as some authors state that there was no positive interaction between mannoproteins and color compounds and that the interaction between mannoproteins and tannins results in a decrease of wine tannin content due to the precipitation of tannin and mannoprotein [112,120,131,132], thus being responsible for a decrease in wine color intensity or lower filterability [133], whereas, others state that mannoproteins appear to stabilize anthocyanin-derived pigments, from a colloidal point of view, avoiding their aggregation and further precipitation [134]. The study of the exhaustive pigment composition of wines has shown that the addition of mannoproteins can stabilize type-A vitisins and other derivative pigments [134].

5. Final Remarks

Wines are complex and evolve physiochemically and sensorially through time. Most of the wine aroma and flavor compounds are produced or released during wine fermentation due to microbial activity of *Saccharomyces*, non-*Saccharomyces* yeast genera, and lactic acid bacteria. A true challenge for winemakers is the selection of a "fit-for-purpose" microbial starter culture or culture strains that can have a crucial role in optimizing flavor, aroma, and color of wines, among other sensory properties.

Co-inoculation involving *S. cerevisiae* and non-*Saccharomyces* yeasts species may result in the death or loss of variability of non-*Saccharomyces*, once *S. cerevisiae* dominates the fermentation and is stress-resistant to the inhibitory ethanol effect. So, several authors suggest that the sequential inoculation (non-*Saccharomyces* followed by *S. cerevisiae*) is a better technique than the mixed culture, allowing a higher expression of the metabolism of non-*Saccharomyces*. Nevertheless, the ratio of inoculation (non-*Saccharomyces* vs. *Saccharomyces*) must be taken into account, especially if the wine should present a special and desirable characteristic such as the expression of a peculiar aroma-flavor, or even the inhibition of the production of a specific family of compounds like, for instance, higher alcohols.

Concerning the MLF, the co-inoculation process has been adopted by some winemakers, in warm climates, where high concentrations of ethanol can inhibit lactic acid bacteria (LAB) growth. Co-inoculation of musts at the beginning of vinification can also accelerate the process without an excessive increase in volatile acidity. However, winemakers must be aware of the possible interactions between yeasts (*Saccharomyces* and non-*Saccharomyces*) and LAB during wine processing. LAB feed on dead and lysed yeast cells but some non-*Saccharomyces* may delay (*M. pulcherrima*) or inhibit (*T. delbrueckii*) bacterial growth, thus inhibiting the occurrence of MLF.

Regarding wine color characteristics, especially red wine, *Saccharomyces* and non-*Saccharomyces* yeasts can directly or indirectly contribute to wine intensity and tonality, by increasing the formation of stable pigment precursors (vinyl phenols, acetaldehyde, and pyruvic acid) and by modifying the pH due to organic acid metabolism. Pyruvic acid is necessary for vitisin synthesis; these important pigments contribute more to wine color parameters than unmodified anthocyanins. Concerning acetaldehyde, this compound has several negative impacts (one on health, the other as a potent binder of SO_2), so its formation, although beneficial to wine color, should be avoided. Also, it is important to choose the right time for promoting MLF, either spontaneously or by inoculation of a commercial LAB strain, to prevent consumption of pyruvic acid by lactic acid bacteria and to promote vitisin synthesis.

Finally, mannoproteins may have a positive effect on sensory perception of red wine, reducing astringency and bitterness and encouraging aroma revelation and odor complexity, but further studies are necessary in order to unravel the possible stabilization mechanism and the relationship between *Saccharomyces* and non-*Saccharomyces* mannoprotein characteristics and their ability to stabilize wine color.

So, in conclusion, knowledge and control of yeast and bacteria can help winemakers enhance the sensory quality of their wines for flavor and color.

Funding: We appreciate the financial support provided to the Research Unit in Vila Real [grant number UID/QUI/00616/2019] by FCT-Portugal and COMPETE

Acknowledgments: The author wants to acknowledge Interreg Program for the financial support of the Project IBERPHENOL, Project Number 0377_IBERPHENOL_6_E, co-financed by European Regional Development Fund (ERDF) through POCTEP 2014-2020.

Conflicts of Interest: The author declares no conflict of interest

References

1. Spence, C. On the psychological impact of food colour. *Flavour* **2015**, *4*, 2044–7248. [CrossRef]
2. Mattes, R.D. Fat Taste in Humans: Is It a Primary. In *Fat Detection: Taste, Texture, and Post Ingestive Effects*; Montmayeur, J.P., le Coutre, J., Eds.; CRC Press: Boca Raton, FL, USA, 2010; pp. 167–193.
3. Melis, M.; Tomassini Barbarossa, I. Taste Perception of Sweet, Sour, Salty, Bitter, and Umami and Changes Due to l-Arginine Supplementation, as a Function of Genetic Ability to Taste 6-n-Propylthiouracil. *Nutrients* **2017**, *9*, 541. [CrossRef] [PubMed]
4. Chaudhari, N.; Roper, S.D. The cell biology of taste. *J. Cell Biol.* **2010**, *190*, 285–296. [CrossRef] [PubMed]
5. Raghu, M.A. Study to Explore the Effects of Sound Vibrations on Consciousness. *Int. J. Soc. Work Hum. Serv. Pract.* **2018**, *6*, 75–88. Available online: http://www.hrpub.org/download/20180730/IJRH2-19290514.pdf (accessed on 15 September 2019).

6. Barham, P.; Skibsted, L.H.; Bredie, W.L.P.; Frøst, M.B.; Møller, P.; Risbo, J.; Snitkjær, P.; Mortensen, L.M. Molecular Gastronomy: A New Emerging Scientific Discipline. *Chem. Rev.* **2010**, *110*, 2313–2365. [CrossRef]
7. Feher, J. The Chemical senses. In *Quantitative Human Physiology, an Introduction*, 2nd ed.; Feher, J., Ed.; Academic Press: London, UK, 2017; pp. 427–439.
8. Swiegers, J.; Bartowsky, E.; Henschke, P.; Pretorius, I. Yeast and bacterial modulation of wine aroma and flavour. *Aust. J. Grape Wine Res.* **2005**, *11*, 139–173. [CrossRef]
9. Bloem, A.; Bertrand, A.; Lonvaud-Funel, A.; de Revel, G. Vanillin production from simple phenols by wine-associated lactic acid bacteria. *Lett. Appl. Microbiol.* **2007**, *44*, 62–67. [CrossRef]
10. Bloem, A.; Lonvaud-Funel, A.; de Revel, G. Hydrolysis of glycosidically bound flavour compounds from oak wood by *Oenococcus Oeni*. *Food Microbiol.* **2008**, *25*, 99–104. [CrossRef]
11. Regodón-Mateos, J.; Pérez-Nevado, F.; Ramírez-Fernández, M. Influence of *Saccharomyces cerevisiae* yeast strain on the major volatile compounds of wine. *Enzyme Microb. Technol.* **2006**, *40*, 151–157. [CrossRef]
12. Vilela, A.; Inês, A.; Cosme, F. Is wine savory? Umami taste in wine. *SDRP J. Food Sci. Technol.* **2016**, *1*, 100–105. [CrossRef]
13. Moreno-Arribas, M.; Polo, M. Winemaking biochemistry and microbiology: Current knowledge and future trends. *Crit. Rev. Food Sci. Nutr.* **2005**, *45*, 265–286. [CrossRef] [PubMed]
14. Victor-Freitas, A.P.; Fernandes, A.; Oliveira, J.; Teixeira, N.; Mateus, N. A review of the current knowledge of red wine colour. *OENO One* **2017**, *51*. [CrossRef]
15. Jordão, A.M.; Vilela, A.; Cosme, F. From Sugar of Grape to Alcohol of Wine: Sensorial Impact of Alcohol in Wine. *Beverages* **2015**, *1*, 292–310. [CrossRef]
16. Belda, I.; Ruiz, J.; Esteban-Fernández, A.; Navascués, E.; Marquina, D.; Santos, A.; Moreno-Arribas, M. Microbial Contribution to Wine Aroma and Its Intended Use for Wine Quality Improvement. *Molecules* **2017**, *22*, 189. [CrossRef]
17. García, V.; Vásquez, H.; Fonseca, F.; Manzanares, P.; Viana, F.; Martínez, C.; Ganga, M. Effects of using mixed wine yeast cultures in the production of chardonnay wines. *Rev. Argent. Microbiol.* **2010**, *42*, 226–229.
18. Jolly, N.P.; Varela, C.; Pretorius, I.S. Not your ordinary yeast: Non-*Saccharomyces* yeasts in wine production uncovered. *FEMS Yeast Res.* **2014**, *14*, 215–237. [CrossRef]
19. Polásková, P.; Herszage, J.; Ebeler, S. Wine flavor: Chemistry in a glass. *Chem. Soc. Rev.* **2008**, *37*, 2478–2489. [CrossRef]
20. Ruiz, J.; Kiene, F.; Belda, I.; Fracassetti, D.; Marquina, D.; Navascués, E.; Calderón, F.; Benito, A.; Rauhut, D.; Santos, A.; et al. Effects on varietal aromas during wine making: A review of the impact of varietal aromas on the flavor of wine. *Appl. Microbiol. Biotechnol.* **2019**, *103*, 7425–7450. [CrossRef]
21. Lanaridis, P.; Salaha, M.J.; Tzourou, I.; Tsoutsouras, E.; Karagiannis, S. Volatile Compounds in Grapes and Wines From Two Muscat Varieties Cultivated In Greek Islands. *J. Int. Sci. Vigne Vin* **2002**, *36*, 39–47. [CrossRef]
22. Ong, P.K.C.; Acree, T.E. Similarities in the Aroma Chemistry of Gewürztraminer Variety Wines and Lychee (*Litchi chinesis* Sonn.) Fruit. *J. Agric. Food Chem.* **1999**, *47*, 665–670. [CrossRef]
23. Arévalo Villena, M.; Úbeda Iranzo, J.; Cordero Otero, R.; Briones Pérez, A. Optimization of a rapid method for studying the cellular location of β-glucosidase activity in wine yeasts. *J. Appl. Microbiol.* **2005**, *99*, 558–564. [CrossRef] [PubMed]
24. Marais, J. Sauvignon blanc Cultivar Aroma - A Review. *S. Afr. J. Enol. Vitic.* **1994**, *15*, 41–45. [CrossRef]
25. Carien, C.; Wessel, J. A comprehensive review on Sauvignon Blanc aroma with a focus on certain positive volatile thiols. *Food Res. Int.* **2012**, *45*, 287–298. [CrossRef]
26. Sacks, G.L.; Gates, M.J.; Ferry, F.X.; Lavin, E.H.; Kurtz, A.J.; Acree, T.E. Sensory Threshold of 1,1,6-Trimethyl-1,2-dihydronaphthalene (TDN) and Concentrations in Young Riesling and Non-Riesling Wines. *J. Agric. Food Chem.* **2012**, *60*, 2998–3004. [CrossRef] [PubMed]
27. Darriet, P.; Thibon, C.; Dubourdieu, D. Aroma and Aroma Precursors in Grape Berry. In *Aroma and Aroma Precursors in Grape Berry*; Hernâni Gerós, M., Manuela, C., Serge, D., Eds.; Bentham Science Publishers: Sharjah, UAE, 2012; pp. 111–136. [CrossRef]
28. Herderich, M.J.; Siebert, T.E.; Parker, M.; Capone, D.L.; Mayr, C.; Zhang, P.; Geffroy, O.; Williamson, P.; Francis, I.L. Synthesis of The Ongoing Works on Rotundone, an Aromatic Compound Responsible of the Peppery Notes in Wines. *Internet J. Enol. Vitic.* **2013**, *6*, 1–6.

29. Claus, H.; Mojsov, K. Enzymes for Wine Fermentation: Current and Perspective Applications. *Fermentation* **2018**, *4*, 52. [CrossRef]
30. Swiegers, J.H.; Capone, D.L.; Pardon, K.H.; Elsey, G.M.; Sefton, M.A.; Francis, I.L.; Pretorius, I.S. Engineering volatile thiol release in *Saccharomyces cerevisiae* for improved wine aroma. *Yeast* **2007**, *24*, 561–574. [CrossRef]
31. Gamero, A.; Belloch, C.; Querol, A. Genomic and transcriptomic analysis of aroma synthesis in two hybrids between *Saccharomyces cerevisiae* and *S. kudriavzevii* in winemaking conditions. *Microb. Cell Fact.* **2015**, *14*, 128. [CrossRef]
32. Ehrlich, F. Über die bedingungen der fuselölbildung und über ihren zusammenhang mit dem eiweissaufbau der hefe. *Ber. Dtsch. Chem. Ges.* **1907**, *40*, 1027–1047. [CrossRef]
33. Neubauer, O.; Fromherz, K. Über den Abbau der Aminosäuren bei der Hefegärung. *Hoppe-Seyler's Z Physiol. Chem.* **1911**, *70*, 326–350. [CrossRef]
34. Hazelwood, L.A.; Daran, J.M.; van Maris, A.J.A.; Pronk, J.T.; Dickinson, J.R. The Ehrlich Pathway for Fusel Alcohol Production: A Century of Research on *Saccharomyces cerevisiae* Metabolism. *Appl. Environ. Microbiol.* **2008**, *74*, 2259–2266. [CrossRef] [PubMed]
35. Parapouli, M.; Sfakianaki, A.; Monokrousos, N.; Perisynakis, A.; Hatziloukas, E. Comparative transcriptional analysis of flavour-biosynthetic genes of a native *Saccharomyces cerevisiae* strain fermenting in its natural must environment, vs. a commercial strain and correlation of the genes' activities with the produced flavour compounds. *J. Biol. Res. Thessalon.* **2019**, *26*. [CrossRef] [PubMed]
36. Gamero, A.; Quintilla, R.; Groenewald, M.; Alkema, W.; Boekhout, T.; Hazelwood, L. High-throughput screening of a large collection of non-conventional yeasts reveals their potential for aroma formation in food fermentation. *Food Microbiol.* **2016**, *60*, 147–159. [CrossRef] [PubMed]
37. Lee, S.-J.; Rathbone, D.; Asimont, S.; Adden, R.; Ebeler, S.E. Dynamic changes in ester formation during chardonnay juice fermentations with different yeast inoculation and initial brix conditions. *Am. J. Enol. Vitic.* **2004**, *55*, 346–354.
38. Swiegers, J.H.; Saerens, S.M.G.; Pretorius, I.S. Novel yeast strains as tools for adjusting the flavour of fermented beverages to market specifications. In *Biotechnology in Flavour Production*, 2nd ed.; Havkin-Frenkel, D., Dudai, N., Eds.; Wiley Online Library: Oxford, UK, 2016; pp. 62–132. [CrossRef]
39. Kruis, A.J.; Levisson, M.; Mars, A.E.; van der Ploeg, M.; Garcés Daza, F.; Ellena, V.; Kengen, S.W.M.; van der Oost, J.; Weusthuis, R.A. Ethyl acetate production by the elusive alcohol acetyltransferase from yeast. *Metab. Eng.* **2017**, *41*, 92–101. [CrossRef] [PubMed]
40. Kruis, A.J.; Brigida, G.; Jonker, T.; Mars, A.E.; van Rijswijck, I.M.H.; Wolkers-Rooijackers Judith, C.M.; Smid, E.J.; Jan, S.; Verstrepen, K.J.; Kengen, S.W.M.; et al. Contribution of Eat1 and other alcohol acyltransferases to ester production in *Saccharomyces cerevisiae*. *Front Microbiol.* **2018**, *9*, 3202. [CrossRef] [PubMed]
41. Querol, A.; Perez-Torrado, R.; Alonso-del-Real, J.; Minebois, R.; Stribny, J.; Oliveira, B.M.; Barrio, E. New trends in the uses of yeasts in oenology. In *Advances in Food and Nutrition Research*; Toldrá, F., Ed.; Elsevier: Cambridge, UK, 2018; pp. 177–210.
42. Mato, I.; Suarez-Luque, S.; Huidobro, J.F. Simple determination of main organic acids in grape juice and wine by using capillary zone electrophoresis with direct UV detection. *Food Chem.* **2007**, *102*, 104–112. [CrossRef]
43. Duan, L.L.; Shi, Y.; Jiang, R.; Yang, Q.; Wang, Y.Q.; Liu, P.T.; Duan, C.Q.; Yan, G.L. Effects of adding unsaturated fatty acids on fatty acid composition of *Saccharomyces cerevisiae* and major volatile compounds in wine. *S. Afr. J. Enol. Vitic.* **2015**, *36*, 285–295. Available online: http://www.scielo.org.za/scielo.php?script=sci_arttext&pid=S2224-79042015000200001&lng=en&tlng=en (accessed on 15 September 2019). [CrossRef]
44. Zhao, P.; Gao, J.; Qian, M.; Li, H. Characterization of the Key Aroma Compounds in Chinese Syrah Wine by Gas Chromatography-Olfactometry-Mass Spectrometry and Aroma Reconstitution Studies. *Molecules* **2017**, *22*, 1045. [CrossRef]
45. Landaud, S.; Helinck, S.; Bonnarme, P. Formation of volatile sulfur compounds and metabolism of methionine and other sulfur compounds in fermented food. *Appl. Microbiol. Biotechnol.* **2008**, *7*, 1191–1205. [CrossRef]
46. Bartowsky, E.J.; Pretorius, I.S. Microbial formation and modification of flavour and off-flavour compounds in wine. In *Biology of Microorganisms on Grapes, in Must and Wine*; König, H., Unden, G., Fröhlich, J., Eds.; Springer: Heidelberg, Germany, 2008; pp. 211–233. [CrossRef]
47. Vermeulen, C.; Gijs, L.; Collin, S. Sensorial contribution and formation pathways of thiols in foods: A review. *Food Rev. Int.* **2005**, *21*, 69–137. [CrossRef]

48. Tominaga, T.; Murat, M.L.; Dubourdieu, D. Development of a method analyzing the volatile thiols involved in the characteristic aroma of wines made from *Vitis vinifera* L. cv. Sauvignon blanc. *J. Agric. Food Chem.* **1998**, *46*, 1044–1048. [CrossRef]
49. Ferreira, D.C.; Hernandes, K.C.; Nicolli, K.P.; Souza-Silva, E.A.; Manfroi, V.; Alcaraz Zini, C.; Elisa Welke, J. Development of a method for determination of target toxic carbonyl compounds in must and wine using HS-SPME-GC/MS-SIM after preliminary GC×GC/TOFMS analyses. *Food Anal. Methods* **2019**, *12*, 108–120. [CrossRef]
50. Lago, L.O.; Nicolli, K.P.; Marques, A.B.; Zini, C.A.; Welke, J.E. Influence of ripeness and maceration of the grapes on levels of furan and carbonyl compounds in wine – Simultaneous quantitative determination and assessment of the exposure risk to these compounds. *Food Chem.* **2017**, *230*, 594–603. [CrossRef]
51. Semchyshyn, H.M. Reactive Carbonyl Species In Vivo: Generation and Dual Biological Effects. *Sci. World J.* **2014**, *2014*. [CrossRef]
52. Bozoudi, D.; Tsaltas, D. Grape Microbiome: Potential and Opportunities as a Source of Starter Cultures. In *Grape and Wine Biotechnology*; InTech: Rijeka, Croatia, 2016.
53. Jemec, P.K.; Cadez, N.; Zagorc, T.; Bubic, V.; Zupec, A.; Raspor, P. Yeast population dynamics in five spontaneous fermentations of Malvasia must. *Food Microbiol.* **2001**, *18*, 247–259. [CrossRef]
54. Combina, M.; Elía Mercado, L.; Catania, C.; Ganga, A.; Martinez, C. Dynamics of indigenous yeast populations during spontaneous fermentation of wines from Mendoza, Argentina. *Int. J. Food Microbiol.* **2005**, *99*, 237–243. [CrossRef]
55. Varela, C.; Siebert, T.; Cozzolino, D.; Rose, L.; McLean, H.; Henschke, P.A. Discovering a chemical basis for differentiating wines made by fermentation with 'wild' indigenous and inoculated yeasts: Role of yeast volatile compounds. *Aust. J. Grape Wine Res.* **2009**, *15*, 238–248. [CrossRef]
56. Restuccia, D.; Loizzo, M.R.; Spizzirri, U.G. Accumulation of Biogenic Amines in Wine: Role of Alcoholic and Malolactic Fermentation. *Fermentation* **2018**, *4*, 6. [CrossRef]
57. Clemente-Jimenez, J.; Mingorance-Cazorla, L.; Martínez-Rodríguez, S.; Las HerasVázquez, F.; Rodríguez-Vico, F. Influence of sequential yeast mixtures on wine fermentation. *Int. J. Food Microbiol.* **2005**, *98*, 301–308. [CrossRef]
58. Gobbi, M.; Comitini, F.; Domizio, P.; Romani, C.; Lencioni, L.; Mannazzu, I.; Ciani, M. *Lachancea thermotolerans* and *Saccharomyces cerevisiae* in simultaneous and sequential co-fermentation: A strategy to enhance acidity and improve the overall quality of wine. *Food Microbiol.* **2013**, *33*, 271–281. [CrossRef] [PubMed]
59. Mateo, J.J.; di Stefano, R. Description of the β-glucosidase activity of wine yeasts. *Food Microbiol.* **1997**, *14*, 583–591. [CrossRef]
60. Van Rensburg, P.; Pretorius, I. Enzymes in winemaking: Harnessing natural catalysts for efficient bio-transformations-a review. *S. Afr. J. Enol. Vitic. Spec. Issue* **2000**, *21*, 52–73.
61. Budić-Leto, I.; Zdunic, G.; Banovic, M.; Kovacevic-Ganic, K.; Tomic-Potrebujes, I.; Lovric, T. Fermentative aroma compounds and sensory descriptors of traditional Croatian dessert wine Prošek from Plavac mali cv. *Food Technol. Biotechnol.* **2010**, *48*, 530–537. [CrossRef]
62. Puertas, B.; Jiménez, M.J.; Cantos-Villar, E.; Cantoral, J.M.; Rodriguez, M. Use of *Torulaspora delbrueckii* and *Saccharomyces cerevisiae* in semi-industrial sequential inoculation to improve quality of Palomino and Chardonnay wines in warm climates. *J. Appl. Microbiol.* **2016**, *122*, 733–746. [CrossRef]
63. García, M.; Arroyo, T.; Crespo, J.; Cabellos, J.M.; Esteve-Zarzoso, B. Use of native non-Saccharomyces strain: A new strategy in D.O. "Vinos de Madrid" (Spain) wines elaboration. *Eur. J. Food Sci. Technol.* **2017**, *5*, 1–31. Available online: https://www.eajournals.org/journals/european-journal-of-food-science-and-technology-ejfst/vol-5-issue-2-april-2017/use-native-non-saccharomyces-strain-new-strategy-d-o-vinos-de-madrid-spain-wines-elaboration/ (accessed on 30 September 2019).
64. Rossouw, D.; Bauer, F.F. Exploring the phenotypic space of non-*Saccharomyces* wine yeast biodiversity. *Food Microbiol.* **2016**, *55*, 32–46. [CrossRef]
65. Wang, C.; Mas, A.; Esteve-Zarzoso, B. Interaction between *Hanseniaspora uvarum* and *Saccharomyces cerevisiae* during alcoholic fermentation. *Int. J. Food Microbiol.* **2015**, *206*, 67–74. [CrossRef]
66. Wang, C.; Mas, A.; Esteve-Zarzoso, B. The interaction between *Saccharomyces cerevisiae* and non-*Saccharomyces* yeast during alcoholic fermentation is species and strain specific. *Front. Microbiol.* **2016**, *7*, 502. [CrossRef]

67. Schaffrath, R.; Meinhard, F.; Klassen, R. Yeast Killer Toxins: Fundamentals and Applications. In *Physiology and Genetics*, 2nd ed.; Anke Schüffler, A., Ed.; Springer International Publishing AG: Cham, Switzerland, 2017. [CrossRef]
68. Velázquez, R.; Zamora, E.; Álvarez, M.L.; Hernández, L.M.; Ramírez, M. Effects of new *Torulaspora delbrueckii* killer yeasts on the must fermentation kinetics and aroma compounds of white table wine. *Front. Microbiol.* **2015**, *6*, 1222. [CrossRef]
69. El-Banna, A.A.; El-Sahn, M.A.; Shehata, M.G. Yeasts Producing Killer Toxins: An Overview. *Alex. J. Food Sci. Technol.* **2011**, *8*, 41–53. [CrossRef]
70. Branco, P.; Francisco, D.; Chambon, C.; Hébraud, M.; Arneborg, N.; Almeida, M.G.; Caldeira, J.; Albergaria, H. Identification of novel GAPDH-derived antimicrobial peptides secreted by *Saccharomyces cerevisiae* and involved in wine microbial interactions. *Appl. Microbiol. Biotechnol.* **2014**, *98*, 843–853. [CrossRef] [PubMed]
71. Loira, I.; Morata, A.; Comuzzo, P.; Callejo, M.J.; González, C.; Calderón, F.; Suárez-Lepe, J.A. Use of *Schizosaccharomyces pombe* and *Torulaspora delbrueckii* strains in mixed and sequential fermentations to improve red wine sensory quality. *Food Res. Int.* **2015**, *76*, 325–333. [CrossRef] [PubMed]
72. Curiel, J.A.; Morales, P.; Gonzalez, R.; Tronchoni, J. Different Non-Saccharomyces Yeast Species Stimulate Nutrient Consumption in *S. cerevisiae* Mixed Cultures. *Front Microbiol.* **2017**, *31*, 2121. [CrossRef]
73. Loira, I.; Morata, A.; Bañuelos, M.A.; Suárez-Lepe, J.A. Isolation, selection, and identification techniques for non-*Saccharomyces* yeasts of oenological interest. In *Biotechnological Progress and Beverage Consumption*; Grumezescu Alexandru, M., Holban Alina, M., Eds.; Elsevier Inc.: Amsterdam, The Netherlands, 2020; pp. 467–521. [CrossRef]
74. Renault, P.; Coulon, J.; de Revel, G.; Barbe, J.C.; Bely, M. Increase of fruity aroma during mixed *T. delbrueckii/S. cerevisiae* wine fermentation is linked to specific esters enhancement. *Int. J. Food Microbiol.* **2015**, *207*, 40–48. [CrossRef]
75. Renault, P.; Coulon, J.; Moine, V.; Thibon, C.; Bely, M. Enhanced 3-sulfanylhexan-1-ol production in sequential mixed fermentation with *Torulaspora delbrueckii/Saccharomyces cerevisiae* reveals a situation of synergistic interaction between two industrial strains. *Front. Microbiol.* **2016**, *7*, 293. [CrossRef]
76. Mosciano, G. Successful flavors: From formulation to QC to applications and beyond. In *Successful Flavors*; Mosciano, G., Ed.; Allured Publishing Corp.: Carol Stream, IL, USA, 2006; p. 240. ISBN 10 1932633197.
77. García, M.; Esteve-Zarzoso, B.; Crespo, J.; Cabellos, J.M.; Arroyo, T. Yeast Monitoring of Wine Mixed or Sequential Fermentations Made by Native Strains from D.O. "Vinos de Madrid" Using Real-Time Quantitative PCR. *Front. Microbiol.* **2017**, *8*. [CrossRef]
78. Loira, I.; Vejarano, R.; Bañuelos, M.A.; Morata, A.; Tesfaye, W.; Uthurry, C.; Villa, A.; Cintora, I.; Suárez-Lepe, J.A. Influence of sequential fermentation with *Torulaspora delbrueckii* and *Saccharomyces cerevisiae* on wine quality. *LWT Food Sci. Technol.* **2014**, *59*, 915–922. [CrossRef]
79. Peinado, R.; Moreno, J.; Medina, M.; Mauricio, J. Changes in volatile compounds and aromatic series in sherry wine with high gluconic acid levels subjected to aging by submerged flor yeast cultures. *Biotechnol. Lett.* **2004**, *26*, 757–762. [CrossRef]
80. Viana, F.; Gil, J.V.; Vallés, S.; Manzanares, P. Increasing the levels of 2-phenylethyl acetate in wine through the use of a mixed culture of *Hanseniaspora osmophila* and *Saccharomyces cerevisiae*. *Int. J. Food Microbiol.* **2009**, *135*, 68–74. [CrossRef]
81. Berbegal, C.; Spano, G.; Tristezza, M.; Grieco, F.; Capozzi, V. Microbial Resources and Innovation in the Wine Production Sector. *S. Afr. J. Enol. Vitic.* **2017**, *38*, 156–166. [CrossRef]
82. Alexandre, H.; Costello, P.J.; Remize, F.; Guzzo, J.; Guilloux-Benatier, M. *Saccharomyces cerevisiae-Oenococcus oeni* interactions in wine: Current knowledge and perspectives. *Int. J. Food Microbiol.* **2004**, *93*, 141–154. [CrossRef] [PubMed]
83. Nehme, N.; Mathieu, F.; Taillandier, P. Impact of the co-culture of *Saccharomyces cerevisiae-Oenococcus oeni* on malolactic fermentation performance and partial characterization of a yeast-derived inhibitory peptidic fraction. *Food Microbiol.* **2010**, *27*, 150–157. [CrossRef] [PubMed]
84. Zapparoli, G.; Tosi, E.; Azzolini, M.; Vagnoli, P.; Krieger, S. Bacterial inoculation strategies for the achievement of malolactic fermentation in high-alcohol wines. *S. Afr. J. Enol. Vitic.* **2009**, *30*, 49–55. [CrossRef]
85. Abrahamse, C.E.; Bartowsky, E.J. Timing of malolactic fermentation inoculation in Shiraz grape must and wine: Influence on chemical composition. *World J. Microbiol. Biotechnol.* **2012**, *28*, 255–265. [CrossRef]

86. Knoll, C.; Fritsch, S.; Schnell, S.; Grossmann, M.; Krieger-Weber, S.; Du Toit, M.; Rauhut, D. Impact of different malolactic fermentation inoculation scenarios on Riesling wine aroma. *World J. Microbiol. Biotechnol.* **2012**, *28*, 1143–1153. [CrossRef]
87. Antalick, G.; Perello, M.; de Revel, G. Co-inoculation with Yeast and LAB Under Winery Conditions: Modification of the Aromatic Profile of Merlot Wines. *S. Afr. J. Enol. Vitic.* **2013**, *34*, 223–232. [CrossRef]
88. Cañas, P.M.; Romero, E.G.; Pérez-Martín, F.; Seseña, S.; Palop, M.L. Sequential inoculation *versus* co-inoculation in Cabernet Franc wine fermentation. *Food Sci Technol Int.* **2015**, *21*, 203–212. [CrossRef]
89. Berbegal, C.; Borruso, L.; Fragasso, M.; Tufariello, M.; Russo, P.; Brusetti, L.; Spano, G.; Capozzi, V. A Metagenomic-Based Approach for the Characterization of Bacterial Diversity Associated with Spontaneous Malolactic Fermentations in Wine. *Int. J. Mol. Sci.* **2019**, *120*, 3980. [CrossRef]
90. Kántor, A.; Hutková, J.; Petrová, J.; Hleba, L.; Kačániová, M. Antimicrobial activity of pulcherrimin pigment produced by *Metschnikowia pulcherrima* against various yeast species. *J. Microbiol. Biotech. Food Sci.* **2015**, *5*, 282–285. [CrossRef]
91. Delgado-Vargas, F.; Jiménez-Aparicio, A.R.; Paredes-Lopez, O. Natural Pigments: Carotenoids, Anthocyanins, and Betalains — Characteristics, Biosynthesis, Processing, and Stability. *Crit. Rev. Food Sci. Nut.* **2000**, *40*, 173–289. [CrossRef] [PubMed]
92. Fei, H.; Na-Na, L.; Lin, M.; Qiu-Hong, P.; Jun, W.; Malcolm, J.R.; Chang-Qing, D. Anthocyanins and their variation in red wines. II. Anthocyanin derived pigments and their color evolution. *Molecules* **2012**, *17*, 1483–1519. [CrossRef]
93. Morata, A.; Gómez-Cordovés, M.C.; Colomo, B.; Suáez, J.A. Pyruvic Ácid and acetaldehyde production by different strains of *Saccharomyces cerevisiae*: Relationship with vitisin A and B formation in red wines. *J. Agric. Food Chem.* **2003**, *51*, 7402–7409. [CrossRef] [PubMed]
94. Morata, A.; Calderón, F.; Gonzalez, C.; Gómez-Cordovés, M.C.; Suarez, C.J.A. Formation of the highly stable pyranoanthocyanins (vitisins A and B) in red wines by the addition of pyruvic acid and acetaldehyde. *Food Chem.* **2007**, *100*, 1144–1152. [CrossRef]
95. Schwarz, M.; Wabnitz, T.C.; Winterhalter, P. Pathway Leading to the Formation of Anthocyanin–Vinylphenol Adducts and Related Pigments in Red Wines. *J. Agric. Food Chem.* **2003**, *51*, 3682–3687. [CrossRef]
96. Fulcrand, H.; Benabdeljalil, C.; Rigaud, J.; Chenyier, V.; Moutounet, M. A new class of wine pigments generated by reaction between pyruvic acid and grape anthocyanins. *Phytochemistry* **1998**, *47*, 1401–1407. [CrossRef]
97. Victor-Freitas, A.P.; Mateus, N. Formation of pyranoanthocyanins in red wines: A new and diverse class of anthocyanin derivatives. *Anal. Bioanal. Chem.* **2011**, *401*, 1463–1473. [CrossRef]
98. Morata, A.; Loira, I.; Suárez-Lepe, J.A. Influence of Yeasts in Wine Colour. In *Grape and Wine Biotechnology*; Morata, A., Loira, I., Eds.; IntechOpen: London, UK, 2016; pp. 285–305. [CrossRef]
99. Ruta, L.L.; Farcasanu, I.C. Anthocyanins and Anthocyanin-Derived Products in Yeast-Fermented Beverages. *Antioxidants (Basel)* **2019**, *8*, 182. [CrossRef]
100. Asenstorfer, R.E.; Markides, A.J.; Iland, P.G.; Jones, G.P. Formation of vitisin A during red wine fermentation and maturation. *Aust. J. Grape Wine Res.* **2003**, *9*, 40–46. [CrossRef]
101. Osborne, J.P.; Orduña, R.M.; Pilone, G.J.; Liu, S.Q. Acetaldehyde metabolism by wine lactic acid bacteria. *FEMS Microbiol. Lett.* **2000**, *191*, 51–55. [CrossRef]
102. He, J.R.; Oliveira, J.; Silva, A.M.-S.; Mateus, N.; Victor-Freitas, A.P. Oxovitisins: A New Class of Neutral Pyranone-anthocyanin Derivatives in Red Wines. *J. Agric. Food Chem.* **2010**, *58*, 8814–8819. [CrossRef] [PubMed]
103. Carew, A.L.; Smith, P.; Close, D.C.; Curtin, C.; Dambergs, R.G. Yeast Effects on Pinot noir Wine Phenolics, Color, and Tannin Composition. *J. Agric. Food Chem.* **2013**, *61*, 9892–9898. [CrossRef] [PubMed]
104. Kennedy, J.A. Wine colour. In *Woodhead Publishing Series in Food Science, Technology and Nutrition*; Reynolds, A.G., Ed.; Woodhead Publishing: Sawston, UK, 2010; pp. 73–104. [CrossRef]
105. Minnaar, P.P.; Ntushelo, N.; Ngqumba, Z.; van Breda, V.; Jolly, N.P. Effect of *Torulaspora delbrueckii* yeast on the anthocyanin and flavanol concentrations of Cabernet franc and Pinotage wines. *S. Afr. J. Enol. Vitic.* **2015**, *36*, 50–58. Available online: http://www.scielo.org.za/scielo.php?script=sci_arttext&pid=S2224-79042015000100013&lng=en&tlng=en (accessed on 25 September 2019). [CrossRef]

106. Pacheco, A.; Santos, J.; Chaves, S.; Almeida, J.; Leão, C.; Sousa, M.J. The Emerging Role of the Yeast Torulaspora delbrueckii. In *Bread and Wine Production: Using Genetic Manipulation to Study Molecular Basis of Physiological Responses, Structure and Function of Food Engineering*; Eissa, A.A., Ed.; IntechOpen: London, UK, 2012; pp. 339–370. [CrossRef]
107. Medina, K.; Boido, E.; Fariña, L.; Dellacassa, E.; Carrau, F. Non-*Saccharomyces* and *Saccharomyces* strains co-fermentation increases acetaldehyde accumulation: Effect on anthocyanin-derived pigments in Tannat red wines. *Yeast* **2016**, *33*, 339–343. [CrossRef] [PubMed]
108. Benito, Á.; Calderón, F.; Benito, S. The Combined Use of *Schizosaccharomyces pombe* and *Lachancea thermotolerans*-Effect on the Anthocyanin Wine Composition. *Molecules* **2017**, *22*. [CrossRef]
109. Benito, S.; Morata, A.; Palomero, F.; Gonzalez, M.; Suárez-Lepe, J. Formation of vinylphenolic pyranoanthocyanins by *Saccharomyces cerevisiae* and *Pichia guillermondii* in red wines produced following different fermentation strategies. *Food Chem.* **2011**, *124*, 15–23. [CrossRef]
110. Strauss, M.; Jolly, N.; Lambrechts, M.; Van Rensburg, P. Screening for the production of extracellular hydrolytic enzymes by non-*Saccharomyces* wine yeasts. *J. Appl. Microbiol.* **2001**, *91*, 182–190. [CrossRef]
111. Klis, F.M.; Mol, P.; Hellingwerf, K.; Brul, S. Dynamics of cell wall structure in *Saccharomyces cerevisiae*. *FEMS Microbiol. Rev.* **2002**, *26*, 239–256. [CrossRef]
112. Rodrigues, A.; Ricardo-Da-Silva, J.M.; Lucas, C.; Laureano, O. Effect of commercial mannoproteins on wine colour and tannins stability. *Food Chem.* **2012**, *131*, 907–914. [CrossRef]
113. Doco, T.; Quellec, N.; Moutounet, M. Polysaccharide patterns during the ageing of Carignan noir red wines. *Am. J. Enol. Vitic.* **1999**, *50*, 25–32.
114. Chalier, P.; Angot, B.; Delteil, D.; Doco, T.; Gunata, Z. Interactions between aroma compounds and whole mannoprotein isolated from *Saccharomyces cerevisiae* strains. *Food Chem.* **2007**, *100*, 22–30. [CrossRef]
115. Dupin, I.; Mc Kinnon, B.M.; Ryan, C.; Boulay, M.; Markides, A.J.; Jones, G.P.; Williams, P.J.; Waters, E.J. *Saccharomyces cerevisiae*, mannoproteins that protect wine from protein haze: Their release during fermentation and lees contact and a proposal mechanism of action. *J. Agric. Food Chem.* **2000**, *48*, 3098–3105. [CrossRef] [PubMed]
116. Vidal, S.; Francis, L.; Williams, P.; Kwitkowski, M.; Gawel, R.; Cheynier, V.; Waters, E. The mouth-feel properties of polysaccharides and anthocyanins in a wine like medium. *Food Chem.* **2004**, *85*, 519–525. [CrossRef]
117. Riou, V.; Vernhet, A.; Doco, T.; Moutonnet, M. Aggregation of grape seed tannins in model wine – effect of wine polysaccharides. *Food Hydrocol.* **2002**, *16*, 17–23. [CrossRef]
118. Juega, M.; Nunez, Y.P.; Carrascosa, A.V.; Martinez-Rodriguez, A.J. Influence of yeast mannoproteins in the aroma improvement of white wines. *J Food Sci.* **2012**, *77*, M499–M504. [CrossRef] [PubMed]
119. Ribéreau-Gayon, P.; Dubourdieu, D.; Donèche, B.; Lonvaud, A. *Handbook of Enology, Volume 1: The Microbiology of Wine and Vinifications*; John Wiley and Sons: Hoboken, NJ, USA, 2006; Volume 1.
120. Guadalupe, Z.; Palacios, A.; Ayestarán, B. Maceration enzymes and mannoproteins: A possible strategy to increase colloidal stability and colour extraction in red wines. *J. Agric. Food Chem.* **2007**, *55*, 4854–4862. [CrossRef]
121. Costa, G.P.; Nicolli, K.P.; Welke, J.E.; Manfroi, V.; Zini, C.A. Volatile Profile of Sparkling Wines Produced with the Addition of Mannoproteins or Lees before Second Fermentation Performed with Free and Immobilized Yeasts. *J. Braz. Chem. Soc.* **2018**, *29*, 1866–1875. [CrossRef]
122. Doco, T.; Vuchot, P.; Cheynier, V.; Moutounet, M. Structural modification of wine arabinogalactans during aging on lees. *Am. J. Enol. Vitic.* **2003**, *54*, 150–157.
123. Lubbers, S.; Voilley, A.; Feuillat, M.; Charpentier, C. Influence of mannoproteins from yeast on the aroma intensity of a model wine. *LWT-Food Sci. Technol.* **1994**, *27*, 108–114. [CrossRef]
124. Braschi, G.; Ricci, A.; Grazia, L.; Versari, A.; Patrignani, F.; Lanciotti, R. Mannoprotein Content and Volatile Molecule Profiles of Trebbiano Wines Obtained by *Saccharomyces cerevisiae* and *Saccharomyces bayanus* Strains. *Fermentation* **2019**, *5*, 66. [CrossRef]
125. Comuzzo, P.; Tat, L.; Tonizzo, A.; Battistutta, F. Yeast derivatives (extracts and autolysates) in winemaking: Release of volatile compounds and effects on wine aroma volatility. *Food Chem.* **2006**, *99*, 217–230. [CrossRef]
126. Nieto-Rojo, R.; Ancín-Azpilicueta, C.; Garrido, J.J. Sorption of 4-ethylguaiacol and 4-ethylphenol on yeast cell walls, using a synthetic wine. *Food Chem.* **2014**, *152*, 399–406. [CrossRef] [PubMed]

127. Domizio, P.; Liu, Y.; Bisson, L.F.; Barile, D. Use of non-Saccharomyces wine yeasts as novel sources of mannoproteins in wine. *Food Microbiol.* **2014**, *43*, 5–15. [CrossRef] [PubMed]
128. Pérez-Través, L.; Querol, A.; Pérez-Torrado, R. Increased mannoprotein content in wines produced by *Saccharomyces kudriavzevii* × *Saccharomyces cerevisiae* hybrids. *Int. J. Food Microbiol.* **2016**, *237*, 35–38. [CrossRef]
129. Escot, S.; Feuillat, M.; Dulau, L.; Charpentier, C. Release of polysaccharides by yeasts and the influence of released polysaccharides on colour stability and wine astringency. *Aust. J. Grape Wine Res.* **2001**, *7*, 153–159. [CrossRef]
130. Poncet-Legrand, C.; Doco, T.; Williams, P.; Vernhet, A. Inhibition of grape seed tannin aggregation by wine mannoproteins: Effect of polysaccharide molecular weight. *Am. J. Enol. Vitic.* **2007**, *58*, 87–91.
131. Guadalupe, Z.; Ayestarán, B. Effect of commercial mannoprotein addition on polysaccharide, polyphenolic, and colour composition in red wines. *J. Agric. Food Chem.* **2008**, *56*, 9022–9029. [CrossRef]
132. Guadalupe, Z.; Martínez, L.; Ayestarán, B. Yeast mannoproteins in red winemaking. Effect on polysaccharide, polyphenolic and colour composition. *Am. J. Enol. Vitic.* **2010**, *61*, 191–200.
133. Morata, A.; Gomez-Cordoves, M.C.; Suberviola, J.; Bartolome, B.; Colomo, B.; Suarez, J.A. Adsorption of anthocyanins by yeast cell walls during the fermentation of red wines. *J. Agric. Food Chem.* **2003**, *51*, 4084–4088. [CrossRef]
134. Alcalde-Eon, C.; García-Estévez, I.; Puente, V.; Rivas-Gonzalo, J.C.; Escribano-Bailón, M.T. Color Stabilization of Red Wines. A Chemical and Colloidal Approach. *J. Agric. Food Chem.* **2014**, *62*, 6984–6994. [CrossRef]

© 2020 by the author. Licensee MDPI, Basel, Switzerland. This article is an open access article distributed under the terms and conditions of the Creative Commons Attribution (CC BY) license (http://creativecommons.org/licenses/by/4.0/).

Article

Co-Existence of Inoculated Yeast and Lactic Acid Bacteria and Their Impact on the Aroma Profile and Sensory Traits of Tempranillo Red Wine

Pedro Miguel Izquierdo-Cañas [1,2], María Ríos-Carrasco [3], Esteban García-Romero [1], Adela Mena-Morales [1], José María Heras-Manso [4] and Gustavo Cordero-Bueso [3,*]

[1] Instituto Regional de Investigación y Desarrollo Agroalimentario y Forestal de Castilla-La Mancha (IRIAF), IVICAM, Ctra. Toledo-Albacete s/n., 13700 Tomelloso, Ciudad Real, Spain; pmizquierdo@jccm.es (P.M.I.-C.); estebang@jccm.es (E.G.-R.); amenam@jccm.es (A.M.-M.)
[2] Parque Científico y Tecnológico de Castilla-La Mancha, Paseo de la Innovación 1, 02006 Albacete, Spain
[3] Department of Biomedicine, Biotechnology and Public Health, University of Cádiz, Puerto Real, 11003 Cádiz, Spain; maria.rioscarrasco@alum.uca.es
[4] Lallemand Península Ibérica, Tomás Edison 4, 28521 Madrid; jmheras@lallemand.com
* Correspondence: gustavo.cordero@uca.es

Received: 29 November 2019; Accepted: 23 January 2020; Published: 25 January 2020

Abstract: This study investigates the effects of simultaneous inoculation of a selected *Saccharomyces cerevisiae* yeast strain with two different commercial strains of wine bacteria *Oenococcus oeni* at the beginning of the alcoholic fermentation on the kinetics of malolactic fermentation (MLF), wine chemical composition, and organoleptic characteristics in comparison with spontaneous MLF in Tempranillo grape must from Castilla-La Mancha (Spain). Evolution of MLF was assessed by the periodic analysis of L-malic acid through the enzymatic method, and most common physiochemical parameters and sensory traits were evaluated using a standardized sensory analysis. The samples were analyzed by GC/MS in SCAN mode using a Trace GC gas chromatograph and a DSQII quadrupole mass analyzer. Co-inoculation reduced the overall fermentation time by up to 2 weeks leading to a lower increase in volatile acidity. The fermentation-derived wine volatiles profile was distinct between the co-inoculated wines and spontaneous MLF and was influenced by the selected wine bacteria used in co-inoculation. Co-inoculation allows MLF to develop under reductive conditions and results in wines with very few lactic and buttery flavors, which is related to the impact of specific compounds like 2,3-butanedione. This compound has been also confirmed as being dependent on the wine bacteria used.

Keywords: Simultaneous inoculation; Alcoholic fermentation; Malolactic fermentation; *Sacccharomyces cerevisiae*; *Oenococcus oeni*; PN4[TM]; Omega[TM]; Aroma profile

---

1. Introduction

Malolactic fermentation (MLF) occurs in wine as the result of the metabolic activity of wine lactic acid bacteria. It can take place after the alcoholic fermentation, normally carried out by yeasts, but it can occur earlier. MLF reduces wine acidity by the conversion of dicarboxylic L-malic acid (malate) into monocarboxylic L-lactic acid (lactate) and carbon dioxide, which implies a modification in wine flavor and its microbiological stability, both of which are considered to be beneficial effects for wine quality [1,2]. Sensory studies carried out in the 1980s by Davis et al. [3] showed that malolactic activity of bacteria imparts recognizable changes to the flavor characteristics of wine. These modifications come from the biotransformation and enzymatic processing of grape nutrients into flavor-active compounds [4].

Several studies show that the combination of selected wine yeasts and bacteria strains has different sensory effects in wines due to the production of certain impact metabolites. They are synthesized in different amounts by these organisms according to the grape variety and bacterial strain, giving us the opportunity to determine if their contribution to aroma and flavor depends on the species or strains [5–7]. Research performed by Pozo-Bayón et al. revealed different malolactic behavior for *Oenococcus oeni* and *Lactobacillus plantarum*, concluding that MLF by both species may contribute to wine quality by modifying the concentration of some of the amino acids and aroma compounds of wine because of metabolic differences [8]. However, according to other previous studies, it was proved that the influence of lactic acid bacteria on wine sensory attributes is strain specific. Experiments carried out using different strains of *O. oeni* showed significant changes in aroma and flavor compounds, such as the modifications in fruity characters in red wines or the greater complexity acquired in Sauvignon Blanc wines [9–11].

Lactic acid bacteria can also produce undesirable aroma and flavor compounds, such as volatile phenols or biogenic amines, both related with quality loss and health problems for humans [12,13]. This problem could be solved with a custom selection of bacterial strains, according to the type of wine and its organoleptic properties, which would be modulated depending on the searched sensory profile and consumer preferences [10,14]. Additionally, the use of selected strains of wine bacteria allows for better control of the timeframe of L-malic acid degradation [15,16].

Once the wine bacterial strains are selected, another relevant parameter to study is the timing of the bacteria starter addition and the number of cells and viability in the wine after inoculation, since these can influence the sensory profile of wine. Three different strategies are being investigated: (i) inoculation at any time during alcoholic fermentation often performed simultaneously with yeast (co-inoculation); (ii) inoculation at pressing; and (iii) inoculation after the alcoholic fermentation (sequential inoculation) [17]. Co-inoculation of yeast and bacteria during fermentation shows several benefits when compared to the other strategies, as it shortens vinification times, enhances the wine organoleptic characteristics and reduces the probability of microbial spoilage [2]. Co-inoculation also influences the volatile chemical composition and the potential sensory attributes of the wine, often giving fruitier notes, as opposed to butter or nuts when the MLF begins after the end of the alcoholic fermentation [18].

The aim of the present study is to determine the effect of the coexistence of inoculated commercial yeast and novel selected lactic acid bacteria strains and its impact on the aroma profile and sensory traits of Tempranillo red wine, by analyzing chemical, physical, and sensory parameters.

2. Materials and Methods

2.1. Microorganisms

The yeast strain selected to perform alcoholic fermentation was Viacell C-58® (Lallemand Inc., Montreal, Quebec, Canada). This *Saccharomyces cerevisiae* yeast strain was isolated in Germany and used according to the manufacturer's recommendations. It has the ability to release polysaccharides to the wine, stabilizing color in red wines, and decreasing astringency of green tannins. Furthermore, it stimulates the development of fruity aromas that persist after barrel aging. This yeast strain also shows a high adaptation to low temperatures, which makes it suitable for cold pre-fermentative maceration.

The commercial wine lactic acid bacteria (LAB) used were *O. oeni* PN4™ (Lallemand Inc., Montreal, Quebec, Canada) and *O. oeni* Omega™ (Lallemand Inc., Montreal, Quebec, Canada). LAB strain PN4™ was isolated and selected at the Edmund Mach Institute (San Michele all'Adige, Trentino, Italy). This LAB stands out for being a robust strain, which showed the ability to complete malolactic fermentation in red and white wines under adverse conditions of pH, alcohol, SO_2, and temperature. In red wines, PN4™ seems to contribute to spicy notes. LAB strain *O. oeni* Omega™ was isolated and selected in the Languedoc region in France by the Institut Françaíse de la Vigne et du Vin (IFV). According to the previous studies made by the manufacturer, it stands out for its ability to finish rapidly malolactic fermentation in a wide range of applications. This LAB endures low pH and high alcohol levels. It is effective in red, rose, and white wines. It complements a fresh and direct fruit style and helps to stabilize color due to its slow degradation of acetaldehyde. Both LAB had been described by the manufacturer (Lallemand Inc., Montreal, Quebec, Canada) to be highly suitable for the use as components of a mixed yeasts/bacteria co-inoculum.

2.2. Vinifications

Tempranillo grape-berries were harvested at the optimum technological maturity (23–25 °Brix) during the 2018 vintage in the Castilla-La Mancha winemaking area. Once grapes were received in the cellar, they were crushed and treated with 4 g/hL of SO_2, and then a dose of 2 g/hL of IOC enzyme (Lallemand Inc., Montreal, Quebec, Canada) was added to favor the color extraction. The initial chemical composition of must was: °Brix 23.50; total acidity 5.35 g/L; pH 3.85, and L-malic acid 1.95 g/L. Must with crushed grapes was homogeneously divided after crushing into six stainless steel vessels with a capacity of 100 L. Two of the vessels were used as control (AF1), in which AF means Alcoholic Fermentation and no lactic acid bacteria strains were added, and after alcoholic fermentation with the selected *S. cerevisiae*, wines carried out malolactic fermentation spontaneously. The four remaining deposits were inoculated with commercial *O. oeni* lactic acid bacteria strains 24 h after yeast inoculation, two of them with PN4™ (AF2) and the other two with Omega™ (AF3). The obtained wines were cold-stabilized at −3 °C for 25 days, and afterwards they were filtered and bottled. Once they were bottled, wines were sensorily and chemically analyzed.

2.3. Microbiological Control

Determination of microbial population was carried out as stated by Izquierdo-Cañas et al. [16]. The viable count of yeasts was performed by plating samples in malt extract agar media (Cultimed, Barcelona, Spain), and incubated at 28 °C for 48 h. In order to quantify lactic acid bacteria, samples were plated onto MLO Agar (MLOA, *O. oeni* Medium, Scharlab, Barcelona, Spain) supplemented with 10% *v/v* tomato juice, 50 mg/L sodium azide, and 100 mg/mL of cycloheximide. Plates were incubated under anaerobic conditions (Gas Pack System, Oxoid Ltd, Basingstoke, UK) at 30 °C for 5 days. The results were expressed in colony forming units per mL (CFU/mL) of wine.

The implantation of the inoculated lactic acid bacteria was assessed by the random selection of isolated colonies from MLOA plates at different fermentation times after inoculation, when L-malic acid concentration had decreased to 50% and when L-malic acid content was below 0.2 g/L.

2.4. FML Progress and Physicochemical and Spectrophotometric Analysis

FML evolution was determined by the periodic analysis of L-malic acid content, using the enzymatic method established by the International Organization of Vine and Wine (OIV) [19]. Most common physiochemical parameters, such as alcoholic strength, pH, total acidity, total and free SO_2, glucose and fructose content, and Folin–Ciocalteu index to determine the total phenolic content of wines were also measured following the recommendations of the official methods of the OIV [20]. Colorimetric parameters were measured at pH 3.6 by a CIE-Lab* color system following the protocol proposed by Ayala et al. [21]. Total anthocyanin content was measured by the sulfur dioxide bleaching method according to Riberèau-Gayon and Stonestree [22]. Concentration of different carboxylic acids (malic, lactic and citric acid) and glycerol content were determined by an isocratic HPLC system that was set up with a column block heater and refractive index detector. The mobile phase was 8 mmol/L H_2SO_4, 0.6 mL/min, and set at 75 °C in an Aminex HPX-87H 300 × 7.8 mm column. Finally, tannin content in wines was quantified by precipitation with methylcellulose following the protocol proposed by Sarneckis et al. [23].

2.5. Volatile Compound Analysis

Volatile compound extraction was performed according to the protocol described by Ibarz et al. [24]. SPE cartridges (Li-Chrolut EN, Merck, 0.2 g of phase, Darmstadt, Germany) and 4-nonanol (0.1 g/L) were used as internal standards. The extracts obtained were concentrated to a final volume of 150 µL by distillation in a Vigreux column and then under a nitrogen stream. Samples were kept at −20 °C until further analysis. A Focus-GC system coupled to a mass spectrometer ISQ with electron impact-ionization source and quadrupole analyzer with an auto-sampler Tri-Plus (Thermo-Quest, Waltham, MA, USA) was used to assess the free volatile composition of wines. We used a BP21 column (SGE, Ringwood, Australia) with a 50 m × 0.32 mm internal diameter and 0.25 µm thick of Free Fatty Acid Phase (FFAP). The chromatographic conditions were as follows: carrier helium gas (1.4 mL min^{-1}, split 1/57); injector temperature, 220 °C; and oven temperature, 40 °C for 15 min, 2 °C min^{-1} to 100 °C, 1 °C/min to 150 °C, 4 °C/min to 210 °C, and 55 min at 210 °C. The detector conditions were as follows: impact energy, 70 eV; electron multiplier voltage, 1250 V; ion source temperature, 250 °C; and mass scanning range, 40–250 amu. Volatile compounds were detected by chromatographic retention times and mass spectra, using commercial products as standards. Quantification of volatile compounds was carried out by analyzing the characteristic m/z fragment for each compound using the internal standard method. Concentrations of non-available volatile compounds were expressed at µg/L or mg/L as 4-nonanol equivalents obtained by normalizing the peak area to that of the internal standard and multiplying by the concentration of the internal standard.

2.6. Sensory Analysis

Descriptive and triangular tests were first performed to determine the differences among treatments. Fifteen tasters were selected to carry out a descriptive sensory analysis according to the Sensory Profile method, based on a scale from 0 (absence of the descriptor) to 5 (maximum intensity), and established by the ISO standard [25]. Selected attributes were color intensity, purplish-red color, aroma of red fruits, aroma of ripe red fruits, aroma of raisins, floral bouquet, aroma of spices, milky aromas, fresh aroma, astringent flavor, noble mature tannins in the mouth, and lengthy finish in the mouth.

2.7. Statistical Analysis

Significant differences among samples were determined for each chemical compound by analysis of variance (ANOVA) and paired Student's t-test ($p = 0.05$) for each chemical compound. A cobweb graph was performed to determine significant differences of the attributes after sensory analysis. Statistical analysis was performed with 20.0 SPSS (IBM Inc. Chicago, IL, USA) for Windows statistical package.

3. Results

3.1. Evolution of Microbial Populations and Fermentation Kinetics

The progress of alcoholic fermentation of the commercial *S. cerevisiae* Viacell C-58® strain was evaluated for each vessel by daily density measurement under the established experimental conditions: spontaneous malolactic fermentation, co-inoculation with *O. oeni* PN4™, and co-inoculation with *O. oeni* Omega™. After seven days, since density was always below 993 g/L, alcoholic fermentation was considered to be finished with similar results in all wines (Figure 1).

Although the presence of native LABs could have an adverse effect in non-sterile must fermentation, our results showed the favorable effect on the start-up period of inoculating pure LAB cultures. The presence of native lactic acid bacteria in the first stages of alcoholic fermentation did not affect the progressive decrease of density in wine (Figure 1). In fact, in the spontaneous malolactic fermentation (control), it was observed that population of *S. cerevisiae* was affected by the presence of native lactic acid bacteria, maintaining an average concentration of 10^6 cells/mL when must density decreased, until LAB took part in malolactic fermentation (Figure 1). In the co-inoculated vinification with the *O. oeni* PN4™ strain, the development of the bacterial population showed a good adaptation to fermentation conditions, following fermentative kinetics similar to the corresponding kinetics of Viacell C-58® yeast. Conversely, in the co-inoculated fermentation with Omega™, a different behavior was observed in presence of Viacell C-58® during the first days of the alcoholic fermentation, reaching concentrations between 10^7 and 10^8 cells/mL, while it reached similar populations to PN4™ during the rest of the alcoholic fermentation (Figure 1).

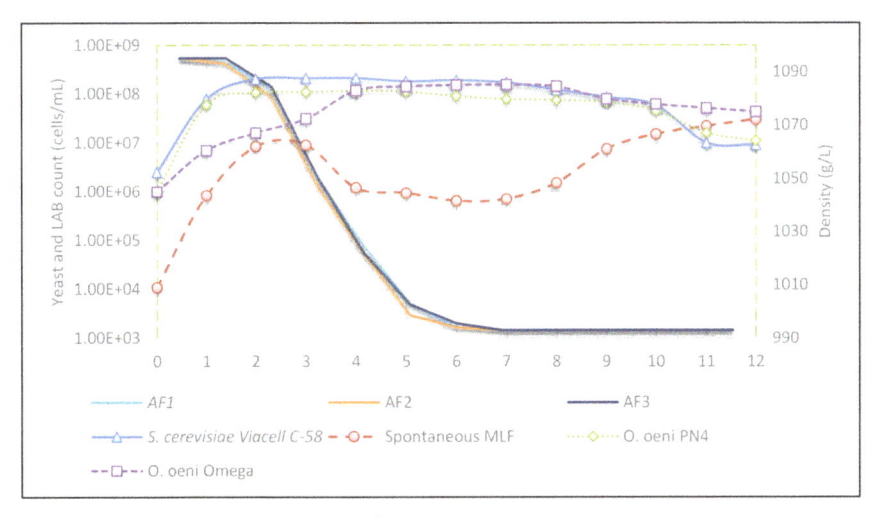

Figure 1. Progress of the yeast Viacell C-58® (Δ) and lactic acid bacteria population (dashed lines) and must density (continuous lines) during the three wine alcoholic fermentations denoted as (AF1, AF2, and AF3) and performed with Tempranillo musts using the following inoculation strategies: no lactic acid bacteria (LAB) inoculation (●); inoculation with LAB PN4™ (♦); and inoculation with LAB Omega™ (■). Data represented correspond to mean values obtained from duplicate fermentations; in all cases standard deviation was <1.5. MLF means malolactic fermentation.

3.2. Malolactic Fermentation

Malolactic fermentation dynamics were controlled by the daily determination of L-malic and L-lactic acid content. In co-inoculated fermentation with PN4™ and Omega™ strains, the behavior was similar during the first days, but variations in L-malic acid consumption appeared in the final stages of the alcoholic fermentation (Figure 2). The strain PN4™ could completely consume L-malic acid in 10 days in the presence of *S. cerevisiae* starters, while the Omega™ strain did the same two days later. Regarding the spontaneous malolactic fermentation, the consumption of L-malic acid was started from day 12, finishing MLF in 26 days (Figure 2). Again, this fact confirms that lactic acid bacteria co-existed with *S. cerevisiae* Viacell C-58® until L-malic acid was almost finished. It is important to highlight that the L-malic acid consumption was reduced between 14 and 16 days when *O. oeni* PN4™ and Omega™ strains were inoculated compared to the spontaneous MLF.

3.3. Determination of Physic-Chemical Parameters of Fermentations and Color

Despite observing significant differences when comparing residual sugars in both co-inoculated and control vinifications, in all cases must sugars were completely consumed, and dry wines were obtained. Regarding acetic acid levels, they decreased to 0.28 g/L for PN4™ and 0.26 g/L for Omega™ values, which were lower than those obtained for spontaneous MLF wine, 0.35 g/L (Table 1).

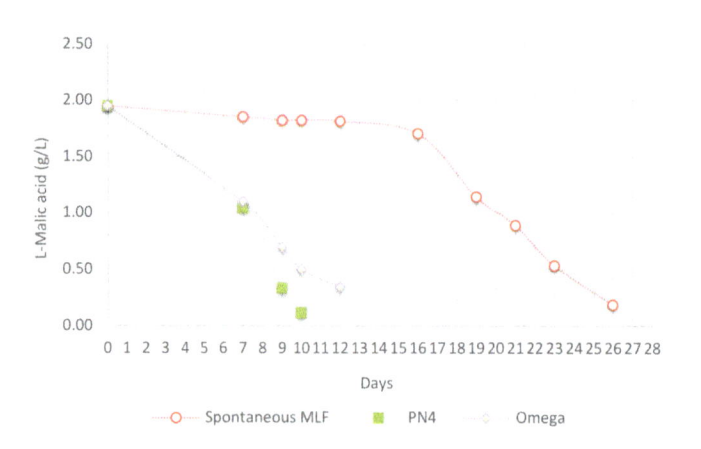

Figure 2. Progression of L-malic acid consumption (g/L) during vinification of Tempranillo must in the two samples inoculated with LAB PN4™ (♦) and with LAB Omega™ (■) at the beginning of the alcoholic fermentation and the uninoculated one (●). Data represented correspond to mean values obtained from duplicate fermentations; in all cases standard deviation was <1.5.

Table 1. Physicochemical parameters, organic acid concentrations, and color values of wines. Data represented correspond to mean and standard deviation values obtained from duplicate vinifications. Characters a, b, and c mean significant differences at $p < 0.05$.

	Spontaneous MLF	PN4™	Omega™
Alcoholic strength (% v/v)	13.34 ± 0.66 [a]	13.45 ± 0.15 [a]	13.52 ± 0.23 [a]
Total acidity (g/L)	3.49 ± 0.03 [a]	3.62 ± 0.18 [a]	3.63 ± 0.18 [a]
pH	4.07 ± 0.08 [a]	4.04 ± 0.07 [a]	4.01 ± 0.03 [a]
Volatile acidity (g/L)	0.35 ± 0.01 [b]	0.28 ± 0.03 [a]	0.26 ± 0.00 [a]
L-malic acid (g/L)	0.07 ± 0.04 [a]	0.05 ± 0.01 [a]	0.08 ± 0.02 [a]
L-lactic acid (g/L)	1.27 ± 0.00 [c]	1.22 ± 0.02 [b]	1.12 ± 0.06 [a]
Citric acid (g/L)	0.07 ± 0.03 [a]	0.20 ± 0.04 [b]	0.19 ± 0.04 [b]
Glycerol (g/L)	9.69 ± 0.93 [a]	9.91 ± 0.26 [a]	9.85 ± 0.28 [a]
L*	17.13 ± 3.29 [a]	15.34 ± 0.36 [a]	19.26 ± 2.23 [a]
a*	43.39 ± 1.31 [a]	43.46 ± 0.68 [a]	44.38 ± 0.09 [a]
b*	22.13 ± 0.84 [a]	21.28 ± 0.18 [a]	20.99 ± 0.82 [a]
Color intensity	6.44 ± 0.83 [a]	6.79 ± 0.32 [a]	5.89 ± 0.55 [a]
Color tone	0.83 ± 0.00 [b]	0.76 ± 0.04 [a]	0.79 ± 0.04 [ab]
Total anthocyanins (mg/L)	446.00 ± 39.60 [a]	432.00 ± 33.94 [a]	426.50 ± 27.58 [a]
Catechins (mg/L)	694.05 ± 67.39 [a]	730.10 ± 107.48 [a]	614.60 ± 55.58 [a]
Tannins (mg/L)	0.61 ± 0.12 [a]	0.64 ± 0.15 [a]	0.51 ± 0.30 [a]
Folin–Ciocalteu	33.00 ± 4.24 [a]	34.50 ± 3.54 [a]	31.00 ± 1.41 [a]

There were no significant differences in volatile acidity concentrations between the two co-inoculated wines, while the spontaneous MLF showed a higher volatile acidity (Table 1). Total acidity, glycerol, and pH were similar in all cases, which indicates that co-inoculation with commercial lactic acid bacteria did not affect in a negative way the chemical properties of wine (Table 1). L-lactic acid values differed for the three experimental conditions, and co-inoculated wine with the LAB Omega™ showed the lowest values and the spontaneous MLF the highest. On the other hand, a decrease in citric acid content was observed in the spontaneous malolactic fermentation (Table 1). This fact indicates that citric acid degradation depends on the LAB strain used. In this study, the commercial strains degraded less citric acid and, as a consequence, the volatile acidity produced was lower than that of the non-co-inoculated one (Table 1). It is well known that during MLF, wine discoloration can take place. However, in this work, it was proved that there were no important differences in the spectral characteristics of wine color, and intensity was similar in all cases. Co-inoculation did not decrease wine color. No significant differences were observed in the rest of analyzed parameters (Table 1).

3.4. Volatile Compound Analysis

Aromatic profiles from the three different wines are shown in Table 2. Metabolism from Viacell C-58® and the native and commercial lactic acid bacteria decisively influenced the volatile composition of wine, including secondary aromas related to MLF. It is appreciated that esters and alcohols were significantly higher, and fatty acid content was lower in wines made by co-inoculation with commercial LAB strains compares to that made by spontaneous MLF, used as a control.

Ethyl acetate content was significantly different between the three fermentation strategies (Table 2). Fermentations carried out with the *O. oeni* Omega™ strain showed a higher isoamyl acetate and hexyl acetate content, which are remarkable compounds in Tempranillo wines when compared to the other two wines. The lower content of 2-phenyl acetate was lower in the wine co-inoculated with the *O. oeni* PN4™ strain (Table 2). Regarding the ethyl lactate, the LAB strain Omega™ contributed to lower levels of lactic aromas, while the isoamyl acetate concentration was higher (Table 2).

Ethyl hexanoate is responsible for the fruity notes in red wines, and it was observed that its content was significantly higher in those wines made with *O. oeni* PN4™ and Omega™, if compared to spontaneous MLF (Table 2). The same happened with 1-hexanol, which contributed to herbaceous notes (Table 2). It should also be emphasized that ethyl butyrate was significantly higher in wines made with Omega™ (Table 2). No significant differences in the ethyl octanoate and ethyl decanoate contents were found between the three wine trials (Table 2).

Carboxylic acids are responsible for fatty, rancid, and buttery notes (Table 2). It was shown that co-inoculated wines had lower levels of hexanoic acid, octanoic acid, and decanoic acid with respect to the spontaneous MLF, highlighting that *O. oeni* PN4™ notably reduced butyric acid and hexanoic acid content.

Taking into account that acetaldehyde content decreased during MLF, a decline was observed in wines with spontaneous MLF, in contrast with LAB co-inoculated wines, where PN4™ showed the highest level of the compound. The 2,3-butanodione is a decisive compound in MLF, since it is mainly derived from lactic acid bacteria metabolism. High concentrations of this metabolite give heavy milky notes. Therefore, it is interesting to obtain it in low concentrations, as it will contribute to wine smoothness. High quantities of this compound were produced in spontaneous MLF wine in contrast to the co-inoculated ones. This fact can be very promising for wines produced by co-inoculation of commercial LAB since they have less lactic notes (octanoic and decanoic acid, 2,3-butanodione), which avoids the masking of fruity notes that give personality to Tempranillo wines made with the co-inoculation of *O. oeni* PN4™ and Omega™ strains.

3.5. Sensory Analysis

Regarding the sensory profile of wines and according to the tasters, differences in organoleptic properties were observed (Figure 3). Wines made by spontaneous MLF showed higher aromas of raisins, milky aromas, and color intensity. In contrast, wines developed by the inoculation of commercial lactic acid bacteria highlighted the fresh aroma and lengthy finish flavor and ripe red fruit character and less milky aromas (Figure 3).

Table 2. Data (Mean ± S.D.) of volatile composition related to the uninoculated fermentations (S) and inoculated fermentations with two commercial LAB strains (PN4™ and Omega™). Characters a, b, and c mean significant differences at $p \leq 0.05$.

	Spontaneous MLF	PN4™	Omega™
Ethyl acetate (mg/L)	80.00 ± 15.89 [a]	95.18 ± 0.13 [b]	108.65 ± 2.93 [c]
Isoamyl acetate (mg/L)	5.85 ± 0.76 [a]	6.13 ± 0.10 [a]	8.85 ± 0.89 [b]
2-Phenylethyl acetate (µg/L)	182.61 ± 7.24 [b]	141.89 ± 7.89 [a]	170.63 ± 31.80 [b]
Hexyl acetate (µg/L)	13.54 ± 1.77 [a]	13.75 ± 0.84 [a]	21.21 ± 7.18 [b]
Ethyl lactate (mg/L)	145.86 ± 19.96 [b]	137.78 ± 6.11 [b]	117.54 ± 1.24 [a]
Ethyl butyrate (µg/L)	632.50 ± 106.77 [a]	687.50 ± 30.41 [a]	896.00 ± 94.75 [b]
Ethyl hexanoate (µg/L)	739.82 ± 26.96 [a]	790.88 ± 57.99 [ab]	866.78 ± 85.64 [b]
Ethyl octanoate (µg/L)	716.64 ± 49.17 [a]	677.01 ± 90.65 [a]	785.66 ± 85.41 [a]
Ethyl decanoate (µg/L)	97.26 ± 8.89 [a]	90.29 ± 14.61 [a]	100.97 ± 10.01 [a]
1-propanol (mg/L)	62.83 ± 8.25 [a]	54.74 ± 5.25 [a]	54.55 ± 0.51 [a]
Isobutanol (mg/L)	30.48 ± 3.71 [a]	28.96 ± 0.51 [a]	30.65 ± 1.20 [a]
1-butanol (mg/L)	1.46 ± 0.30 [a]	1.65 ± 0.12 [a]	1.46 ± 0.23 [a]
Isoamylic alcohols (mg/L)	342.50 ± 36.63 [a]	332.19 ± 1.25 [a]	327.92 ± 10.87 [a]
1-octanol (µg/L)	13.82 ± 2.45 [a]	16.14 ± 2.27 [a]	17.33 ± 1.77 [a]
3-methyl-1-pentanol (µg/L)	167.16 ± 42.26 [a]	212.66 ± 6.96 [c]	181.06 ± 1.16 [ab]
3-ethoxy-1-propanol (µg/L)	53.47 ± 13.10 [a]	48.15 ± 3.02 [a]	45.46 ± 2.95 [a]
1-hexanol (mg/L)	3.24 ± 0.12 [a]	3.46 ± 0.16 [b]	3.53 ± 0.08 [b]
t-3-hexenol (µg/L)	74.70 ± 0.47 [a]	75.24 ± 9.86 [a]	74.74 ± 0.86 [a]
c-3-hexen-1-ol (µg/L)	513.62 ± 88.94	516.16 ± 21.29	560.66 ± 4.26
2-phenylethanol (mg/L)	23.68 ± 2.79 [a]	26.07 ± 1.58 [a]	22.54 ± 1.49 [a]
Isobutiric acid (µg/L)	460.53 ± 49.70 [a]	428.63 ± 51.80 [a]	426.93 ± 10.52 [a]
Butyric acid (µg/L)	570.86 ± 1.69 [b]	436.89 ± 2.96 [a]	513.28 ± 91.90 [b]
Isovaleric acid (mg/L)	1.43 ± 0.05 [c]	1.31 ± 0.08 [b]	1.22 ± 0.06 [a]
Valeric acid (µg/L)	13.68 ± 3.89 [b]	7.71 ± 2.71 [a]	6.68 ± 4.72 [a]
Hexanoic acid (mg/L)	8.29 ± 0.46 [b]	7.28 ± 0.57 [a]	7.80 ± 0.34 [ab]
Octanoic acid (mg/L)	7.25 ± 1.65 [b]	5.02 ± 0.08 [a]	5.41 ± 0.16 [a]
Decanoic acid (mg/L)	3.78 ± 0.94 [b]	2.08 ± 0.10 [a]	1.92 ± 0.17 [a]
γ-nonalactone (µg/L)	6.35 ± 0.53 [a]	5.39 ± 0.56 [a]	6.14 ± 0.55 [a]
δ-dodecalactone (µg/L)	9.58 ± 0.67 [a]	6.89 ± 1.05 [a]	7.56 ± 3.70 [a]
δ-ethoxycarbonyl-δ-butyrolactone (µg/L)	321.46 ± 39.16 [ab]	275.65 ± 26.25 [a]	355.95 ± 31.85 [b]
Damascenone (µg/L)	6.03 ± 0.81 [a]	6.27 ± 0.23 [a]	5.74 ± 0.97 [a]
3-oxo-α-ionol (µg/L)	55.82 ± 0.60 [a]	49.64 ± 7.01 [a]	53.82 ± 4.76 [a]
Acetaldehyde (mg/L)	9.28 ± 0.05 [a]	10.87 ± 0.96 [b]	14.00 ± 1.16 [c]
2.3 Butanodione (mg/L)	18.35 ± 1.67 [c]	14.77 ± 1.61 [b]	8.55 ± 0.48 [a]
1-octen-3-ona (µg/L)	14.25 ± 0.59 [a]	14.02 ± 1.84 [a]	14.19 ± 2.08 [a]
3-methyl-thio-propanol (µg/L)	277.33 ± 85.07 [a]	311.44 ± 66.18 [a]	211.11 ± 10.75 [a]
Furaneol (µg/L)	12.32 ± 1.26 [b]	5.94 ± 3.73 [a]	10.39 ± 2.52 [b]
Guaiacol (µg/L)	12.39 ± 1.65 [a]	18.44 ± 0.37 [b]	19.88 ± 5.18 [b]

Table 2. Cont.

	Spontaneous MLF	PN4™	Omega™
Eugenol (µg/L)	3.97 ± 0.36 [b]	2.52 ± 0.79 [a]	2.71 ± 0.13 [a]
Phenol (µg/L)	8.33 ± 0.98 [a]	9.41 ± 0.66 [a]	10.24 ± 3.49 [a]
4-metil-2, 6-ditercbutil-fenol (µg/L)	95.91 ± 29.03 [b]	66.28 ± 8.35 [a]	63.06 ± 0.07 [a]

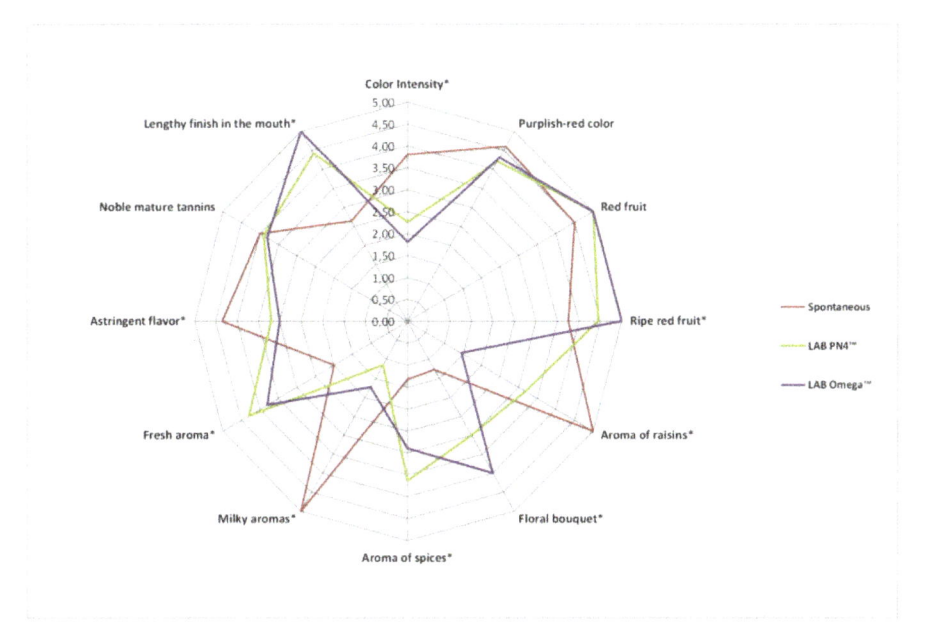

Figure 3. This figure shows the polar coordinate (cobweb) graph of mean sensory score ratings of "color intensity", "purplish-red color", "red fruit", "ripe red fruit", "aroma of raisins", "floral bouquet", "aroma of spices", "milky aroma", "fresh aroma", "astringent flavor", "noble mature tannins", and "lengthy finish in the mouth" for the spontaneous Tempranillo wine and the PN4™ and Omega™ LAB strains. In sensorial variables indicated with an asterisk (*) a difference between trials was verified for $p \leq 0.05$.

Panelists concluded that wines fermented with LAB reached more fruity and floral flavors, and spontaneous wines were less fruity and had stronger astringency and milky, raisin flavors. LAB Omega™ had strongly fruity and floral aromas, and PN4 in spices and fresh aromas in the Tempranillo wines. Spontaneous LAB produced the least ripe red fruit, floral, spices, and fresh aromas wines in this comparison. These wines tended to have milky, raisin, and astringent flavors. Statistical analysis revealed that the descriptors that the descriptors differed significantly in non-inoculated Tempranillo wines from those ones, which were inoculated,) suggesting that the different LAB strains deeply affected the flavor compounds of the wines (Figure 3).

4. Discussion

Under normal conditions, spontaneous malolactic fermentation (MLF) takes place once alcoholic fermentation has finished. This fact is due to a competition for must nutrients between yeast and lactic acid bacteria (LAB) naturally occurring in grapes. *S. cerevisiae* develops rapidly and consumes the essential nutrients for lactic acid bacteria growth, in addition to the production of ethanol and other byproducts that avoid their proliferation in the first stages of fermentation [26,27]. However, as alcoholic fermentation finishes, yeast suffer what is called lysis or autolysis, where they free intracellular nutrients that will be used as nitrogen and carbon sources to those lactic acid bacteria that resisted the ethanol concentration of the final wine [27]. Then, lactic acid bacteria population will exponentially proliferate to carry out MLF. It is possible to reproduce this natural phenomenon in controlled conditions. When winemakers use selected or commercial yeast starters to perform alcoholic fermentation, generally they inoculate lactic acid bacteria in sequential steps reproducing the succession that would take place naturally.

Other winemakers decide to co-inoculate LAB with fermentative yeast at the beginning of fermentation or just before alcoholic fermentation finishes [27–29]. This technique is gaining popularity because not only MLF is assured, but also carries recognized advantages by enologists and winemakers. One of the most important benefits of simultaneous inoculation of yeast/LAB is the reduction of total time of fermentation [18]. Jussier et al. [30] observed a significant reduction of MLF time in Chardonnay wines at pH of 3.53 and 13% ethanol (v/v) in co-inoculation when compared to sequential AF/MLF. This study corroborates this assertion at a pilot scale and is consistent with previous investigations performed at the laboratory scale [29,31–34].

Reduction of MLF times can be explained by specific interactions that occur between S. cerevisiae and O. oeni during AF and MLF, when the chosen strategy is the co-inoculation of both starter cultures [35]. In the experiments of this study, Viacell C-58® was used as fermentative yeast and two O. oeni (PN4™ and Omega™) as commercial LAB as a mixed co-inoculum of yeast/bacteria. Yeast and LAB strains had been described by the producers to be highly suitable for the use as component of a mixed yeast/bacteria co-inoculum. It can be observed that interactions between the yeast and the commercial co-inoculated bacteria are different from those that take place in spontaneous fermentation. The concentration of the inoculated yeast was about 10^6 CFU/mL in all cases for the commercial LAB, while the LAB initial population of spontaneous elaborations was 10^4 CFU/mL, a little higher than that reported in red wine musts [36]. MLF also occurred when AF was finished, but the duration of the total process was substantially superior than in the case of the two starter culture inoculations, due to the adaptation of bacterial starter to the must environment from the beginning of AF. Furthermore, viability of S. cerevisiae starter did not get influenced during the simultaneous progress of AF and MLF, as the yeast exponential growth phase did not decrease before reaching the stationary phase [32–37]. These evidences are consistent with those obtained in previous studies in Cabernet Franc, Tempranillo, and Merlot wines with other LAB and yeast strains [16,38] and Neroamaro wines [29].

This study also confirmed that MLF can take place in the presence of fermentable sugars without a significant increasing of acetic acid, in contrast with data obtained in other studies where the content of acetic acid increased in yeast and LAB co-inoculations [12,39,40]. This variability in the reduction of volatile acidity in co-inoculated wines with PN4™ and Omega™ strains could depend on the yeast strain used, on the LAB starter used, or both. This result shows consistency with those obtained by Tristezza et al. [29] in Neroamaro wines and du Plessis et al. [41] in Shiraz wines. In fact, in the vinification experiments, approximately 50% of the malic acid was consumed at the 7th day of the AF, and the total consumption of malic acid took place after the AF (days 7–12). This result reconfirmed a previous study of this research group in which an improvement of organoleptic properties of Cabernet Franc, Merlot, and Tempranillo wines was observed [16,38].

Regarding the color, no significant differences were observed in the spectral properties of color in co-inoculated MLF when compared to the control, but in contrast, lower shades were found in LAB co-inoculated wines. According to Burns and Osborne [42], MLF can affect red wine color independently of pH change. The species O. oeni could change the concentration of phenolic and non-phenolic compounds involved in red wine color development independently of pH change. As stated before, there were no strain specific differences in the color, but the tonality could be related to the O. oeni strain used.

The obtained results not only confirm that the correct selection of yeast and LAB strains for the performing of MLF by co-inoculation has a positive influence in the reduction of fermentation times and volatile acidity production, but also in the aromatic composition of wine. In fact, it was clearly proved the considerable effect of co-inoculation of the commercial yeast with two different LAB strains on the organoleptic properties of Tempranillo wines obtained, when compared with those obtained by spontaneous MLF. Recent investigations have highlighted the variation in aromatic profile in wine produced by different LAB inoculation processes [16,18,38,39]. Our data suggest that coexistence of yeast *Saccharomyces* with the lactic acid bacterial strain Omega™ increases some ester levels, improving the fruity aroma according to the sensory analysis, as showed by Tristezza et al [29]. Nine esters were identified and quantified, and wines produced by co-inoculation contained higher concentrations of diethyl and monoethylsuccinate, ethyl lactate, 2-phenyl acetate, and ethyl fatty acids esters [2,29].

In general, the co-inoculation of both LAB strains gave as a result a significant change in the ester profile of wine, those of ethyl fatty acid being the most representative. This suggests that coexistence of yeast and lactic acid bacteria stimulates the formation of medium chain fatty acids and, consequently, the concentration of fatty acid esters varies in wine [29,38,43]. On the other hand, the presence of 2,3-butanodiol shows that, in the case of co-inoculation, bacteria were able to degrade diacetyl, a derived-MLF compound with high impact on wine organoleptic properties [18,39]. When this compound appears in high concentrations in wine, it can affect in a negative way its aroma, conferring buttery notes that interfere with fruity aromas [44]. Consequently, applying the co-inoculation technique, it is possible to reduce unwanted buttery and lactic notes, allowing the fruity ones to prevail, bringing out the characteristic varietal aroma of Tempranillo wines. Moreover, after sensory analysis, the wines from non-inoculated LAB fermentations had the majority of high scores for these descriptors, while those fermented with LAB had the lowest scores.

It is well known that the period from the end of the AF to the beginning of the FML is specifically conducive to the development of *Brettanomyces/Dekkera*. Early wine inoculation with LAB, immediately after AF or co-inoculation, has been proved a simple and effective method to prevent *Brettanomyces* development and the production of off-flavors due to high concentrations of ethylphenols [45]. This study showed that phenol content in wine elaborated by co-inoculation of LAB starters was significantly lower than those ones by spontaneous MLF. Co-inoculation reduced the overall fermentation time by up to 2 weeks leading to a lower increase in volatile acidity and an increase of the aromatic quality of wines with more varietal and fruity notes.

Co-inoculation allows MLF to develop under reductive conditions and results in wine with very few milky and buttery flavors, related to the impact of specific compounds like 2,3-butanedione. This compound has also confirmed as being dependent on the wine bacteria used.

Author Contributions: P.M.I.-C., E.G.-R., and A.M.-M. contributed to the conception and design of experiments; P.M.I.-C., M.R.-C., G.C.-B, and J.M.H.-M. performed the experiments; P.M.I.-C., G.C.-B., E.G.-R., and A.M.-M. analyzed the data; P.M.I.-C., M.R.-C. and G.C.-B. contributed to writing the paper. All authors critically revised the manuscript before submission. All authors have read and agreed to the published version of the manuscript.

Funding: This research received no external funding

Acknowledgments: The authors thank J.J. Armario for critically revising the manuscript and the English grammar.

Conflicts of Interest: The authors declare no conflict of interest.

References

1. Davis, C.R.; Wibowo, D.; Eschenbruch, R.; Lee, T.H.; Fleet, G.H. Practical implications of malolactic fermentation: A review. *Am. J. Enol. Viticult.* **1985**, *36*, 290–301.
2. Versari, A.; Patrizi, C.; Parpinello, G.P.; Mattioli, A.U.; Pasini, L.; Meglioli, M.; Longhini, G. Effect of co-inoculation with yeast and bacteria on chemical and sensory characteristics of commercial Cabernet Franc red wine from Switzerland. *J. Chem. Technol. Biot.* **2016**. [CrossRef]
3. Davis, C.R.; Wibowo, D.; Fleet, G.H.; Lee, T.H. Properties of wine lactic acid bacteria: Their potential enological significance. *Am. Soc. Enol. Viticult.* **1988**, *39*, 137–142.
4. Styger, G.; Prior, B.; Bauer, F.F. Wine flavor and aroma. *J. Ind. Microbiol. Biot.* **2011**. [CrossRef] [PubMed]

5. Bokulich, N.A.; Collins, T.S.; Masarweh, C.; Allen, G.; Heymann, H.; Ebeler, S.E.; Mills, D.A. Associations among wine grape microbiome, metabolome, and fermentation behavior suggest microbial contribution to regional wine characteristics. *mBio* **2016**, *7*, e00631-16. [CrossRef] [PubMed]
6. Belda, I.; Ruiz, J.; Esteban-Fernández, A.; Navascués, E.; Marquina, D.; Santos, A.; Moreno-Arribas, M.V. Microbial contribution to Wine aroma and its intended use for Wine quality improvement. *Molecules* **2017**, 189. [CrossRef] [PubMed]
7. Bartowsky, E.J.; Borneman, A.R. Genomic variations of *Oenococcus oeni* strains and the potential to impact on malolactic fermentation and aroma compounds in wine. *Appl. Microbiol. Biot.* **2011**, *92*, 441–447. [CrossRef]
8. Pozo-Bayón, M.; G.-Alegría, E.; Polo, M.C.; Tenorio, C.; Martín-Álvarez, P.J.; Calvo De La Banda, M.T.; Ruiz-Larrea, F.; Moreno-Arribas, M.V. Wine volatile and amino acid composition after malolactic fermentation: Effect of *Oenococcus oeni* and *Lactobacillus plantarum* starter cultures. *J. Agr. Food Chem.* **2005**, *53*, 8729–8735.
9. Cappello, M.S.; Zapparoli, G.; Logrieco, A.; Bartowsky, E.J. Linking wine lactic acid bacteria diversity with wine aroma and flavour. *Int. J. Food Microbiol.* **2017**. [CrossRef]
10. De Revel, G.; Martin, N.; Pripis-Nicolau, L.; Lonvaud-Funel, A.; Bertrand, A. Contribution to the knowledge of malolactic fermentation influence on wine aroma. *J. Agr. Food Chem.* **1999**, *47*, 4003–4008. [CrossRef]
11. Malherbe, S.; Menichelli, E.; du Toit, M.; Tredoux, A.; Muller, N.; Næs, T.; Nieuwoudt, H. The relationships between consumer liking, sensory and chemical attributes of *Vitis vinifera* L. cv. Pinotage wines elaborated with different Oenococcus oeni starter cultures. *J. Sci. Food Agr.* **2013**. [CrossRef] [PubMed]
12. Liu, S. Malolactic fermentation in wine—beyond deacidification. *J. Appl. Microbiol.* **2002**, *92*, 589–601. [CrossRef]
13. Guo, Y.-Y.; Yang, Y.-P.; Peng, Q.; & Han, Y. Biogenic amines in wine: A review. *Int. J. Food Sci. Tech.* **2015**, *50*, 1523–1532. [CrossRef]
14. Inês, A.; Falco, V. Lactic Acid Bacteria Contribution to Wine Quality and Safety. In *Generation of Aromas and Flavours*; Vilela, A., Ed.; InTechOpen: London, UK, 2018. [CrossRef]
15. Gao, C.; Fleet, G.H. The degradation of malic acid by high-density cell suspensions of *Leuconostoc oenos*. *J. Appl. Bacteriol.* **1994**. [CrossRef]
16. Izquierdo-Cañas, P.M.I.; Pérez-Martín, F.; Romero, E.G.; Prieto, S.S.; de los Herreros, M.L.P. Influence of inoculation time of an autochthonous selected malolactic bacterium on volatile and sensory profile of Tempranillo and Merlot wines. *Int. J. Food Microbiol.* **2012**. [CrossRef]
17. Fugelsang, K.C.; Edwards, C.G. *Wine microbiology. Practical Applications and Procedures, in Lactic Acid Bacteria*, 2nd ed.; Springer: New York, NY, USA, 2007; pp. 29–41.
18. Abrahamse, C.E.; Bartowsky, E.J. Timing of malolactic fermentation inoculation in Shiraz grape must and wine: Influence on chemical composition. *W J. Microb Biot.* **1007**. [CrossRef] [PubMed]
19. OIV. *Compendium of international methods of analysis–OIV Chromatic Characteristics Method OIV-MA-AS2-07B: R2009*; OIV: Paris, France, 2009; Available online: http://www.oiv.int/oiv/files/6-Domainesscientifiques/6-4Methodesdanalyses/6-4-1/EN/OIV-MA-AS2-07B.pdf.
20. OIV. *Compendium of international methods of wine and must analysis*; International Organization of Vine and Wine: Paris, France.
21. Ayala, F.; Echávarri, J.F.; Negueruela, A.I. A new simplified method for measuring the color of wines. I. Red and Rosé wines. *Am. J. Enol. Viticult.* **1997**, *48*, 357–363.
22. Ribéreau-Gayon, P.; Stonestreet, E. Le dosage des anthocyanes dans les vins rouges. *Bull. Soc. Chim.* **1965**, *9*, 2649–2652.
23. Sarneckis, C.J.; Dambergs, R.G.; Jones, P.; Mercurio, M.; Herderich, M.J.; Smith, P.A. Quantification of condensed tannins by precipitation with methyl-cellulose: Development and validation of an optimized tool for grape and wine analysis. *Aust. J. Grape Wine Res.* **2006**, *12*, 39–42. [CrossRef]
24. Ibarz, M.; Ferreira, V.; Hernández-Orte, P.; Loscos, N.; Cacho, J. Optimization and evaluation of a procedure for the gas chromatographic-mass spectrometric analysis of the aromas generated by fast acid hydrolysis of flavors precursors extracted from grapes. *J. Chromatogr. A.* **2006**, *1116*, 217–229. [CrossRef]
25. ISO Standard 11035. In *Sensory analysis. Identification and selection of descriptors for stablishing a sensory profile*; ISO: Geneva, Switzerland, 1994.
26. Lucio, O.; Pardo, I.; Krieger-Weber, S.; Heras, J.M.; Ferrer, S. Selection of *Lactobacillus* strains to induce biological acidification in low acidity wines. *LWT Food Sci Tech.* **2016**. [CrossRef]
27. Pardo, I.; Ferrer, S. Yeast-Bacteria Co-inoculation. *Red Wine Technol.* **2019**. [CrossRef]

28. Lasik, M. The application of malolactic fermentation process to create good-quality grape wine produced in cool-climate countries: A review. *Eur. Food Res. Technol.* **2013**. [CrossRef]
29. Tristezza, M.; di Feo, L.; Tufariello, M.; Grieco, F.; Capozzi, V.; Spano, G.; Mita, G.; Grieco, F. Simultaneous inoculation of yeasts and lactic acid bacteria: Effects on fermentation dynamics and chemical composition of Negroamaro wine. *LWT Food Sci. Tech.* **2016**. [CrossRef]
30. Jussier, D.; Morneau, A.D.; De Orduña, R.M. Effect of simultaneous inoculation with yeast and bacteria on fermentation kinetics and key wine parameters of cool-climate Chardonnay. *Appl. Environ. Microb.* **2006**. [CrossRef]
31. Rosi, I.; Fia, G.; Canuti, V. Influence of different pH values and inoculation time on the growth and malolactic activity of a strain of *Oenococcus oeni*. *Aust. J. Grape. Wine Res.* **2003**. [CrossRef]
32. Massera, A.; Soria, A.; Catania, C.; Krieger, S.; Combina, M. Simultaneous inoculation of Malbec (*Vitis Vinifera*) musts with yeast and bacteria: Effects on fermentation performance, sensory and sanitary attributes of wines. *Food Technol. Biotech.* **2009**.
33. Zapparoli, G.; Tosi, E.; Azzolini, M.; Vagnoli, P.; Krieger, S. Bacterial inoculation strategies for the achievement of Malolactic fermentation in high-alcohol wines. *S. Afr J. Enol. Vitic.* **2009**. [CrossRef]
34. Antalick, G.; Perello, M.C.; De Revel, G. Co-inoculation with yeast and LAB under winery conditions: Modification of the aromatic profile of merlot wines. *S. Afr. J. Enol. Vitic.* **2013**. [CrossRef]
35. Alexandre, H.; Costello, P.J.; Remize, F.; Guzzo, J.; Guilloux-Benatier, M. *Saccharomyces cerevisiae-Oenococcus oeni* interactions in wine: Current knowledge and perspectives. *Int. J. Food Microbiol.* **2004**. [CrossRef]
36. Ribéreau-Gayon, P.; Glories, Y.; Maujean, A.; Dubourdieu, D. *Handbook of Enology, The Chemistry of Wine: Stabilization and Treatments: Second Edition*, 2nd ed.; Wiley: Hoboken, NJ, USA, 2006. [CrossRef]
37. Tristezza, M.; Fantastico, L.; Vetrano, C.; Bleve, G.; Corallo, D.; Mita, G.; Grieco, F. Molecular and technological characterization of *Saccharomyces cerevisiae* strains isolated from natural fermentation of susumaniello grape must in Apulia, Southern Italy. *Int. J. Microbiol.* **2014**. [CrossRef] [PubMed]
38. Izquierdo-Cañas, P.M.; Romero, E.G.; Pérez-Martín, F.; Seseña, S.; Palop, M.L. Sequential inoculation versus co-inoculation in Cabernet Franc wine fermentation. *Food Sci. Technol. Int.* **2015**. [CrossRef]
39. Knoll, C.; Fritsch, S.; Schnell, S.; Grossmann, M.; Krieger-Weber, S.; du Toit, M.; Rauhut, D. Impact of different malolactic fermentation inoculation scenarios on Riesling wine aroma. *World J. Microb. Biot.* **2012**. [CrossRef] [PubMed]
40. Garofalo, C.; El Khoury, M.; Lucas, P.; Bely, M.; Russo, P.; Spano, G.; Capozzi, V. Autochthonous starter cultures and indigenous grape variety for regional wine production. *J. Appl. Microbiol.* **2015**. [CrossRef]
41. Du Plessis, H.; Du Toit, M.; Nieuwoudt, H.; Van Der Rijst, M.; Kidd, M.; Jolly, N. Effect of *Saccharomyces*, non-*Saccharomyces* yeasts and malolactic fermentation strategies on fermentation kinetics and flavor of Shiraz wines. *Fermentation* **2017**, 64. [CrossRef]
42. Burns, T.R.; Osborne, J.P. Impact of malolactic fermentation on the color and color stability of pinot noir and Merlot wine. *Am. J. Enol. Viticult.* **2013**. [CrossRef]
43. Fang, Y.; Qian, M. Aroma compounds in Oregon Pinot Noir wine determined by aroma extract dilution analysis (AEDA). *Flavour Frag. J.* **2005**. [CrossRef]
44. Bartowsky, E.J.; Henschke, P.A. The "buttery" attribute of wine - Diacetyl - Desirability, spoilage and beyond. *Int J. Food Microbiol.* **2004**. [CrossRef]
45. Gerbaux, V.; Briffox, C.; Dumont, A.; Krieger, S. Infuence of inoculation with Malolactic bacteria on volatile phenols in wines. *Am. J. Enol. Viticult.* **2009**, *60*, 233–235.

© 2020 by the authors. Licensee MDPI, Basel, Switzerland. This article is an open access article distributed under the terms and conditions of the Creative Commons Attribution (CC BY) license (http://creativecommons.org/licenses/by/4.0/).

fermentation

Article

The Use of CRISPR-Cas9 Genome Editing to Determine the Importance of Glycerol Uptake in Wine Yeast During Icewine Fermentation

Jared Muysson [1], Laurianne Miller [2], Robert Allie [1] and Debra L. Inglis [1,2,3],*

[1] Centre for Biotechnology, Brock University, St. Catharines, ON L2S 3A1, Canada; jared.muysson@gmail.com (J.M.); ra18rs@brocku.ca (R.A.)
[2] Department of Biological Sciences, Brock University, St. Catharines, ON L2S 3A1, Canada; 22laurianne@gmail.com
[3] Cool Climate Oenology and Viticulture Institute, Brock University, St. Catharines, ON L2S 3A1, Canada
* Correspondence: dinglis@brocku.ca; Tel.: +1-905-688-5550 (ext. 3828)

Received: 11 October 2019; Accepted: 28 October 2019; Published: 30 October 2019

Abstract: The high concentration of sugars in Icewine juice causes formidable stress for the fermenting *Saccharomyces cerevisiae*, causing cells to lose water and shrink in size. Yeast can combat this stress by increasing the internal concentration of glycerol by activating the high osmolarity glycerol response to synthesize glycerol and by actively transporting glycerol into the cell from the environment. The H$^+$/glycerol symporter, Stl1p, has been previously characterized as being glucose repressed and inactivated, despite osmotic stress induction. To further investigate the role of Stl1p in Icewine fermentations, we developed a rapid single plasmid CRISPR-Cas9-based genome editing method to construct a strain of the common Icewine yeast, *S. cerevisiae* K1-V1116, that lacks *STL1*. In an Icewine fermentation, the ΔSTL1 strain had reduced fermentation performance, and elevated glycerol and acetic acid production compared to the parent. These results demonstrate that glycerol uptake by Stl1p has a significant role during osmotically challenging Icewine fermentations in K1-V1116 despite potential glucose downregulation.

Keywords: Icewine; *Saccharomyces cerevisiae*; hyperosmotic stress; CRISPR-Cas9; glycerol transport; STL1

1. Introduction

Icewine is a signature sweet dessert wine that is fermented from the juice of naturally frozen grapes. Since much of the water in the grape is frozen at harvest, the pressed Icewine juice is highly concentrated in solutes such as sugars, organic acids, and nitrogenous compounds. For Icewine juice produced in the Niagara Peninsula of Ontario, Canada, the average sugar concentration for Vidal Icewine juice was reported at 39.3 Brix for 298 Icewine juice samples (approximately 450 g/L sugar) [1]. The high sugar concentration presents a significant osmotic stress for the fermenting *Saccharomyces cerevisiae*. High osmotic stress, commonly referred to as hyperosmotic stress, causes prolonged fermentation times, less than ideal alcohol yields, reduced biomass growth, and elevated levels of undesired metabolites [2,3]. The undesired metabolites, namely acetate, acetaldehyde, and ethyl acetate, indirectly arise because of the increased synthesis of glycerol, the primary compound responsible for the adaptation to hyperosmotic stress [4]. Synthesis of glycerol is a two-step process that involves the reduction and dephosphorylation of the glycolytic intermediate, dihydroxyacetone phosphate (DHAP). Glycerol synthesis is not redox neutral, as one equivalent of NADH (nicotinamide adenine dinucleotide (NAD$^+$) + hydrogen (H)) is required per synthesized glycerol, resulting in the production of NAD$^+$ [5,6]. Therefore, a second reaction is required to reduce the generated NAD$^+$ back to NADH. The NADH generating step of glycolysis, the oxidation of glyceraldehyde 3-phosphate by glyceraldehyde 3-phosphate dehydrogenase, is balanced by coupling this step to alcoholic fermentation, where acetaldehyde is reduced to ethanol by alcohol dehydrogenase, regenerating NAD$^+$ and making this a redox neutral process. Excess oxidation of NADH through glycerol synthesis can lead to a significant shift in the NAD$^+$/NADH ratio [7]. Due to the importance of NADH for the reduction of acetaldehyde to ethanol, reduced levels of NADH relative to NAD$^+$ causes insufficient acetaldehyde reduction, thus leading to elevated levels of acetaldehyde, and subsequently acetate, due to acetaldehyde oxidation by aldehyde dehydrogenases [7–9].

In addition to glycerol synthesis, yeast can also actively uptake glycerol from the extracellular environment [10]. One such transporter, Stl1p, is a plasma membrane H$^+$/glycerol symporter. The proton-glycerol symport mechanism of Stl1p allows for extracellular glycerol to be imported against a concentration gradient, at the cost of H$^+$ influx into the cell [11]. As the name *STL* (sugar transporter like) suggests, the protein is one of 34 sugar permeases in yeast and is part of the major facilitator superfamily (MFS) [12]. In laboratory yeast strains of *S. cerevisiae*, *STL1* is reported as upregulated by salt-induced osmotic stress and repressed and inactivated by glucose, but glucose repression was overcome at high temperatures [11,13]. Previous gene expression analysis on the short-term hyperosmotic stress response in wine yeast to grape must inoculation revealed that *STL1* was highly expressed between the first 10 and 30 minutes, before rapid downregulation to near undetectable levels [14]. Another gene expression analysis study found that *STL1* expression upregulation was very low in *S. cerevisiae* wine yeast throughout table wine fermentations and in glucose-containing stress conditions [15]. Conversely, we have previously reported that *STL1* is the most highly upregulated gene in *S. cerevisiae* K1-V1116 during osmotically challenging Icewine fermentations, with a 25-fold increase relative to that measured in table wine fermentation being seen on the fifth day of fermentation [16].

While gene expression analysis is a useful tool for identifying potential gene targets associated with a given phenotype, further investigation is often required to confirm protein function and phenotypic impact. Genetic modification is one such tool that can be used for gene function elucidation. Given the hyperosmotic conditions of Icewine, and the differences between laboratory yeast strains and wine yeast strains, constructing engineered wine yeast strains would be optimal for studying Icewine hyperosmotic stress condition responses [17–19]. While *S. cerevisiae* is amendable with existing genome editing techniques, primarily through the chromosomal integration of selectable markers in haploid auxotrophic strains, applying these techniques to industrially relevant diploid and polyploid strains, like wine yeasts, can be challenging and time-consuming [20,21]. Luckily, with the advent of CRISPR-Cas9 genome editing methods, yeast strain engineering has become rapid, efficient, and multiplexed [22,23]. At the time of this work, there has only been one other published study on the use of CRISPR-Cas9 for genome editing in wine yeasts. The work of Vigentini et al. (2017) used a double plasmid CRISPR-Cas9 method to eliminate *CAN1*, an arginine permease, in the popular wine yeast strains, EC-1118 and AWR1796, thus lowering urea production [24].

In this work, we developed a rapid and effective CRISPR-Cas9 genome editing method for the wine yeast *S. cerevisiae* strain, K1-V1116, which is commonly used for Icewine production. Using a single linearized CRISPR-Cas9 plasmid, that is repaired in vivo with a linear DNA fragment containing the target sequence of the sgRNA, we were able to construct a K1-V1116 strain featuring the complete removal of the *STL1* open reading frame. Subsequent Icewine fermentations were conducted with K1-V1116 and K1-V1116 Δ*STL1* strain. The reduced fermentation performance, along with increased glycerol and acetic acid production per gram of sugar consumed, of the Δ*STL1* strain, provides further evidence that Stl1p contributes to the osmotolerance of K1-V1116 during Icewine fermentations, despite the high concentration of glucose.

2. Materials and Methods

2.1. Yeast Strains and Media

The commercial wine yeast used in this work was *S. cerevisiae* K1-V1116 (Lallemand, Montreal, QC, Canada). K1-V1116 strains were propagated using YPD (0.5% (w/v) Yeast Extract, 1% (w/v) Peptone, and 2% (w/v) Dextrose), unless otherwise stated. For solid media, 1% (w/v) agar was added. For selection media geneticin (G418) (Teknova, Hollister, CA, USA) was supplemented on YPD plates at 200 µg mL^{-1}.

2.2. Description of CRISPR-Cas9 Method

The CRISPR-Cas9 method used in this work was built upon the work by the Tom Ellis lab [25]. The plasmid used, pWS173, contains cassettes for the yeast expression of both Cas9 and sgRNA (Figure 1). The yeast-optimized Cas9 contains a nuclear localization sequence and is driven by the PGK1 promoter. The optimized sgRNA cassette contains a yeast tRNA and 5' HDV ribozyme and is flanked by 500 bp junk DNA homology arms. In pWS173, the sgRNA cassette contains a BsmBI-flanked GFP dropout for the 20 bp protospacer sequence required for targeted endonuclease activity. The method relies on the repair of a BsmBI linearized pWS173 with a 1.1 kb linear DNA fragment, which has the sgRNA cassette, with a 20 bp protospacer sequence, inserted within. The sgRNA repair DNA is generated using pWS173 as a template, with two PCR amplicons containing primer overhangs that encode for the protospacer sequence and the terminal sequence of the adjacent fragment. The two protospacer encoding amplicons are then assembled using overlap extension PCR, as described in Section 2.3. When the linearized plasmid, sgRNA repair DNA, and donor DNA are transformed into yeast, the homology arms on the sgRNA repair DNA and pWS173 are recognized and the plasmid is circularized (Figure 2a). The circularization of the plasmid restores its functionality, offering G418 resistance and expression of CRISPR-Cas9 and the active sgRNA species (Figure 2b).

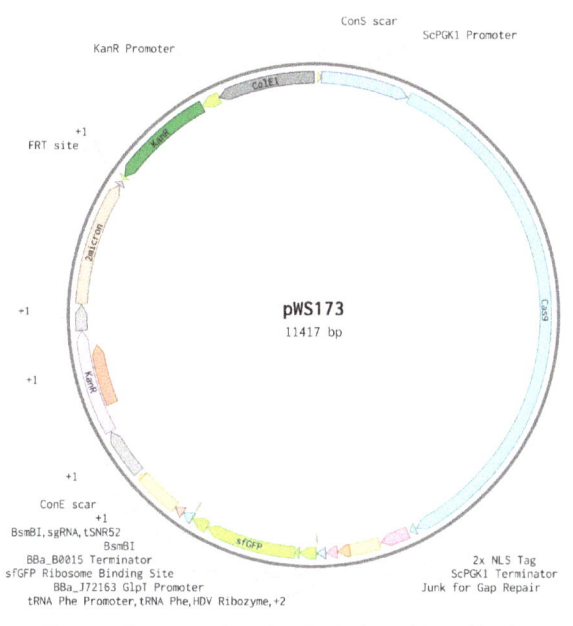

Figure 1. Representation of pWS173 plasmid (Benchling).

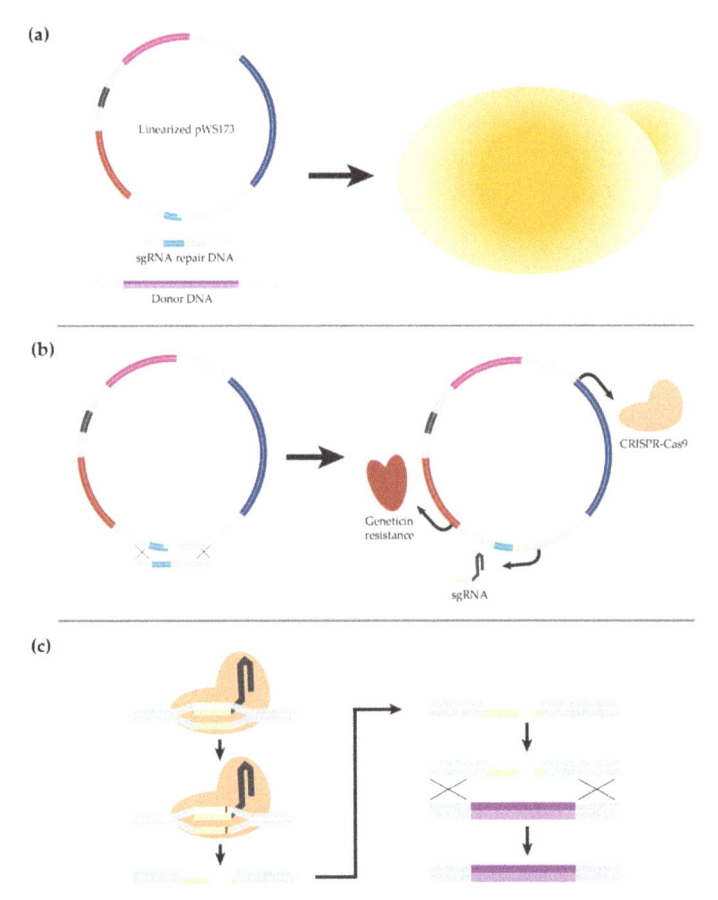

Figure 2. Schematic of DNA components which are transformed in yeast (**a**) and in vivo repair of pWS173, and subsequent expression of CRISPR-Cas9, sgRNA, and G418 resistance cassettes (**b**). The CRISPR-Cas9 ribonucleoprotein recognizes the genomic target sequence and cleaves the site, resulting in a double-stranded DNA break in the genomic DNA. Subsequent homology directed repair with the exogenous donor DNA results in successful genome editing (**c**).

2.3. Cloning

The plasmid used in this work, pWS173, was a gift from Tom Ellis (Addgene plasmid # 90960). Plasmids were propagated in *NEB Stable™ Competent Escherichia coli* (New England Biolabs), under the selection of LB (1% (*w/v*) Tryptone, 0.5% (*w/v*) Yeast Extract, 1% (*w/v*) Sodium Chloride) supplemented with 50 µg mL^{-1} kanamycin (Tocris Biosciences, Bristol, UK). The plasmid was isolated using QIAprep Spin Miniprep (Qiagen, Hilden, Germany), linearized using BsmBI (New England Biolabs, Ipswich, MA, USA) and used without gel extraction. The sequences from EC-1118 [26] were used in place of K1-V1116 for all construct and primer design purposes, since the sequence of EC-1118 was available on the Ensembl genome browser. Benchling (San Francisco, CA, USA) was used for plasmid and construct design. The target sequence for CRISPR-Cas9 sgRNA, used in K1-V1116, was designed using the Benchling CRISPR tool, with the default parameters for the NGG PAM single guide RNA for Cas9. The R64-1-1 *S. cerevisiae* reference genome was used for off-target analysis. The target sequence used can be found in Table A1. Donor DNA and sgRNA repair DNA was synthesized using 2-part overlap extension PCR, using Q5 polymerase (New England Biolabs, Ipswich, MA, USA) and the primers in Table A2. An initial annealing temperature of 55 °C was used for most PCR reactions, with the temperature adjusted empirically to resolve low yield reactions. An extension time of 15 second per kb and 35 cycles was used for all PCRs. For overlap extension PCR, 0.5 µL of the PCR mix each fragment was used as templates in a 50 µL reaction. Genomic DNA was isolated and purified from K1-V1116 using the Fungi/Yeast Genomic DNA Isolation Kit (Norgen Biotek, Thorold, ON, Canada). PCR amplified DNA was purified using QIAquick Gel Extraction Kit (Qiagen, Hilden, Germany). DNA was quantified using a NanoDrop Lite Spectrophotometer (ThermoFisher, Waltham, MA, USA). PCR amplicons were visualized using 1% agarose-TAE (40 mM Tris, 20 mM acetate, and 1 mM EDTA), containing 0.003% (v/v) RedSafe nucleic acid staining solution (iNtRON Biotechnology, Seongnam, Korea).

2.4. Strain Construction

The *S. cerevisiae* strain, K1-V1116, was transformed using a modified electroporation protocol, which was largely based off the protocol described by DiCarlo et al. (2013) [27]. A colony of K1-V1116 was inoculated into 5.5 mL of YPD. The culture was then incubated overnight at 30 °C in a roller drum. The following day, the culture was centrifuged at 8000 G for 1 min in a microcentrifuge. The media was removed, and the pellets were resuspended in 250 µL of room-temperature water and consolidated into a single tube. The cells were washed twice more, using 1 mL of water, and then resuspended in 1.5 mL of 500 mM lithium acetate with 5 mM of dithiothreitol (DTT) before being incubated for 30 min at 30 °C in a roller drum. Following conditioning, the cells were collected by centrifugation and washed twice with 1.5 mL of ice-cold electroporation buffer (1 M sorbitol and 1 mM calcium chloride). The final cell pellet was then resuspended in 950 µL of ice-cold electroporation buffer and kept on ice. For each transformation, 100 µL of cells were mixed with linearized pWS173 plasmid, sgRNA repair DNA, and donor DNA. The mixture was then transferred to an ice cold 0.2 cm gap electroporation cuvette and electroporated at 2.5 kV, 25 µF, and 200 Ω. The electroporated cells were recovered in 900 µL of room-temperature recovery media (0.5 M sorbitol and 0.5x YPD) and incubated for 4 h at 30 °C in a roller drum. After recovery, 50 µL of culture was plated onto YPD, supplemented with 200 µg mL^{-1} G418 (geneticin) and incubated at 30 °C for 24 to 48 h. The remaining recovered culture was saved at 4 °C and plated again if required. Integration events were confirmed with colony PCR using Q5 polymerase (New England Biolabs, Ipswich, MA, USA). To obtain colony PCR DNA, a single colony was mixed into 25 µL of 20 mM sodium hydroxide and boiled at 90 °C for 10 min. PCR products were visualized using agarose gel electrophoresis, as described in Section 2.2.

2.5. Icewine Juice

Vidal Icewine juice was previously acquired from Huebel Grape Estates in Niagara-on-the-Lake, Ontario, Canada and stored at −35 °C until use. The Icewine juice was thawed at 7 °C for 24 h, before being racked off and filtered through a series of coarse, medium, and fine pore size pad filters using the Bueno Vino Mini Jet filter (Vineco, St. Catharines, ON, Canada). The juice was then filter sterilized using 0.45 µm membrane cartridge (Millipore, Etobicoke, ON, Canada). The final sterile filtered Icewine juice (40.2 Brix) was used for Icewine fermentations. The sterile filtered Icewine juice contained 447 g L^{-1} sugar, pH 3.79, titratable acidity of 7.0 g L^{-1} tartaric acid, glycerol at 8.67 g L^{-1}, acetic acid at 0.06 g L^{-1} and yeast assimilable nitrogen (YAN) at 441 mg N L^{-1}.

2.6. Permissive Condition Growth Kinetics Characterization

Permissive condition growth kinetics characterization was conducted aerobically in YPD liquid media at 30 °C in an orbital shaker. In a sterile bottle with a stir bar, 250 mL of YPD was added, along with 450 mg of yeast cells from an agar plate, for each strain. This culture was stirred for 5 min on a stir plate and then was sampled into 75 mL aliquots in 250 mL Erlenmeyer flasks with sponge stoppers. The flasks were then placed in an orbital incubator set to 120 RPM and 30° C, and were sampled every two hours. Optical density at 600 nm (OD$_{600}$) values were obtained using a Genesys 10S UV-Vis Spectrometer (Thermo Scientific, Waltham, MA, USA) set to 600 nm in absorbance mode, with methacrylate cuvettes. Dilutions were made with sterile YPD as needed, to stay under a reading of 1.0. All readings were done in triplicate.

2.7. Icewine Fermentations and Samplings

The K1-V1116 strains used for fermentations were prepared as starter cultures. The starter culture consisted of filter sterilized Icewine juice diluted to 10 Brix, with an additional 2 g L^{-1} of diammonium phosphate (DAP). From YPD plates, four freshly cultured yeast colonies were collected and used to inoculate 100 mL of starter culture media. The culture was incubated until the cell concentration reached 2×10^8 cells mL^{-1}, or approximately 17 h, using a shaker table incubator set to 30 °C.

The starter culture was used to inoculate 1 L of Icewine juice to a final cell density of approximately 1×10^7 cells mL^{-1}. The fermentations were incubated at 17 °C for 34 days. Samples were taken every day between days 0 and 6, every other day between days 8 and 16, and every three days between days 19 and 34. Before sampling, the fermentations were stirred for 5 min to ensure homogeneity, after which 2 mL of the culture was taken. After centrifugation, the supernatant of the sample was transferred to a clean tube and stored at −35 °C for later metabolite analysis. The remaining cell pellet was used for total and viable cell counts using methylene blue on a hemocytometer. Larger 50 mL samples were taken from the initial juice and final wine and stored at −35 °C for later analysis. Colony PCR was conducted on the strains at the beginning and end of fermentation to verify the genotype. The fermentations were conducted in triplicate for each strain.

2.8. Metabolite Analysis

Soluble solids were determined by ABBE bench top refractometer (model 10450; American Optical, Buffalo, NY, USA). Acidity was determined by pH measurement using a sympHony pH meter (model B10P; VWR, Mississauga, ON, Canada), and titratable acidity by titration against 0.1 mol L^{-1} NaOH, to an endpoint of pH 8.2 [28]. Glucose, fructose, glycerol, acetic acid, amino nitrogen, and ammonia nitrogen were measured using Megazyme assay kits (K-FRUGL, K-ACET, K-GCROL, K-PANOPA, K-AMIAR; Megazyme International Ireland, Ltd., Bray, Co. Wicklow, Ireland). Ethanol was determined by gas chromatography (Model 6890; Agilent Technologies Inc., Palo Alto, CA, USA) equipped with a flame ionization detector (FID), split/split-less injector, and Chemstation software (version E.02.00.493). Separations were carried out with a DB®-WAX (30 m, 0.25 mm, 0.25 µm) GC column (122-7032 model; Agilent Technologies, Santa Clara, CA, USA) with helium as the carrier gas, at a flow rate of 1.5 mL min^{-1}. Metabolite production during fermentation was calculated by the difference in the respective metabolite concentration measured between the time zero point (immediately after inoculation) and at each sampling time point throughout fermentations. Normalized metabolite production was determined by dividing the final metabolite production by the final sugar consumed.

2.9. Statistical Analysis

XLSTAT-Pro by Addinsoft (New York, NY, USA) was used for statistical analysis. Analysis of variance (ANOVA) with mean separation by Tukey's Honest Significant Difference ($p < 0.05$) and Student's *t*-Test ($p < 0.05$, $p < 0.01$, $p < 0.001$) were used to evaluate differences between variables.

3. Results

3.1. Development of a CRISPR-Cas9 Genome Editing Protocol for K1-V1116

3.1.1. K1-V1116 Drug Sensitivity Testing

To assess the selection potential of Geneticin (G418) for the isolation of successful transformants, the wild type resistance of K1-V1116 needed to be determined. Given the proven effectiveness of G418 in previous studies [23,27], a rapid approach was used to determine wild type sensitivity: 200 µL of stationary phase K1-V1116 cultured in YPD was plated onto a series of YPD plates with G418 (140,

280, 560, 710, and 850 µg mL^{-1}) and incubated at 32 °C. After three days, only one colony appeared on the lowest concentration plate. As such, YPD with 200 µg mL^{-1} G418 was selected as the future selection media.

3.1.2. Initial K1-V1116 Transformations

The following step was to assess the G418 positive selection functionality of the intact and linearized pWS173, as well as validate the electroporation protocol developed. As the goal of this work was to develop a rapid and facile genome editing method, testing the use of stationary phase cultures as the source of cells for electroporation was desired. The use of stationary phase cultures would eliminate the need to monitor the OD$_{600}$, allow for more schedule flexibility and reduce the hands-on time by over 5 h on the day of transformation. While convenient, the previously reported cost of using stationary phase yeast is reduced transformation efficiency, thus we needed to determine if using stationary phase culture would be a viable strategy for future strain construction [29,30]. Three YPD cultures of K1-V1116 were incubated at 32 °C in a roller drum until log phase (OD$_{600}$ of 1.8), 18 h of growth, and 40 h of growth, until prepared and transformed with either 300 ng of non-linearized pWS173 or 300 ng of linearized pWS173 plus 1 µg of sgRNA repair DNA, which would produce a truncated, non-functional sgRNA (Figure 3).

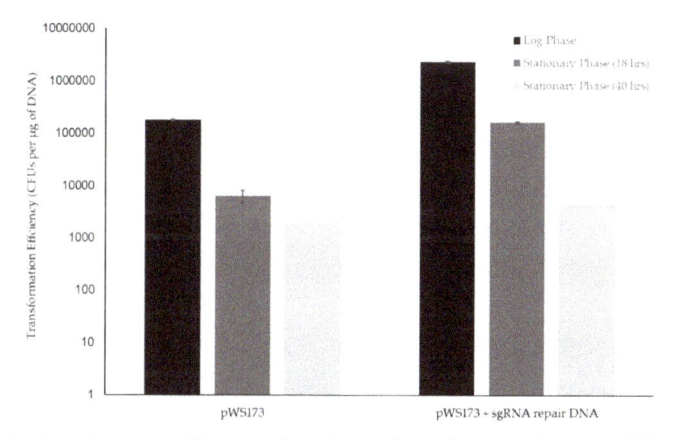

Figure 3. Log transformation efficiency of log phase culture, 18-hour culture, and 40-hour culture, with 300 ng of pWS173 or 300 ng of linearized pWS173, with 1 µg of non-functional sgRNA repair DNA. Single transformations were plated in triplicate on YPD with 200 µg mL^{-1} G418. Transformation efficiency represents the average ± standard deviation.

There was a difference in transformation efficiency between the log phase culture, 18-hour culture, and 40-hour culture. For the repaired pWS173 plasmid, the transformation efficiency of the log phase culture was 14x greater than the 18-hour culture, and 541x greater than the 40-culture. Despite this difference, the 18-hour culture still achieved a transformation efficiency of 1.6 × 10^5 ± 5.1 × 10^3 CFUs per µg of DNA. The same trend of transformation efficiency between cell culture phases was shown using the intact pWS173. The transformation efficiency measured at each growth phase of the repaired pWS173 appears greater than that observed with the intact pWS173, although we cannot say if this difference is significant due to the experimental design. The transformation efficiency seen in the 18-hour culture was deemed acceptable for future strain construction given the isogenic goal of this work.

3.1.3. Construction of a K1-V1116 ΔSTL1 Strain

Upon validation of the electroporation protocol and the G418 positive selection of the pWS173, K1-V1116 ΔSTL1 was constructed. For this, the sgRNA repair DNA that contained a target sequence within the *STL1* open reading frame was synthesized (Figure 4a). The donor DNA for homology-directed repair contained 500 bp homology arms, and would result in the deletion of the entire *STL1* open reading frame (Figure 4b). K1-V1116 was transformed with 525 ng of linearized pWS173, 1 µg of sgRNA repair DNA, containing a target in *STL1*, and 1.3 µg of donor DNA. The transformation efficiency was 932 ± 62 CFUs per µg of DNA. Of four selected transformants, all four had the genotype which corresponded with the intended integration event (Figure 4c).

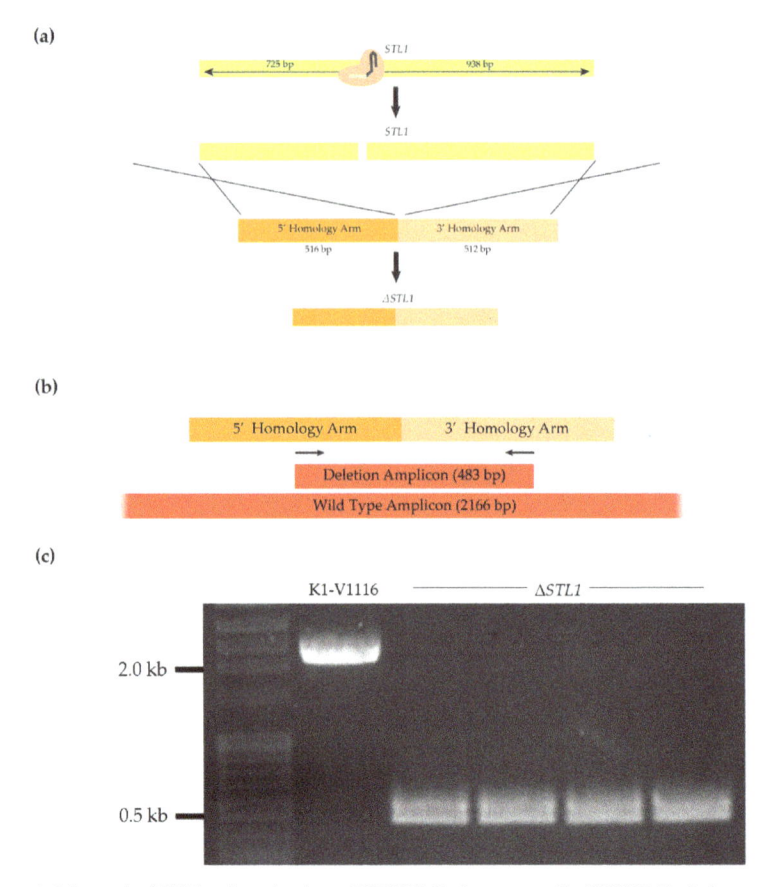

Figure 4. Schematic of *STL1* and mechanism of CRISPR-Cas9 genome edit. CRISPR-Cas9 cleaves at the depicted location, resulting in homology directed repair with the Δ*STL1* donor DNA (**a**). PCR genotype scheme used to validate the integration (**b**). Gel image of *STL1* genotype of K1-V1116 in lane 2 and K1-V1116 Δ*STL1* transformants in lanes 3 to 6 (**c**).

Further testing of an additional 15 transformants revealed that one isolate was a Δ*STL1* heterozygote with an additional two transformants maintaining a wild type genotype, thus resulting in an overall homozygous editing efficiency of 84%. Eight transformants were streaked to single colonies on YPD with G418, followed by two more rounds on YPD, in order to eliminate pWS173. The final K1-V1116 Δ*STL1* strain used for further testing was negative for pWS173.

3.2. ΔSTL1 Growth Characterization in Permissive Conditions and Icewine Fermentations

3.2.1. ΔSTL1 Growth Kinetics in Permissive Conditions

To ensure that the modification to the Δ*STL1* strain did not cause any unintended permissive growth defects, its growth kinetics were characterized in permissive conditions. As *STL1* is reported as having remarkably low expression during conditions containing glucose, especially in absence of significant osmotic stress, this strain was expected to have a similar growth kinetic profile to the parent strain in YPD [11]. The YPD OD$_{600}$ growth curve of K1-V1116 and K1-V1116 Δ*STL1* did reveal similar growth kinetics between the two strains (Figure 5). Interestingly, K1-V1116 Δ*STL1* consistently reached a slightly higher stationary phase OD$_{600}$ value compared to K1-V1116.

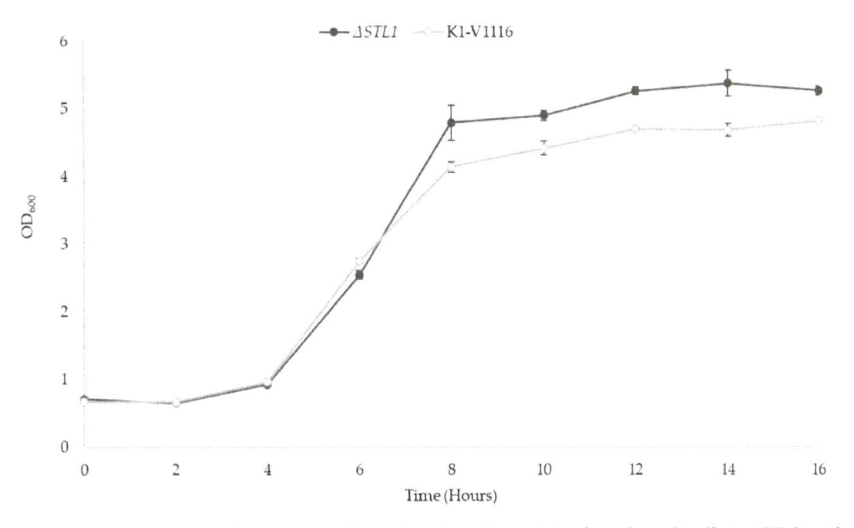

Figure 5. Growth kinetics of K1-V1116 Δ*STL1* (●) and K1-V1116 (○) cultured aerobically in YPD liquid media at 30 °C. Cultures were setup and measured in triplicate. Samples were taken every 2 h, with each point representing a corrected OD$_{600}$ average ± standard deviation.

3.2.2. ΔSTL1 Icewine Fermentation

The fermentation performance of the K1-V1116 Δ*STL1* compared to K1-V1116 was compared in Icewine juice. Within the first 10 days of the fermentation, K1-V1116 Δ*STL1* began to lag in sugar consumption compared to K1-V1116 (Figure 6). The sluggish performance continued for the remainder of the fermentation, with sugar consumption continuing even as K1-V1116 reached a plateau at around day 25.

The total cell concentrations of strains on day 7 revealed that K1-V1116 Δ*STL1* had only reached around 3.5×10^7 cells mL^{-1}, versus the 5.5×10^7 cells mL^{-1} of K1-V1116 (Figure 7a). The reduction in viable cells seen in K1-V1116 later in the fermentation was not as prominent with K1-V1116 Δ*STL1* (Figure 7b).

The average total amount of glycerol and acetic acid produced throughout fermentation by K1-V1116 Δ*STL1* is lower than K1-V1116 but K1-V1116 Δ*STL1* produces more glycerol and acetic acid per gram of sugar consumed (Figure 8). The differences in glycerol production appear in the first few days of fermentation, while the differences in acetic acid production appear after day 5.

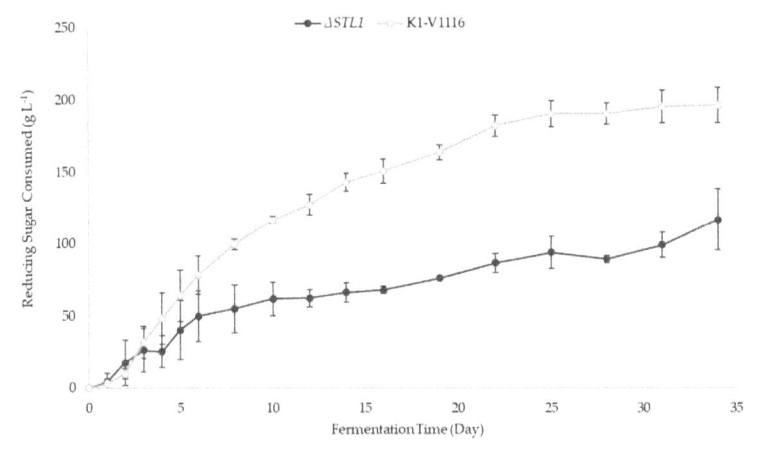

Figure 6. Reducing sugar consumed during fermentation. Fermentations of K1-V1116 Δ*STL1* (●) and K1-V1116 (○) were performed in triplicate, with samples from each trial being analyzed in duplicate. Sugar values represent the average ± standard deviation.

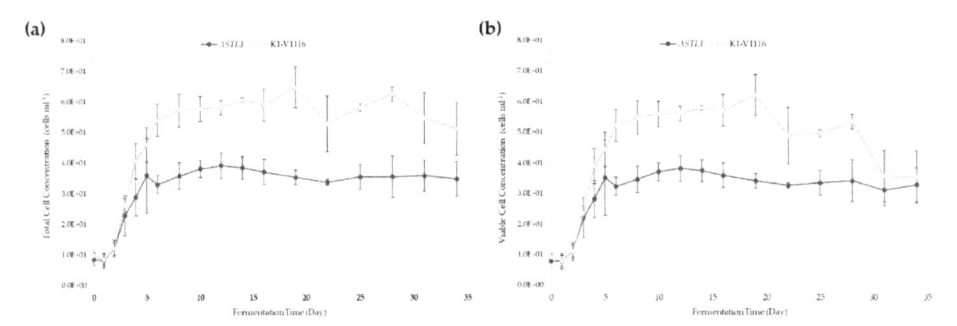

Figure 7. Fermentation total cell concentration (**a**) and viable cell concentration (**b**) of K1-V1116 Δ*STL1* (●) and K1-V1116 (○). Fermentations were performed in triplicate, with counts being determined in duplicate. Cell concentrations represent the average ± standard deviation.

The final wine parameters (Table 1) demonstrate the significant differences between K1-V1116 and K1-V1116 Δ*STL1*. As a result of the difference in reducing sugar consumption, K1-V1116 Icewine had 80% more alcohol than that of K1-V1116 Δ*STL1*. The differences in total glycerol and acetic acid were not statistically significant, despite the differences in the final alcohol and reducing sugar levels.

Table 1. Icewine juice and final wine chemical composition. Data represent the mean value ± standard deviation from triplicate fermentations, with each sample tested in duplicate. Lowercase letters within the same parameter indicate differences between treatments as tested by ANOVA, with mean separation by Tukey's Honest Significant Difference (HSD$_{0.05}$).

Parameter	Icewine Juice	Wine from K1-V1116	Wine from K1-V1116 Δ*STL1*
Reducing Sugars (g L^{-1})	447 ± 19 a	251 ± 11 c	333 ± 19 b
pH	3.79 ± 0.04 b	3.90 ± 0.02 a	3.84 ± 0.08 a,b
Titratable Acidity (g L^{-1})	7.0 ± 0.1 b	8.6 ± 0.2 a	8.6 ± 0.3 a
Yeast Assimilable Nitrogen (mg N L^{-1})	441 ± 2 a	345 ± 13 c	396 ± 14 b
Glycerol (g L^{-1})	8.67 ± 0.32 b	17.42 ± 1.12 a	16.36 ± 0.32 a
Acetic Acid (g L^{-1})	0.06 ± 0.01 b	1.48 ± 0.07 a	1.39 ± 0.05 a
Ethanol (% v/v)	1.3 ± 0.0 c	12.3 ± 0.6 a	6.7 ± 0.6 b

The final differences in normalized glycerol and acetic acid production were significant, despite the insignificant differences in the total glycerol and acetic acid levels (Table 2). K1-V1116 Δ*STL1* produced nearly 49% more glycerol and 71% more acetic acid per gram of sugar consumed.

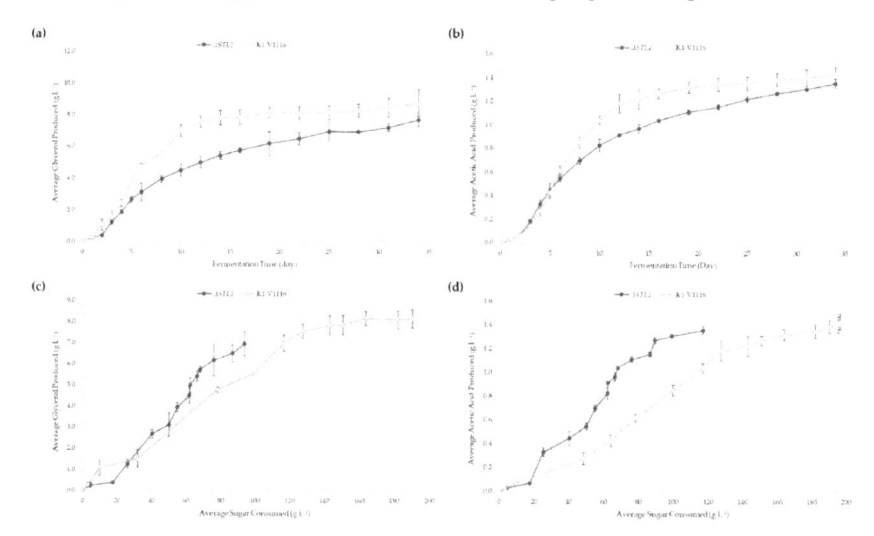

Figure 8. Average glycerol produced (**a**) and average acetic acid produced (**b**) over the time course of the fermentation. Average glycerol produced (**c**) and average acetic acid produced (**d**) versus sugar consumed over the course of the fermentation. Includes K1-V1116 Δ*STL1* (●) and K1-V1116 (○). Values are the averages of triplicate fermentations, with each sample tested in duplicate for metabolite analysis.

Table 2. Consumption of juice components and production of yeast metabolites.

Parameter	K1-V1116	K1-V1116 Δ*STL1*
Reducing Sugar consumed (g L^{-1})	196 ± 13 **	117 ± 21 **
Yeast Assimilable Nitrogen consumed (mg N L^{-1})	92 ± 17 *	44 ± 12 *
Glycerol produced (g L^{-1})	8.76 ± 0.84	7.68 ± 0.44
Acetic Acid produced (g L^{-1})	1.42 ± 0.07	1.35 ± 0.07
Normalized Glycerol production (mg glycerol per g sugar consumed)	45 ± 3 *	67 ± 12 *
Normalized Acetic Acid production (mg acetic acid per g sugar consumed)	7 ± 0 **	12 ± 2 **

Production refers to the difference between the initial Icewine juice and the final wine of each strain. Data represent the mean value ± standard deviation of duplicate measurements per sample (three winemaking replicates per treatment). Asterisks (* $p < 0.05$, ** $p < 0.01$) indicate significant differences between yeast strains (Student's *t*-Test).

4. Discussion

The CRISPR-Cas9 genome editing method developed was effective in modifying the commercial wine yeast strain, K1-V1116, with various protocol optimization strategies. The use of a single plasmid which contains all required elements, except for the protospacer sequence, in conjunction with overlap extension PCR generated sgRNA repair DNA, allows for rapid testing of other loci. While the use of primers and in vitro DNA assembly can lead to mutations, the resulting 1.1 kb sgRNA repair DNA can be rapidly sequenced with Sanger sequencing. Alternatively, commonly used sgRNA repair DNA fragments could be cloned into an *E. coli* vector for sequence verification and maintenance. The use of overlap extension PCR to generate the Δ*STL1* donor DNA was fitting for an edit which is not sensitive to single-nucleotide polymorphisms and indels, but such a cloning method is not appropriate for a sequence sensitive edit, like a gene overexpression cassette. Future work that ventures into sensitive edit types will need to rely on other high-fidelity and verifiable cloning methods.

The initial antibiotic resistance testing of K1-V1116 demonstrated that the strain is no more resistant than previously used strains. We believe the rapid test was sufficient, as notable background growth was not encountered for the remaining transformations. The initial transformation trials, with various K1-V1116 cultures as the source for electrocompetent cells, reinforced the previous findings that use of stationary phase cells was many orders of magnitude less efficient than log phase cells [29,30]. For the isogenic goal of this work, the use of stationary phase yeast for electroporation resulted in acceptable transformation efficiencies and greatly reduced the work required for transformation. Whether future work utilizes stationary phase cells will depend on the goal of the project. Conveniently, the use of the repaired pWS173 and sgRNA repair DNA appears to be more effective than a larger circularized plasmid. This observation indicates that the homologous recombination in K1-V1116 is active and potentially comparable to that of many laboratory *S. cerevisiae* strains [31]. The editing efficiency seen with the Δ*STL1* edit, 84%, was high enough for practical uses, and is within the ranges seen across many CRISPR-Cas9 methods [22].

The Icewine fermentation trials conducted with K1-V1116 Δ*STL1* demonstrated the serious consequences that removing the Stl1p transporter has on fermentation performance. The sluggish fermentation, as indicated by the drastically reduced sugar consumption and total cell count, indicates that cell growth was inhibited by the deletion. The higher level of normalized glycerol production suggests that, since glycerol uptake is disrupted, the cell has an increased reliance on endogenous glycerol synthesis to satisfy demand under the hyperosmotic conditions. With the increased glycerol production per sugar consumed, the normalized acetic acid levels increased as well. Given the current understanding of glycerol synthesis and the NAD^+/NADH ratio during Icewine fermentations, the increased glycerol demand correlates with higher acetic acid levels [4,8,16]. The exact cause of the reduced fermentation performance still needs to be determined. The increased demand on glycerol synthesis may further increase the NAD^+/NADH ratio and disrupt the metabolic flux of the many reactions which rely on NAD^+/NADH [32,33]. The inability to meet the glycerol demand could result in cell shrinkage, thus leading to the crowding of macromolecules and the slowing of cellular processes [34]. The accumulation of the toxic fermentation intermediate, acetaldehyde, may also contribute to the sluggish fermentation [35]. The exact cause of this phenomenon appears to not be acutely toxic to the cells, as the viable cell concentration of K1-V1116 Δ*STL1* was relatively stable throughout stationary phase. The maintained viability, despite the reduced performance, is not abnormal for other known causes of sluggish fermentations [36].

These results, in conjunction with previous findings that *STL1* was the most highly upregulated gene during Icewine fermentations, strongly suggest that Stl1p has a role in maintaining internal glycerol levels during the hyperosmotic stress conditions of an Icewine fermentation [4,16]. The previous studies on *STL1* expression reveal that expression is low when glucose is present, even in the relatively stressful environments of table wine fermentations [11,13–15]. Our findings are similar to recent gene expression analysis research in sugarcane bioethanol strains, which showed that *STL1* glucose repression was overcome in 30 and 35 °Brix conditions [37]. Future work should investigate whether overcoming glucose repression and inactivation at high sugar concentrations is a shared attribute between strains, or if certain strains have evolved the ability to overcome it. Considering many other yeast species rely on glycerol uptake to satisfy osmotic stress glycerol demand, it appears natural for *S. cerevisiae* strains like K1-V1116 to rely more on glycerol uptake during hyperosmotic stress conditions [38,39]. Furthermore, the role of initial glycerol levels in Icewine juice needs to be determined, as reduced starting glycerol levels may negatively impact the effectiveness of Stl1p-mediated osmotolerance.

5. Conclusions

Our work demonstrates that CRISPR-Cas9 genome editing is an effective tool for studying the role of glycerol uptake by Stl1p in the commercial wine yeast, K1-V1116. The CRISPR-Cas9 genome editing method developed allows for rapid strain engineering method by employing time- and cost-saving strategies typically only used with laboratory yeast strains. We found that K1-V1116 is not challenging to engineer despite the lack of genome sequence information and oenological nature. The use of the engineered K1-V1116-lacking *STL1* in Icewine fermentations generated significant data, which support the idea that Stl1p contributes to osmotolerance during the harsh hyperosmotic conditions.

Author Contributions: J.M. and D.L.I. conceived and designed the experiments; J.M. performed the genome editing experiments; R.A. performed the yeast growth curves in YPD, filter sterilized the juice and determined the chemical analysis of the sterile-filtered Icewine juice; L.M. conducted the Icewine fermentations and measured the yeast metabolites and final wine composition; J.M. and D.L.I. contributed to the writing of the manuscript and J.M., D.L.I. and R.A. contributed to the review and editing of the manuscript.

Funding: This project was funded by grants from the Natural Sciences and Engineering Research Council of Canada (NSERC 238872-2012 and NSERC DDG-2018-00007).

Acknowledgments: We would like to thank the Tom Ellis lab for sharing pWS173 and for assembling an in-depth guide on using their CRISPR/Cas9 method. We appreciate the additional technical support provided by Sudarsana Poojari.

Conflicts of Interest: The authors declare no conflict of interest.

Appendix A

Table A1. CRISPR-Cas9 Target sequence used for STL1 deletion.

Protospacer Sequence	Protospacer Adjacent Motif
GGTAGGAACACTAGACGACG	CGG

Table A2. Primers used in this work.

Primer Name	Sequence
pWS173 sgRNA FW	ATCAACAACAGAGGACATATGCCCTACCTCCATG
pWS173 sgRNA RV	ATCCTGCACTCATCTACTACCCGCATCCC
STL1 sgRNA FW	GCTATTTCTAGCTCTAAAACCGTCGTCTAGTGTTCCTACC AAAGTCCCATTCGCCACCCG
STL1 sgRNA RV	CGGGTGGCGAATGGGACTTTGGTAGGAACACTAGA CGACGGTTTTAGAGCTAGAAATAGC
No target sgRNA FW	GCCTTATTTTAACTTGCTATTTCTAGCTCTAAAACAAAGT CCCATTCGCCACCCG
No target sgRNA RV	GCTGGGCAACACCTTCGGGTGGCGAATGGGACTTTGTTTT AGAGCTAGAAATAGC
STL1 5′ homology arm FW	CGCGGGGTATTCGGCGCTACG
STL1 5′ homology arm RV	GTGTATGAATCAACCCTCAAAATTTGCTTTATCGTGGTCTAA AACTTTCTATGTTCTAC
STL1 3′ homology arm FW	TAAATAAAAAATAGAAAAGTAGAACATAGAAAGTTTTAGACC ACGATAAAGCAAATTTTG
STL1 3′ homology arm RV	GCAGTGTTCGAAAGTTGGTGGCTTTTTGTATGTTGC
STL1 genotyping FW	CAGAATTGCAGTTCAGGAGTAGTCACAC
STL1 genotyping RV	CGCTCCAATGTAATAACACGTCCACGTTATC
pWS173 KanR genotyping FW	GGATGGCAAGATCCTGGTATCGG
pWS173 KanR genotyping RV	ACTGAATCGGGAGGATCTGCTG

References

1. Inglis, D.L.; Pickering, G.P. Vintning on thin ice—The making of Canada's iconic dessert wine. In *The World of Niagara Wine*; Ripmeester, M., Mackintosh, P., Fullerton, C., Eds.; Wilfrid Laurier University Press: Waterloo, ON, Canada, 2013; pp. 229–248.
2. Kontkanen, D.; Inglis, D.L.; Pickering, G.J.; Reynolds, A. Effect of Yeast Inoculation Rate, Acclimatization, and Nutrient Addition on Icewine Fermentation. *Am. J. Enol. Vitic.* **2004**, *55*, 363–370.
3. Pigeau, G.M.; Inglis, D.L. Upregulation of ALD3 and GPD1 in *Saccharomyces cerevisiae* during Icewine fermentation. *J. Appl. Microbiol.* **2005**, *99*, 112–125. [CrossRef] [PubMed]
4. Yang, F.; Heit, C.; Inglis, D.L. Cytosolic Redox Status of Wine Yeast (*Saccharomyces cerevisiae*) under Hyperosmotic Stress during Icewine Fermentation. *Fermentation* **2017**, *3*, 61. [CrossRef]
5. Björkqvist, S.; Ansell, R.; Adler, L.; Lidén, G. Physiological response to anaerobicity of glycerol-3-phosphate dehydrogenase mutants of *Saccharomyces cerevisiae*. *Appl. Environ. Microbiol.* **1997**, *63*, 128–132.
6. Bakker, B.M.; Overkamp, K.M.; van Maris, A.J.A.; Kötter, P.; Luttik, M.A.H.; van Dijken, J.P.; Pronk, J.T. Stoichiometry and compartmentation of NADH metabolism in *Saccharomyces cerevisiae*. *FEMS Microbiol. Rev.* **2001**, *25*, 15–37. [CrossRef]
7. Cronwright, G.R.; Rohwer, J.M.; Prior, B.A. Metabolic Control Analysis of Glycerol Synthesis in *Saccharomyces cerevisiae*. *Appl. Environ. Microbiol.* **2002**, *68*, 4448–4456. [CrossRef]
8. Blomberg, A.; Adler, L. Roles of glycerol and glycerol-3-phosphate dehydrogenase (NAD +) in acquired osmotolerance of *Saccharomyces cerevisiae*. *J. Bacteriol.* **1989**, *171*, 1087–1092. [CrossRef]
9. Aranda, A.; del Olmo, M.L. Response to acetaldehyde stress in the yeast *Saccharomyces cerevisiae* involves a strain-dependent regulation of several ALD genes and is mediated by the general stress response pathway. *Yeast* **2003**, *20*, 747–759. [CrossRef]
10. Hohmann, S. An integrated view on a eukaryotic osmoregulation system. *Curr. Genet.* **2015**, *61*, 373–382. [CrossRef]
11. Ferreira, C.; van Voorst, F.; Martins, A.; Neves, L.; Oliveira, R.; Kielland-Brandt, M.C.; Lucas, C.; Brandt, A. A Member of the Sugar Transporter Family, Stl1p Is the Glycerol/H+ Symporter in *Saccharomyces cerevisiae*. *Mol. Biol. Cell* **2005**, *16*, 2068–2076. [CrossRef]
12. Nelissen, B.; De Wachter, R.; Goffeau, A. Classification of all putative permeases and other membrane pluripanners of the major facilitator superfamily encoded by the complete genome of *Saccharomyces cerevisiae*. *FEMS Microbiol. Rev.* **1997**, *21*, 113–134. [CrossRef] [PubMed]
13. Ferreira, C.; Lucas, C. Glucose repression over *Saccharomyces cerevisiae* glycerol/H+ symporter gene STL1 is overcome by high temperature. *FEBS Lett.* **2007**, *581*, 1923–1927. [CrossRef] [PubMed]
14. Noti, O.; Vaudano, E.; Pessione, E.; Garcia-Moruno, E. Short-term response of different *Saccharomyces cerevisiae* strains to hyperosmotic stress caused by inoculation in grape must: RT-qPCR study and metabolite analysis. *Food Microbiol.* **2015**, *52*, 49–58. [CrossRef] [PubMed]
15. Pérez-Torrado, R.; Oliveira, B.M.; Zemančíková, J.; Sychrová, H.; Querol, A. Alternative Glycerol Balance Strategies among *Saccharomyces* Species in Response to Winemaking Stress. *Front. Microbiol.* **2016**, *7*. [CrossRef]
16. Heit, C.; Martin, S.J.; Yang, F.; Inglis, D.L. Osmoadaptation of wine yeast (*Saccharomyces cerevisiae*) during Icewine fermentation leads to high levels of acetic acid. *J. Appl. Microbiol.* **2018**, *124*, 1506–1520. [CrossRef] [PubMed]
17. Hauser, N.C.; Fellenberg, K.; Gil, R.; Bastuck, S.; Hoheisel, J.D.; Pérez-Ortín, J.E. Whole Genome Analysis of a Wine Yeast Strain. *Comp. Funct. Genomics* **2001**, *2*, 69–79. [CrossRef] [PubMed]
18. Jiménez-Martí, E.; Gomar-Alba, M.; Palacios, A.; Ortiz-Julien, A.; del Olmo, M. Towards an understanding of the adaptation of wine yeasts to must: Relevance of the osmotic stress response. *Appl. Microbiol. Biotechnol.* **2011**, *89*, 1551–1561. [CrossRef]
19. Jiménez-Martí, E.; Zuzuarregui, A.; Gomar-Alba, M.; Gutiérrez, D.; Gil, C.; del Olmo, M. Molecular response of *Saccharomyces cerevisiae* wine and laboratory strains to high sugar stress conditions. *Int. J. Food Microbiol.* **2011**, *145*, 211–220. [CrossRef]
20. Pronk, J.T. Auxotrophic Yeast Strains in Fundamental and Applied Research. *Appl. Environ. Microbiol.* **2002**, *68*, 2095–2100. [CrossRef]

21. Le Borgne, S. Genetic Engineering of Industrial Strains of *Saccharomyces cerevisiae*. In *Recombinant Gene Expression*; Lorence, A., Ed.; Springer: Berlin/Heidelberg, Germany, 2012; pp. 451–465. [CrossRef]
22. Stovicek, V.; Holkenbrink, C.; Borodina, I. CRISPR/Cas system for yeast genome engineering: Advances and applications. *FEMS Yeast Res.* **2017**, *17*. [CrossRef]
23. Zhang, Y.; Wang, J.; Wang, Z.; Zhang, Y.; Shi, S.; Nielsen, J.; Liu, Z. A gRNA-tRNA array for CRISPR-Cas9 based rapid multiplexed genome editing in *Saccharomyces cerevisiae*. *Nat. Commun.* **2019**, *10*, 1–10. [CrossRef] [PubMed]
24. Vigentini, I.; Gebbia, M.; Belotti, A.; Foschino, R.; Roth, F.P. CRISPR/Cas9 System as a Valuable Genome Editing Tool for Wine Yeasts with Application to Decrease Urea Production. *Front. Microbiol.* **2017**, *8*. [CrossRef] [PubMed]
25. Shaw, W. Quick and Easy CRISPR Engineering in Saccharomyces cerevisiae. 2018. Available online: https://www.benchling.com/2018/11/06/quick-and-easy-crispr-engineering-in-saccharomyces-cerevisiae/ (accessed on 10 November 2018).
26. Novo, M.; Bigey, F.; Beyne, E.; Galeote, V.; Gavory, F.; Mallet, S.; Cambon, B.; Legras, J.-L.; Wincker, P.; Casaregola, S.; et al. Eukaryote-to-eukaryote gene transfer events revealed by the genome sequence of the wine yeast *Saccharomyces cerevisiae* EC1118. *Proc. Natl. Acad. Sci. USA* **2009**, *106*, 16333–16338. [CrossRef]
27. DiCarlo, J.E.; Norville, J.E.; Mali, P.; Rios, X.; Aach, J.; Church, G.M. Genome engineering in *Saccharomyces cerevisiae* using CRISPR-Cas systems. *Nucleic Acids Res.* **2013**, *41*, 4336–4343. [CrossRef]
28. Zoecklein, B.W.; Fugelsang, K.C.; Gump, B.H. Laboratory Procedures. In *Wine Analysis and Production*, 1st ed.; Chapman and Hall, International Thomson Publishing: New York, NY, USA, 1995; pp. 374–378.
29. Grey, M.; Brendel, M. A ten-minute protocol for transforming *Saccharomyces cerevisiae* by electroporation. *Curr. Genet.* **1992**, *22*, 335–336. [CrossRef]
30. Tripp, J.D.; Lilley, J.L.; Wood, W.N.; Lewis, L.K. Enhancement of plasmid DNA transformation efficiencies in early stationary phase yeast cell cultures. *Yeast* **2013**, *30*, 191–200. [CrossRef]
31. Kuijpers, N.G.; Solis-Escalante, D.; Bosman, L.; van den Broek, M.; Pronk, J.T.; Daran, J.-M.; Daran-Lapujade, P. A versatile, efficient strategy for assembly of multi-fragment expression vectors in *Saccharomyces cerevisiae* using 60 bp synthetic recombination sequences. *Microb. Cell Factories* **2013**, *12*, 47. [CrossRef]
32. Förster, J.; Famili, I.; Fu, P.; Palsson, B.Ø.; Nielsen, J. Genome-Scale Reconstruction of the *Saccharomyces cerevisiae* Metabolic Network. *Genome Res.* **2003**, *13*, 244–253. [CrossRef]
33. Nielsen, J. It Is All about Metabolic Fluxes. *J. Bacteriol.* **2003**, *185*, 7031–7035. [CrossRef]
34. Miermont, A.; Waharte, F.; Hu, S.; McClean, M.N.; Bottani, S.; Léon, S.; Hersen, P. Severe osmotic compression triggers a slowdown of intracellular signaling, which can be explained by molecular crowding. *Proc. Natl. Acad. Sci. USA* **2013**, *110*, 5725–5730. [CrossRef]
35. Jones, R.P. Biological principles for the effects of ethanol. *Enzym. Microb. Technol.* **1989**, *11*, 130–153. [CrossRef]
36. Bisson, L.F. Stuck and Sluggish Fermentations. *Am. J. Enol. Vitic.* **1999**, *50*, 107–119.
37. Monteiro, B.; Ferraz, P.; Barroca, M.; da Cruz, S.H.; Collins, T.; Lucas, C. Conditions promoting effective very high gravity sugarcane juice fermentation. *Biotechnol. Biofuels* **2018**, *11*, 251. [CrossRef] [PubMed]
38. Lages, F.; Silva-Graça, M.; Lucas, C. Active glycerol uptake is a mechanism underlying halotolerance in yeasts: A study of 42 species. *Microbiology* **1999**, *145*, 2577–2585. [CrossRef]
39. Dušková, M.; Ferreira, C.; Lucas, C.; Sychrová, H. Two glycerol uptake systems contribute to the high osmotolerance of *Zygosaccharomyces rouxii*. *Mol. Microbiol.* **2015**, *97*, 541–559. [CrossRef]

© 2019 by the authors. Licensee MDPI, Basel, Switzerland. This article is an open access article distributed under the terms and conditions of the Creative Commons Attribution (CC BY) license (http://creativecommons.org/licenses/by/4.0/).

fermentation

Communication

PTR-ToF-MS for the Online Monitoring of Alcoholic Fermentation in Wine: Assessment of VOCs Variability Associated with Different Combinations of *Saccharomyces*/Non-*Saccharomyces* as a Case-Study

Carmen Berbegal [1,2], Iuliia Khomenko [3], Pasquale Russo [1], Giuseppe Spano [1], Mariagiovanna Fragasso [1], Franco Biasioli [3,*] and Vittorio Capozzi [4,*]

[1] Department of the Sciences of Agriculture, Food and Environment, University of Foggia, Via Napoli 25, 71122 Foggia, Italy; carmen.berbegal@uv.es (C.B.); pasquale.russo@unifg.it (P.R.); giuseppe.spano@unifg.it (G.S.); mariagiovanna.fragasso@gmail.com (M.F.)
[2] EnolabERI BioTecMed, Universitat de València, 46100 Valencia, Spain
[3] Department of Food Quality and Nutrition, Research and Innovation Centre, Fondazione Edmund Mach (FEM), via E. Mach 1, 38010 San Michele all'Adige, Italy; iuliia.khomenko@fmach.it
[4] Institute of Sciences of Food Production, National Research Council (CNR), c/o CS-DAT, Via Michele Protano, 71121 Foggia, Italy
* Correspondence: franco.biasioli@fmach.it (F.B.); vittorio.capozzi@ispa.cnr.it (V.C.); Tel.: +39-0461-615-187 (F.B.); +39-0881-630-201 (V.C.)

Received: 31 December 2019; Accepted: 24 May 2020; Published: 26 May 2020

Abstract: The management of the alcoholic fermentation (AF) in wine is crucial to shaping product quality. Numerous variables (e.g., grape varieties, yeast species/strains, technological parameters) can affect the performances of this fermentative bioprocess. The fact that these variables are often interdependent, with a high degree of interaction, leads to a huge 'oenological space' associated with AF that scientists and professionals have explored to obtain the desired quality standards in wine and to promote innovation. This challenge explains the high interest in approaches tested to monitor this bioprocess including those using volatile organic compounds (VOCs) as target molecules. Among direct injection mass spectrometry approaches, no study has proposed an untargeted online investigation of the diversity of volatiles associated with the wine headspace. This communication proposed the first application of proton-transfer reaction-mass spectrometry coupled to a time-of-flight mass analyzer (PTR-ToF-MS) to follow the progress of AF and evaluate the impact of the different variables of wine quality. As a case study, the assessment of VOC variability associated with different combinations of *Saccharomyces*/non-*Saccharomyces* was selected. The different combinations of microbial resources in wine are among the main factors susceptible to influencing the content of VOCs associated with the wine headspaces. In particular, this investigation explored the effect of multiple combinations of two *Saccharomyces* strains and two non-*Saccharomyces* strains (belonging to the species *Metschnikowia pulcherrima* and *Torulaspora delbrueckii*) on the content of VOCs in wine, inoculated both in commercial grape juice and fresh grape must. The results demonstrated the possible exploitation of non-invasive PTR-ToF-MS monitoring to explore, using VOCs as biomarkers, (i) the huge number of variables influencing AF in wine, and (ii) applications of single/mixed starter cultures in wine. Reported preliminary findings underlined the presence of different behaviors on grape juice and on must, respectively, and confirmed differences among the single yeast strains 'volatomes'. It was one of the first studies to include the simultaneous inoculation on two non-*Saccharomyces* species together with a *S. cerevisiae* strain in terms of VOC contribution. Among the other outcomes, evidence suggests that the addition of *M. pulcherrima* to the coupled *S. cerevisiae*/*T. delbrueckii* can modify the global release of volatiles as a function of the characteristics of the fermented matrix.

Keywords: volatile organic compounds; proton-transfer reaction-mass spectrometry; *Saccharomyces cerevisiae*; *Metschnikowia pulcherrima*; *Torulaspora delbrueckii*; wine; flavor

1. Introduction

Wine is the result of alcoholic fermentation (AF) performed by yeasts that convert the sugars present in grape must into ethanol and carbon dioxide. During this fermentation, other chemical changes are produced as a consequence of yeast metabolic activities. Among the chemical changes, a consistent part of volatile organic compounds (VOCs) is released, influencing wine flavor [1,2]. The interest in the monitoring of this bioprocess is high due to (i) the vast number of variables that can affect AF performances, and (ii) the crucial relevant impact of AF on wine quality. Non-separative approaches based on direct injection mass spectrometry (DIMS) have recently emerged as an alternative for the high-throughput and cost-effective quantitative profiling of volatiles in food and beverages [3]. To the best of our knowledge, no study has explored the potential of DIMS techniques to assess online VOC variability in association with alcoholic fermentation in wine [4].

Saccharomyces cerevisiae has a leading role in performing AF in wine [5,6]. However, an increasing interest has been given to non-*Saccharomyces* yeasts as drivers of the differentiation of the quality of final wines [7,8]. Non-*Saccharomyces* yeasts can possess enzymatic activities different from the *S. cerevisiae* enzymatic inventory, catalyzing the synthesis and the release (from non-volatile bound precursors) of VOCs able to modulate aromatic wine complexity [9,10]. Moreover, they may influence other characteristics such as glycerol and mannoprotein content, volatile acidity, color stability, and ethanol levels of wines [11,12]. Usually, as a reason for non-optimal fermentative performances, non-*Saccharomyces* yeasts are used in combination with *S. cerevisiae* strains. Some studies have shown that the strategy of co-inoculating *S. cerevisiae* starter together with selected non-*Saccharomyces* yeasts at high cell density produces wines with distinctive characteristics [13]. The interaction between the different yeast species influences the content of VOCs associated with fermentations [14]. Among the non-*Saccharomyces* species, *Torulaspora delbrueckii*, *Metschnikowia pulcherrima*, *Candida zemplinina*, and *Hanseniaspora uvarum* are mostly cited and have been intensively investigated [9,15–21]. Strains belonging to the species *Lachancea thermotolerans*, *Metschnikowia fructicola*, *Schizosaccharomyces pombe*, *T. delbrueckii*, *Kluyveromyces thermotolerans*, *Pichia kluyveri*, and *M. pulcherrima* are commercialized or have patented applications [16,22]. Belonging to the class of direct injection mass spectrometry (DIMS) approaches, proton transfer reaction mass spectrometry (PTR-MS) is an established method for the rapid, direct, and non-invasive online monitoring of VOCs characterized by short response time and high sensitivity [23]. The coupling of proton transfer ionization with time-of-flight (ToF) mass spectrometers and automated sampling offers several advantages in terms of mass resolution, throughput, and reproducibility [24–26]. This analytical strategy has found several applications in the food field (e.g., [27–31]), with a specific interest in bioprocess monitoring associated with microbial-based processes (e.g., [32–34]). Furthermore, several studies have applied PTR-based approaches to monitor VOC release associated with yeast metabolisms [35], often in association with food matrices [36–38]. In the case of matrices containing ethanol, consistent experimental efforts have been performed to avoid the adverse effects of a high concentration of this alcohol (primary ion depletion and ethanol–ethanol/water clusters formation responsible for the loss of efficiency in the qualitative/quantitative detection) [34,39–42].

In the present work, PTR-ToF-MS was used for the online monitoring of AF in wine and to compare the performance of four (autochthonous and commercial) yeast strains, both in single cultures and in multiple inoculations, using two diverse model matrices as substrates (real grape must and commercial grape juice). This study also aimed to preliminarily explore the interest in PTR-ToF-MS analysis of flavor-related volatile compounds in the control, design, and application of single/mixed starter cultures for wine.

2. Materials and Methods

2.1. Microorganisms and the Determination of Microbial Population

The following microorganisms were used for grape juice and grape must inoculation: the commercially available *Saccharomyces cerevisiae* strain DV10 (Lallemand, Montreal, QC, Canada), autochthonous characterized *S. cerevisiae* I6 strain from the Apulian region (Southern Italy) [43], and the commercially available non-*Saccharomyces* strains *Metschnikowia pulcherrima* FLAVIA (Lallemand, USA) and *Torulaspora delbrueckii* BIODIVA (Lallemand, Montreal, QC, Canada). Yeast starters were purchased in active-dried form. Rehydration procedures were done according to the suppliers' instructions. Starter cultures were prepared by growing pure cultures of the yeast strains separately grown in liquid Yeast Peptone Dextrose (YPD) medium (2% glucose, 2% Bacto peptone, 1% yeast extract) at 28 °C.

The viable count of yeasts during the AF was enumerated on Wallerstein Laboratory (WL) agar medium (Sigma-Aldrich, St. Louis, USA). WL discriminates between the used yeast species by colony morphology and color (*S. cerevisiae* produces large white colonies, whereas non-*Saccharomyces* yeasts produce green colonies on this medium). Plates were incubated at 28 °C for 48 h.

2.2. Micro-Vinifications and Wine Analysis

Starter cultures were prepared by growing strains in YPD medium as described above and then inoculating the strains into commercial red grape juice (Vitafit, Lidl Stiftung & Co., Neckarsulm, Germany) and red grape must from Apulian autochthonous grape varieties (20° Babo; 7.2 g/L total acidity; 3.5 g/L malic acid; pH 3.5; free ammonium 163.5 mg/L). Fermentations were performed inoculating at concentrations of 1×10^6 cfu/mL (colony-forming units per milliliter) of *M. pulcherrima* FLAVIA, 1×10^6 cfu/mL of *T. delbrueckii* BIODIVA and 1×10^4 cfu/mL of *S. cerevisiae* strains (DV10 or I6). Each fermentation experiment was carried out by performing three simultaneous independent repetitions. With these four biotypes, 14 different combinations of strains were carried out (Table 1). Fermentative kinetics from grape must were monitored daily by gravimetric determinations for seven days. With this purpose, samples were weighed daily to follow the weight loss caused by CO_2 production.

Table 1. Microorganisms employed in different grape must/juice fermentations (trials 1–15).

Sample Code	Inoculated Yeast Cultures
1	*S. cerevisiae* DV10
2	*S. cerevisiae* I6
3	*M. pulcherrima* FLAVIA
4	*T. delbrueckii* BIODIVA
5	*S. cerevisiae* DV10 + *S. cerevisiae* I6
6	*S. cerevisiae* DV10 + *M. pulcherrima* FLAVIA
7	*S. cerevisiae* DV10 + *T. delbrueckii* BIODIVA
8	*S. cerevisiae* I6 + *M. pulcherrima* FLAVIA
9	*S. cerevisiae* I6 + *T. delbrueckii* BIODIVA
10	*S. cerevisiae* DV10 + *S. cerevisiae* I6 + *M. pulcherrima* FLAVIA
11	*S. cerevisiae* DV10 + *S. cerevisiae* I6 + *T. delbrueckii* BIODIVA
12	*S. cerevisiae* DV10 + *M. pulcherrima* FLAVIA + *T. delbrueckii* BIODIVA
13	*S. cerevisiae* I6 + *M. pulcherrima* FLAVIA + *T. delbrueckii* BIODIVA
14	*S. cerevisiae* DV10 + *S. cerevisiae* I6 + *M. pulcherrima* FLAVIA + *T. delbrueckii* BIODIVA
15	Uninoculated grape must/juice

2.3. Samples Preparation and PTR-ToF-MS (Proton Transfer Reaction-Time of Flight-Mass Spectrometry) Analysis

Nano-vinifications were performed in the vials using the above yeast combinations (Table 1) in commercial red grape juice and fresh red grape must. Nitrogen flux in the vial headspace assured maintaining the conditions comparable with those present in vinification. While the manufacturer sterilized the commercial grape juice, the must was not treated to reduce microbial presence. When present in the same trial, yeasts were co-inoculated in the juice/must. The resulting AF was monitored for three days. The whole experiment was performed in five replicates. For the measurements, a commercial PTR-ToF-MS 8000 apparatus from Ionicon Analytik GmbH (Innsbruck, Austria) was used in its standard configuration (V mode). The air associated with the headspace of the sample was directly injected in the PTR-MS drift tube. An argon dilution system was applied after headspace sampling. The dilution ratio was one part of headspace to three parts of argon. The argon flow rate was 120 sccm and was controlled by a multigas controller (MKS Instruments, Inc, Andover, MA, USA). Ionization conditions were as follows: 110 °C drift tube temperature, 2.30 mbar drift pressure, and 550 V drift voltage. These conditions led to an E/N ratio of about 140 Td (1 Townsend = 10–17 cm^2 V^{-1} s^{-1}). The inlet line was a PEEK capillary tube (internal diameter 0.04 in.) heated at 110 °C, with a flow set at 40 sccm. The acquisition rate of the instrument was one spectrum per second.

2.4. Data Analysis

Deadtime correction, internal calibration of mass spectral data, and peak extraction were performed according to the procedure described by Cappellin et al. [44,45]. Peak intensity in ppbv was estimated using the formula described by Lindinger et al. [46] using a constant value for the reaction rate coefficient (k = 2.10^{-9} cm^3 s^{-1}). This approach introduces a systematic error for the absolute concentration for each compound that is, in most cases, below 30% and could be accounted for if the actual rate constant coefficient is available [45]. All data detected and recorded by the PTR-ToFMS were processed and analyzed using MATLAB R2017a (MathWorks Inc., Natick, MA, USA) and R (R Foundation for Statistical Computing, Vienna, Austria). Principal component analysis, analysis of variance, and Tukey's post-hoc test were performed to spot the differences in the volatile aroma compounds emitted by the 28 grape must and juice fermentations used in this study.

3. Results

3.1. Alcoholic Fermentation Kinetics and Yeast Dynamics

The kinetics of the 14 fermentations in red grape must were monitored daily for seven days, evaluating the loss of weight due to the production of CO_2 (data not shown). All fermentations were completed in four days except for sample 3 (inoculated with a single culture of *M. pulcherrima* FLAVIA), which was not able to complete the AF. The interactions between *Saccharomyces* spp. and both non-*Saccharomyces* spp. of enological interest, *M. pulcherrima* FLAVIA and *T. delbrueckii* BIODIVA, were investigated in terms of cell density. The differential morphology of the colonies on WL medium allowed us to calculate the proportion of each yeast species in different phases of AF (Figure 1). Only when both *S. cerevisiae* strains were co-inoculated the viable cell count was considered as total *S. cerevisiae* viable cells without distinguishing between DV10 and I6 strains.

Results from plate counting revealed that the maximum cell density of the single cultures was obtained after 48–72 h of the grape must inoculation both for *S. cerevisiae* (with an initial cell population of 1 × 10^4 cfu/mL) and non-*Saccharomyces* yeasts (with an initial cell population of 1 × 10^6 cfu/mL) (Figure 1, Experiments 1–4). In terms of single cultures studied, *M. pulcherrima* FLAVIA reached the lowest cell concentration (slightly more than 1 × 10^7 cfu/mL) after 72 h of inoculation (Figure 1, Experiment 3). Conversely, *T. delbrueckii* BIODIVA, even if with a different profile, achieved a biomass concentration comparable to those of the two *S. cerevisiae* strains. Considering the strain combinations, when two *S. cerevisiae* strains (DV10 and I6) were inoculated simultaneously, the growth behavior was the same as when they were inoculated in a single form and reached the maximum yeast population after 48 h of the inoculation (Figure 1, Experiment 5). Results from the co-inoculation of one strain of *S. cerevisiae* with *M. pulcherrima* FLAVIA (Figure 1 Experiments 6 and 8) showed that in 24 h, the *S. cerevisiae* strains were able to overtake the non-*Saccharomyces* yeast concentration, and in 48–72 h achieved the maximum yeast population. On the other hand, the *M. pulcherrima* FLAVIA population decreased drastically after 72 h of inoculation. Contrary to *M. pulcherrima* FLAVIA, yeast *T. delbrueckii* BIODIVA presented a high cell density when it was co-inoculated with one of the *S. cerevisiae* strains, and with similar cell concentration levels to *S. cerevisiae* (Figure 1, Experiments 7 and 9). In the same way, when the two *S. cerevisiae* strains, DV10 and I6, were co-inoculated with *M. pulcherrima* FLAVIA or with *T. delbrueckii* BIODIVA strains, the growth behavior of the non-*Saccharomyces* strains was the same than when they were inoculated with only one *S. cerevisiae* strain (Figure 1, Experiments 10 and 11). The simultaneous inoculation of one *S. cerevisiae* strain with both, *T. delbrueckii* BIODIVA and *M. pulcherrima* FLAVIA (Figure 1, Experiments 12 and 13), revealed that *S. cerevisiae* strains needed more time to reach the maximum cell concentration than when co-inoculated with only one non-*Saccharomyces* strain, and *T. delbrueckii* BIODIVA presented a higher population than *S. cerevisiae* strains for 48 h.

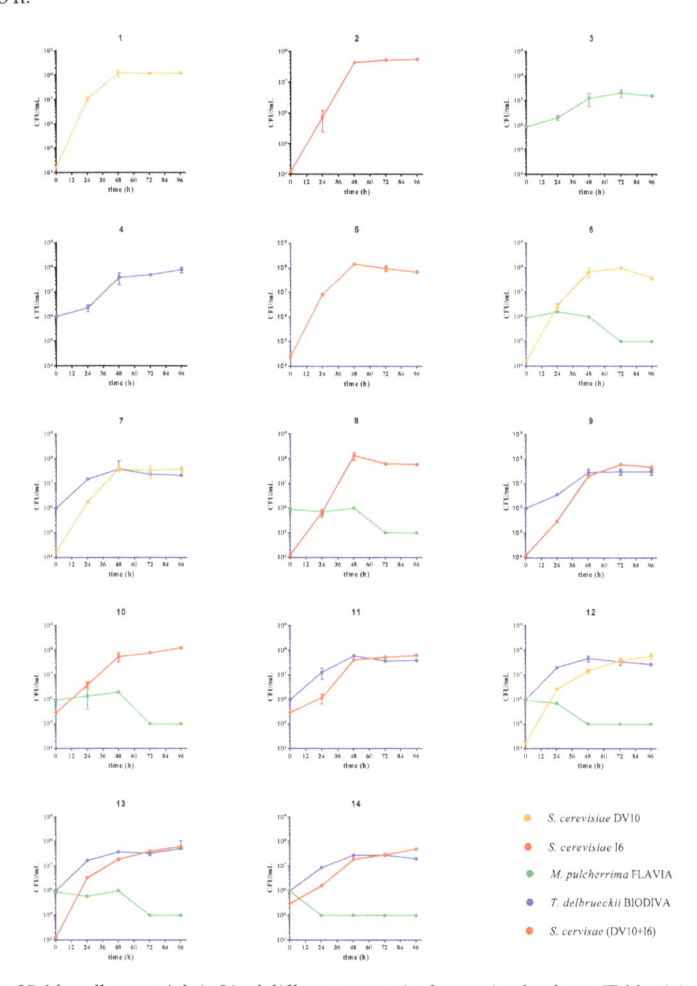

Figure 1. Viable cell count (cfu/mL) of different yeast single or mixed culture (Table 1) inoculated. The cell enumeration was performed on Wallerstein Laboratory agar medium that discriminates *S. cerevisiae* (large white colonies) from non-*Saccharomyces* yeasts (green colonies).

Inoculating the four starter cultures simultaneously (Figure 1, Experiment 14) triggered, as in the previous cases, that *S. cerevisiae* strains required more time to reach the maximum cell population and, the maximum cell concentration was lower than when they were inoculated in a single culture form. Furthermore, the population of *S. cerevisiae* and *T. delbrueckii* BIODIVA presented similar population levels after 48 h of inoculation. Otherwise, the *M. pulcherrima* FLAVIA population decreased from the inoculation time. Overall, biological interactions influenced single yeast growth behavior. Nevertheless, in all of the studied experimental modes, the most significant changes related to yeast population occurred during the first 72 h, which led us to focus on this temporal interval for the online monitoring of VOCs associated with the considered experimental modes using the PTR-ToF-MS technology.

3.2. Automated Monitoring Volatile Organic Compound (VOC) Evolution in Red Grape Must and Juice Fermentation by Different Yeast Mixed Cultures

A preliminary data exploration has been made to visualize the results of the PTR-ToF-MS grape must and juice analysis through a principal component analysis (PCA). The first and second PCA components (Figure 2) accounted for 84% of total variability and showed that the two matrices (grape juice and must) used in this study led to clear changes of VOC release. Differences in the distribution of variances were also observed concerning single yeasts or yeast combinations. It is easy to follow a time-dependent dimension of the phenomena, observing the increasing dimensions of the symbols in Figure 2.

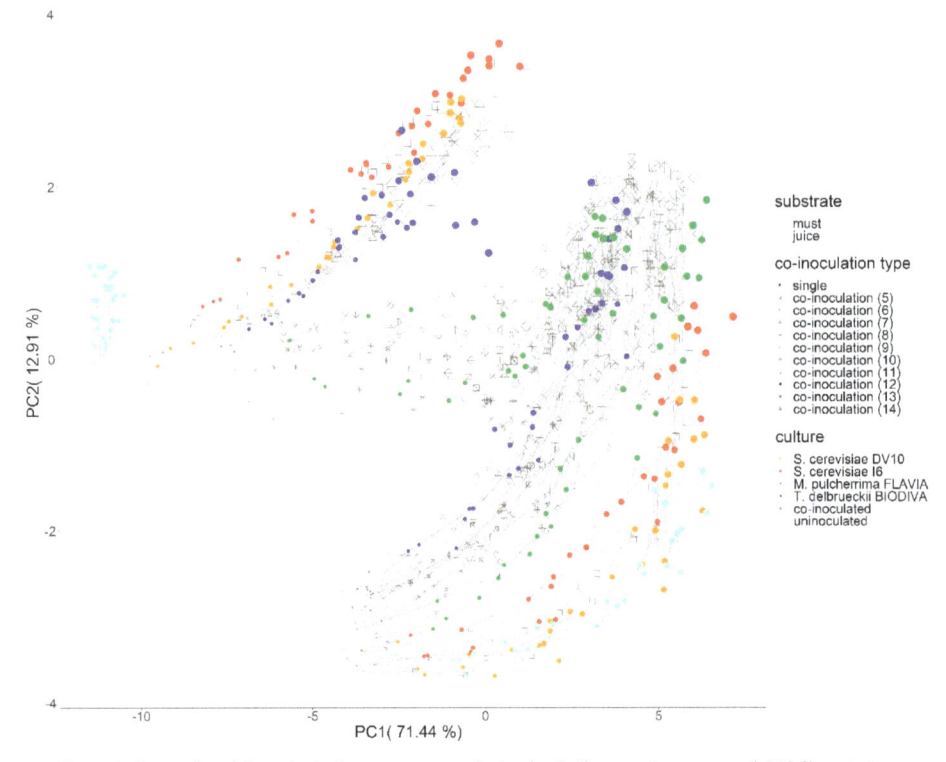

Figure 2. Score plot of the principal component analysis of volatile organic compound (VOC) emission evolution associated with the first three days of AF for each trial tested in this study. Data were logarithmically transformed and centered. Different colors indicate the different yeast managements, medium, and blank samples. The size of the points grew with the time of measurements. For a detailed view of the figure, the original image is included in the Supplementary Materials (Figure S1).

Separations and different evolutions were evident by comparing the matrices 'grape must' and 'grape juice'. In the 'must' assays, it was clear the partition between the trials with only *S. cerevisiae* strains inoculated (sample codes 1, 2, and 5; Table 1) and those that included, in the starter cultures, non-*Saccharomyces* strains (sample codes 3, and 4; Table 1) (Figure 2). Additionally, a diverse behavior was noticeable for the fermentations inoculated with pure cultures of *M. pulcherrima* FLAVIA and *T. delbrueckii* BIODIVA strains, respectively. All the experiments that included both *S. cerevisiae* and non-*Saccharomyces* strains (sample codes 6–14; Table 1) followed a trend that appeared more similar to the non-*Saccharomyces* pure cultures. Concerning the 'juice' experimental plan, an uniform trend was confirmed for the samples inoculated with the *S. cerevisiae* strains (Figure 2). In contrast, the behavior observed for the pure inoculation of *M. pulcherrima* FLAVIA strain (sample code 3; Table 1) was radically different. The pure culture of *T. delbrueckii* BIODIVA strain (sample code 4; Table 1) and all the combinations of *S. cerevisiae*–*T. delbrueckii* (sample codes 7, 9, and 11; Table 1) followed trajectories closer to those of *S. cerevisiae* strains than to the *M. pulcherrima* one. In contrast, all the other trials (sample codes 6, 8, 10, and 12–14; Table 1) observed patterns of evolution similar to *M. pulcherrima*. Concerning these last trials, some samples also included *T. delbrueckii* BIODIVA among the inoculated strains. In the case of PCA, the loading plot (Figure 3) indicates the mass peaks related to the observed evolution of the VOC profile in Figure 2.

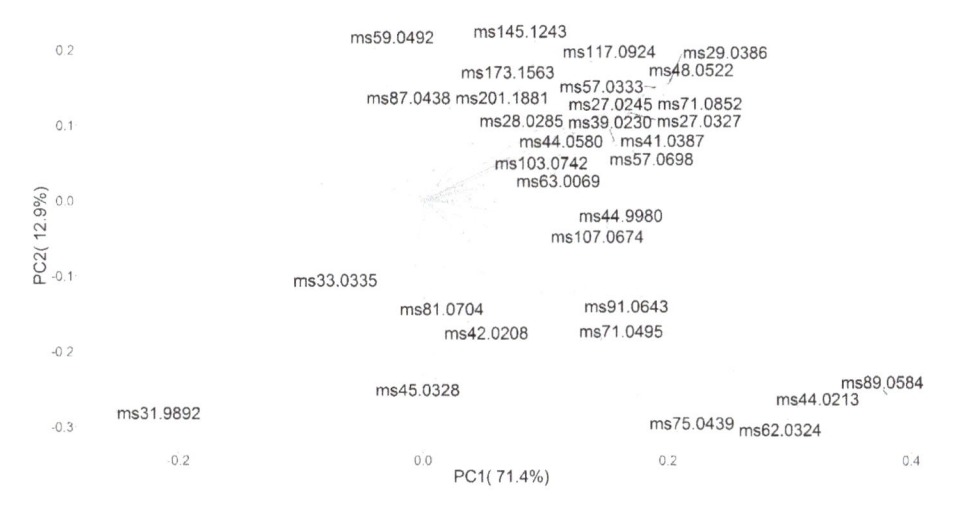

Figure 3. Loading plot of principal component analysis of the mass peaks (ms) related to the observed evolution of VOCs profile in Figure 2. For a detailed view of the figure, the original image is included in the Supplementary Materials (Figure S2).

More than 70 mass peaks were identified among the four yeast commercial starters during online monitoring throughout the three days of fermentation. For each of these mass peaks, it was possible to perform a tentative identification (allowing a possible link of the ion with a given molecule/molecular fragment) and to follow the evolution of the intensity in the time, allowing a direct analytic determination to evaluate the yeast metabolic activity during the progress of AF.

More specifically, differences between the matrices used were observed when the score plot of the PCA analysis on the distribution of variances associated with VOC emission during the first three days of AF was represented separately for each trial (Figure 4). Negligible VOC evolution was evident in uninoculated grape juice and slow evolution in grape must as revealed by the first PCA dimension, which is related to the increase of volatile concentration in the sample headspace (Figure 4, uninoculated trial, experiment 15). Regarding the inoculated yeasts, differences in the VOC emissions were also found. For example, *M. pulcherrima* and *T. delbrueckii* in single culture (Figure 4, experiments

3 and 4) tended to reach a lesser concentration of VOCs in juice, while both *S. cerevisiae* kept producing more volatile compounds with time (Figure 4, experiments 1 and 2). Moreover, this graph confirmed that there were differences between the different yeast combinations inoculated, as they were arranged in the graph according to different patterns. This effect is of particular interest, if we consider that together with the effect of different strains/species combinations we also tested the impact of the increasing microbial diversity of the starter cultures inoculated.

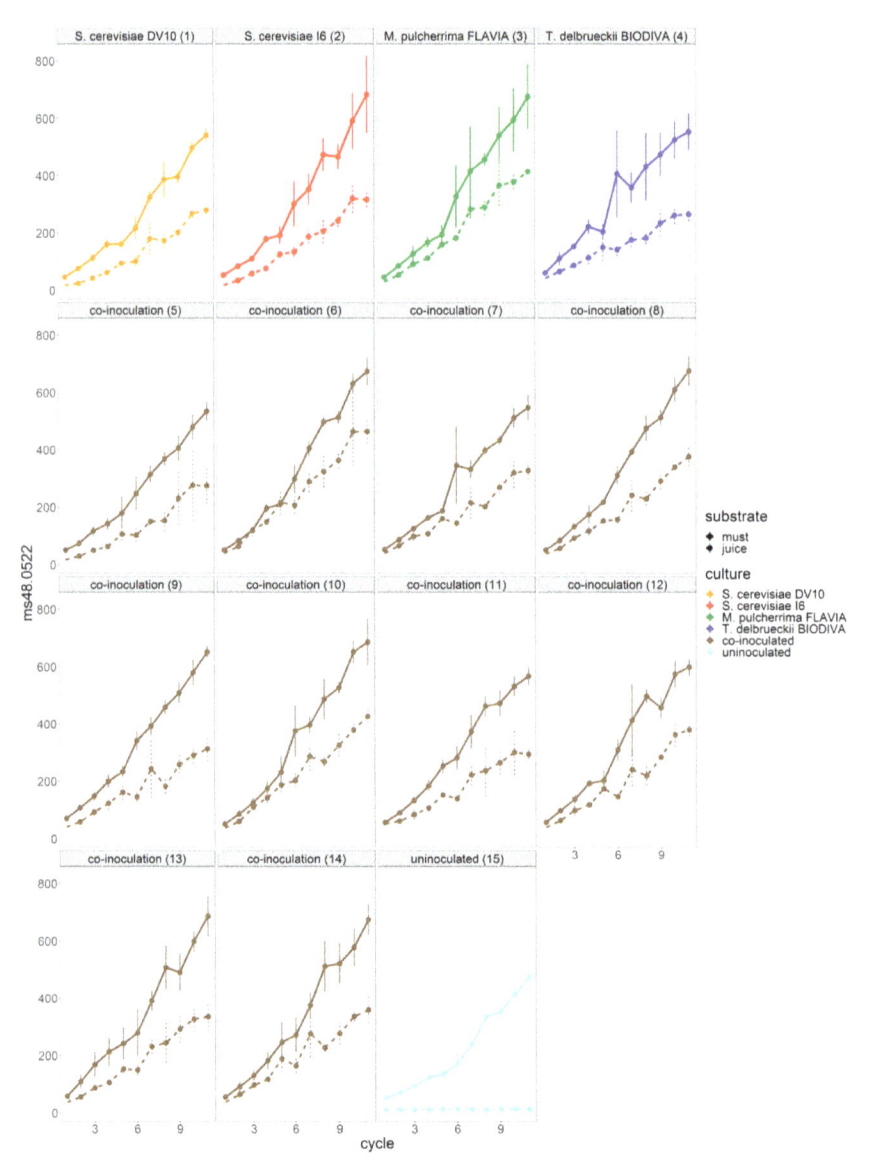

Figure 4. Score plot of principal component analysis of VOC emission evolution associated with the first three days of AF, separately represented for each trial (Table 1). Continuous lines indicate grape must and broken lines indicate grape juice. For a detailed view of the figure, the original image is included in the Supplementary Materials (Figure S3).

4. Discussion

Wine is a peculiar commodity in the agrifood sector in terms of business opportunities and innovative trends [47]. The management of AF deeply affects the optimization of the product quality and the improvement of process sustainability [48–52]. Several variables can influence the performance of AF such as grape variety, yeast species, yeast strain, nutrient availability for the yeasts, temperature of the process, addition of chemical compounds, and technological regimen [49,53–55]. The fact that these variables are often intimately connected leads to a huge 'oenological space' that needs to be explored. This observation explains the high interest in approaches tested to monitor this bioprocess [53,56] including those using VOCs as target molecules [57–61]. Furthermore, it is important to underline that the study of VOC diversity has a dual significance; on one hand, VOC variability is the effect of yeast metabolism, on the other, VOCs represent the molecular basis of the sensory perception of wine tasting [1,62,63]. Among the DIMS techniques, no study has delved into the survey of the untargeted diversity of volatiles associated with the wine headspace in order to (i) monitor online the progress of AF, and (ii) evaluate the impact of the different variables of wine quality [4].

As a case study, the assessment of VOCs variability associated with different combinations of *Saccharomyces*/non-*Saccharomyces* was selected. In fermented beverages such as wine, a relevant field of study deals with the contribution of microbiological resources to the organoleptic and sensory properties of the final product [64,65]. In the winemaking process, some of the most characteristic flavor and aroma components are synthesized by yeasts during the AF [66]. *S. cerevisiae* is the main responsible microorganism of the AF in wine, but nowadays, non-*Saccharomyces* yeasts are used in industry to improve flavor, aroma, and stability [16,22,64,65]. This heterogeneous class of eukaryotic microorganisms detains a wide enzymatic diversity [67–69]. In this light, it appears comprehensible the interest in the formation of volatile compounds by both *Saccharomyces* and non-*Saccharomyces* yeasts, which are important to maximize the sensorial quality of the final products. The present study, in particular, tested the PTR-ToF-MS-based approach recently optimized to compare the performance of different yeasts in cultural media [35]. This technology has been successfully employed in fermented foods and beverages to monitor the effect of different microorganisms responsible for the fermentative process, for instance, to discriminate wines inoculated with different malolactic starters [39], monitor lactic fermentation driven by different yoghurt commercial starter cultures [33], and characterize single commercial yeast starters in bread productions [36]. This study proposed the first application of PTR-ToF-MS for the AF monitoring in wine, demonstrating the high potential of this analytical approach to explore the huge number of variables influencing this bioprocess crucial in winemaking.

Concerning the yeast population kinetics, the respective inoculation of non-*Saccharomyces* and *Saccharomyces* strains to promote/drive alcoholic fermentation in wine were generally performed (i) by inoculating together the strains (generally with a ratio 100:1 in favor of the non-*Saccharomyces* strain) (simultaneous inoculation) or (ii) inoculating the *Saccharomyces* strain with a delay of 24–48 h compared to the non-*Saccharomyces* inoculation (sequential inoculation) [70–72]. Both approaches aimed to maximize the development of non-*Saccharomyces*, concretizing an advantage for these yeasts [73]. The oenological objective is to counteract the fermentative advantage of *S. cerevisiae*, allowing non-*Saccharomyces* yeasts to influence wine quality [74]. The findings reported in the present article suggest that simultaneous inoculation led to good growth/survival for the tested non-*Saccharomyces* in combination with the selected *S. cerevisiae* strain. Cell concentration remained particularly high for *T. delbrueckii*, confirming the ability of this non-*Saccharomyces* yeast to survive at high ethanol concentrations [75]. For *M. pulcherrima*, the evidence was only partially in accordance to that reported by Dutraive et al. [17], who observed an initial decline of this yeast between the second and the third day after the inoculation, but followed by the complete annulment of the population.

The analysis included both commercial grape juice and fresh grape must to test the efficacy of the technique both in model conditions and in the real winemaking conditions. In fact, commercial grape juice, together with synthetic grape must [76] represents a common model medium for the fermentative studies in oenology (e.g., [77,78]). We found an evolution of volatiles during the three days of the study, which was in accordance with the evolution of yeast cell counts carried out during the AF. The results reported from the analysis of 'volatomes' associated with the development of single yeast species depicted different trends that could be coherent with different claimed aromatic properties for three commercialized strains, which received a considerable interest in the scientific literature (*S. cerevisiae* DV10, e.g., [79–83]; *M. pulcherrima* FLAVIA, e.g., [17,19,84–87]; *T. delbrueckii* BIODIVA, e.g., [17,84–88]). The study highlighted a global separation of VOC variability associated with the headspaces of the two tested matrices that can be ascribable to the chemical differences and/or to variable microbiological properties of the two media. Interesting, the behavior VOCs released by *M. pulcherrima* FLAVIA radically changed, shifting from fresh must to commercial juice, meaning that chemical/microbiological determinants of these media can directly or indirectly modulate VOC production by the yeasts. Even if the effect of abiotic and biotic interaction in the wine environment have been extensively investigated [89–91], further studies are needed to understand the biology affecting this phenotype, particularly in light of the huge intraspecific variability in terms of oenological properties within the species *M. pulcherrima* [92]. Intriguingly, variable patterns in must versus juice have also been observed in the trials where *M. pulcherrima* was co-inoculated with *S. cerevisiae* and with *T. delbrueckii*. A few studies have delved into the compatibility of *S. cerevisiae* combined in the same vinification with more than one non-*Saccharomyces* species [69,93]. Except for the coupled *Lachancea thermotolerans* and *Schizosaccharomyces pombe* that (used both in combination but not with also *S. cerevisiae*) has been extensively explored [94–96], only one study has tested the sensory impact (but not the VOC analysis) of this non-*Saccharomyces* multiple inoculation in wine [69]. In fact, usually, the articles considered the impact on volatile diversity of single strains or mixed starters composed of one *S. cerevisiae* strain and one non-*Saccharomyces* species. While the effect of multiple *Saccharomyces* yeast co-inoculations on volatile wine composition has been assessed (yeast inocula differed substantially in volatile thiols and other flavor compounds) [97,98], the interactions among different non-*Saccharomyces* wine yeast species need to be further elucidated [69]. The present findings suggest that the addition of *M. pulcherrima* to the coupled (in the case of the strains we tested) *S. cerevisiae/T. delbrueckii* can modify the global release of volatiles during the AF in wine as a function of the fermented matrix.

The different behaviors of the 'volatomes' associated with the single and mixed cultures showed promising results in terms of variability of the single mass peaks. The study of individual peak mass profiles during the three first days of AF in association with the tested yeast combinations will be the natural follow-up of the present communication. The objective will be focused to elucidate the single mass peaks/molecules responsible for the strain/species-specific differences and the specific yeast interactions/combinations, but also for the selection of candidate 'volatile' markers for the rapid screening of new microbial resources for 'flavoring' starter culture design in wine fermentations. Some findings have corroborated the evidence that the complexity of the microbial starter cultures inoculated can be among the levers capable of improving sensory wine complexity, assuring the safety of the productions (also for the possible exploitation in terms of biocontrol activity) [12,99–101]. It is interesting to underline that the tested strategy could find an application also in testing the interaction of yeast with malolactic bacteria [102–105]. Furthermore, it is important to stress how the proposed exploration of the phenotypic space of yeast activity in oenology can open new research lines for fundamental research in the field of yeast biology [106–108].

5. Conclusions

PTR-ToF-MS, combining high sensitivity/accuracy without neither sample preparation nor sample destruction, allows rapid real-time determination of volatile organic compounds (VOCs). In this paper, preliminary findings on the application of this analytical approach for the online monitoring of alcoholic fermentation in wine is proposed. The study explored different single and multiple inoculation of diverse oenological yeasts both in commercial grape juice and fresh must. The experiment highlighted a variability of the global volatiles in association with (i) the different yeast species, (ii) the different yeast combinations, and (iii) the different fermenting matrices. The evidence demonstrates the potential of PTR-ToF-MS in monitoring experimental variables associated with alcoholic fermentation in wine, opening new opportunities to manage this crucial phase, thus improving the quality of the final products and optimizing the processes.

Supplementary Materials: The following are available online at http://www.mdpi.com/2311-5637/6/2/55/s1, Figure S1: source file for Figure 2. Figure S2: source file for Figure 3. Figure S3: source file for Figure 4.

Author Contributions: Conceptualization, C.B., I.K., P.R., G.S., M.F., F.B., and V.C.; Methodology, C.B., I.K., P.R., G.S., M.F., F.B., and V.C.; Investigation, C.B., I.K., P.R., M.F., F.B., and V.C.; Resources, G.S., F.B., and V.C.; Data curation, C.B., I.K., F.B., and V.C.; Writing—original draft preparation, C.B., I.K., and V.C.; Writing—review and editing, P.R., G.S., M.F., and F.B.; Supervision, G.S. and F.B.; Project administration, F.B. and V.C.; Funding acquisition, F.B. and V.C. All authors have read and agreed to the published version of the manuscript.

Funding: Pasquale Russo is the beneficiary of a grant by MIUR in the framework of 'AIM: Attraction and International Mobility' (PON R&I 2014–2020) (practice code D74I18000190001). The work has been partially supported by the Autonomous Province of Trento (ADP 2020).

Acknowledgments: We would like to thank (i) the three anonymous reviewers for their suggestions and comments, (ii) Francesco De Marzo and Massimo Franchi of the Institute of Sciences of Food Production—CNR for their skilled technical support provided during the realization of this work, and (iii) Sergio Pelosi of the Institute of Sciences of Food Production—CNR for their critical reading of the manuscript.

Conflicts of Interest: The authors declare no conflicts of interest.

References

1. Belda, I.; Ruiz, J.; Esteban-Fernández, A.; Navascués, E.; Marquina, D.; Santos, A.; Moreno-Arribas, M.V. Microbial Contribution to Wine Aroma and Its Intended Use for Wine Quality Improvement. *Mol. Basel Switz.* **2017**, *22*, 189. [CrossRef] [PubMed]
2. Benito, Á.; Calderón, F.; Benito, S. The Influence of Non-*Saccharomyces* Species on Wine Fermentation Quality Parameters. *Fermentation* **2019**, *5*, 54. [CrossRef]
3. Lebrón-Aguilar, R.; Soria, A.C.; Quintanilla-López, J.E. Comprehensive evaluation of direct injection mass spectrometry for the quantitative profiling of volatiles in food samples. *Philos. Trans. R. Soc. Math. Phys. Eng. Sci.* **2016**, *374*, 20150375. [CrossRef]
4. Pinu, F.R. Grape and Wine Metabolomics to Develop New Insights Using Untargeted and Targeted Approaches. *Fermentation* **2018**, *4*, 92. [CrossRef]
5. Tufariello, M.; Maiorano, G.; Rampino, P.; Spano, G.; Grieco, F.; Perrotta, C.; Capozzi, V.; Grieco, F. Selection of an autochthonous yeast starter culture for industrial production of Primitivo "Gioia del Colle" PDO/DOC in Apulia (Southern Italy). *LWT* **2019**, *99*, 188–196. [CrossRef]
6. Garofalo, C.; Tristezza, M.; Grieco, F.; Spano, G.; Capozzi, V. From grape berries to wine: Population dynamics of cultivable yeasts associated to "Nero di Troia" autochthonous grape cultivar. *World J. Microbiol. Biotechnol.* **2016**, *32*, 59. [CrossRef]
7. Capozzi, V.; Garofalo, C.; Chiriatti, M.A.; Grieco, F.; Spano, G. Microbial terroir and food innovation: The case of yeast biodiversity in wine. *Microbiol. Res.* **2015**, *181*, 75–83. [CrossRef] [PubMed]
8. Ciani, M.; Comitini, F.; Mannazzu, I.; Domizio, P. Controlled mixed culture fermentation: A new perspective on the use of non-*Saccharomyces* yeasts in winemaking. *FEMS Yeast Res.* **2010**, *10*, 123–133. [CrossRef] [PubMed]
9. Comitini, F.; Gobbi, M.; Domizio, P.; Romani, C.; Lencioni, L.; Mannazzu, I.; Ciani, M. Selected non-*Saccharomyces* wine yeasts in controlled multistarter fermentations with *Saccharomyces cerevisiae*. *Food Microbiol.* **2011**, *28*, 873–882. [CrossRef]
10. Tronchoni, J.; Curiel, J.A.; Morales, P.; Torres-Pérez, R.; Gonzalez, R. Early transcriptional response to biotic stress in mixed starter fermentations involving *Saccharomyces cerevisiae* and *Torulaspora delbrueckii*. *Int. J. Food Microbiol.* **2017**, *241*, 60–68. [CrossRef]
11. Ciani, M.; Morales, P.; Comitini, F.; Tronchoni, J.; Canonico, L.; Curiel, J.A.; Oro, L.; Rodrigues, A.J.; Gonzalez, R. Non-conventional Yeast Species for Lowering Ethanol Content of Wines. *Front. Microbiol.* **2016**, *7*, 642. [CrossRef] [PubMed]
12. Rossouw, D.; Bauer, F.F. Exploring the phenotypic space of non-*Saccharomyces* wine yeast biodiversity. *Food Microbiol.* **2016**, *55*, 32–46. [CrossRef] [PubMed]

13. Beckner Whitener, M.E.; Stanstrup, J.; Panzeri, V.; Carlin, S.; Divol, B.; Du Toit, M.; Vrhovsek, U. Untangling the wine metabolome by combining untargeted SPME–GCxGC-TOF-MS and sensory analysis to profile Sauvignon blanc co-fermented with seven different yeasts. *Metabolomics* **2016**, *12*, 53. [CrossRef]
14. Lu, Y.; Huang, D.; Lee, P.R.; Liu, S.Q. Assessment of volatile and non-volatile compounds in durian wines fermented with four commercial non-*Saccharomyces* yeasts. *J. Sci. Food Agric.* **2016**, *96*, 1511–1521. [CrossRef]
15. Medina, K.; Boido, E.; Dellacassa, E.; Carrau, F. Growth of non-Saccharomyces yeasts affects nutrient availability for *Saccharomyces cerevisiae* during wine fermentation. *Int. J. Food Microbiol.* **2012**, *157*, 245–250. [CrossRef]
16. Petruzzi, L.; Capozzi, V.; Berbegal, C.; Corbo, M.R.; Bevilacqua, A.; Spano, G.; Sinigaglia, M. Microbial Resources and Enological Significance: Opportunities and Benefits. *Front. Microbiol.* **2017**, *8*, 995. [CrossRef]
17. Dutraive, O.; Benito, S.; Fritsch, S.; Beisert, B.; Patz, C.D.; Rauhut, D. Effect of Sequential Inoculation with Non-*Saccharomyces* and *Saccharomyces* Yeasts on Riesling Wine Chemical Composition. *Fermentation* **2019**, *5*, 79. [CrossRef]
18. Du Plessis, H.; Du Toit, M.; Nieuwoudt, H.; Van der Rijst, M.; Hoff, J.; Jolly, N. Modulation of Wine Flavor using Hanseniaspora uvarum in Combination with Different *Saccharomyces cerevisiae*, Lactic Acid Bacteria Strains and Malolactic Fermentation Strategies. *Fermentation* **2019**, *5*, 64. [CrossRef]
19. Morata, A.; Loira, I.; Escott, C.; del Fresno, J.M.; Bañuelos, M.A.; Suárez-Lepe, J.A. Applications of *Metschnikowia pulcherrima* in Wine Biotechnology. *Fermentation* **2019**, *5*, 63. [CrossRef]
20. Garofalo, C.; Russo, P.; Beneduce, L.; Massa, S.; Spano, G.; Capozzi, V. Non-*Saccharomyces* biodiversity in wine and the 'microbial terroir': A survey on Nero di Troia wine from the Apulian region, Italy. *Ann. Microbiol.* **2016**, *66*, 143–150. [CrossRef]
21. Sgouros, G.; Chalvantzi, I.; Mallouchos, A.; Paraskevopoulos, Y.; Banilas, G.; Nisiotou, A. Biodiversity and Enological Potential of Non-*Saccharomyces* Yeasts from Nemean Vineyards. *Fermentation* **2018**, *4*, 32. [CrossRef]
22. Roudil, L.; Russo, P.; Berbegal, C.; Albertin, W.; Spano, G.; Capozzi, V. Non-*Saccharomyces* Commercial Starter Cultures: Scientific Trends, Recent Patents and Innovation in the Wine Sector. *Recent Pat. Food Nutr. Agric.* **2019**. [CrossRef] [PubMed]
23. Lindinger, W.; Jordan, A. Proton-transfer-reaction mass spectrometry (PTR–MS): On-line monitoring of volatile organic compounds at pptv levels. *Chem. Soc. Rev.* **1998**, *27*, 347–375. [CrossRef]
24. Jordan, A.; Haidacher, S.; Hanel, G.; Hartungen, E.; Märk, L.; Seehauser, H.; Schottkowsky, R.; Sulzer, P.; Märk, T.D. A high resolution and high sensitivity proton-transfer-reaction time-of-flight mass spectrometer (PTR-TOF-MS). *Int. J. Mass Spectrom.* **2009**, *286*, 122–128. [CrossRef]
25. Capozzi, V.; Yener, S.; Khomenko, I.; Farneti, B.; Cappellin, L.; Gasperi, F.; Scampicchio, M.; Biasioli, F. PTR-ToF-MS Coupled with an Automated Sampling System and Tailored Data Analysis for Food Studies: Bioprocess Monitoring, Screening and Nose-space Analysis. *JoVE J. Vis. Exp.* **2017**, e54075. [CrossRef]
26. Cappellin, L.; Loreto, F.; Aprea, E.; Romano, A.; Del Pulgar, J.S.; Gasperi, F.; Biasioli, F. PTR-MS in Italy: A Multipurpose Sensor with Applications in Environmental, Agri-Food and Health Science. *Sensors* **2013**, *13*, 11923–11955. [CrossRef]
27. Pico, J.; Khomenko, I.; Capozzi, V.; Navarini, L.; Bernal, J.; Gómez, M.; Biasioli, F. Analysis of volatile organic compounds in crumb and crust of different baked and toasted gluten-free breads by direct PTR-ToF-MS and fast-GC-PTR-ToF-MS. *J. Mass Spectrom.* **2018**, *53*, 893–902. [CrossRef]
28. Deuscher, Z.; Andriot, I.; Sémon, E.; Repoux, M.; Preys, S.; Roger, J.M.; Boulanger, R.; Labouré, H.; Quéré, J.L.L. Volatile compounds profiling by using proton transfer reaction-time of flight-mass spectrometry (PTR-ToF-MS). The case study of dark chocolates organoleptic differences. *J. Mass Spectrom.* **2019**, *54*, 92–119. [CrossRef]
29. van Ruth, S.M.; Koot, A.; Akkermans, W.; Araghipour, N.; Rozijn, M.; Baltussen, M.; Wisthaler, A.; Märk, T.D.; Frankhuizen, R. Butter and butter oil classification by PTR-MS. *Eur. Food Res. Technol.* **2008**, *227*, 307–317. [CrossRef]
30. Capozzi, V.; Lonzarich, V.; Khomenko, I.; Cappellin, L.; Navarini, L.; Biasioli, F. Unveiling the Molecular Basis of Mascarpone Cheese Aroma: VOCs analysis by SPME-GC/MS and PTR-ToF-MS. *Molecules* **2020**, *25*, 1242. [CrossRef]

31. Makhoul, S.; Yener, S.; Khomenko, I.; Capozzi, V.; Cappellin, L.; Aprea, E.; Scampicchio, M.; Gasperi, F.; Biasioli, F. Rapid non-invasive quality control of semi-finished products for the food industry by direct injection mass spectrometry headspace analysis: The case of milk powder, whey powder and anhydrous milk fat. *J. Mass Spectrom.* **2016**, *51*, 782–791. [CrossRef] [PubMed]
32. Romano, A.; Capozzi, V.; Spano, G.; Biasioli, F. Proton transfer reaction–mass spectrometry: Online and rapid determination of volatile organic compounds of microbial origin. *Appl. Microbiol. Biotechnol.* **2015**, *99*, 3787–3795. [CrossRef] [PubMed]
33. Benozzi, E.; Romano, A.; Capozzi, V.; Makhoul, S.; Cappellin, L.; Khomenko, I.; Aprea, E.; Scampicchio, M.; Spano, G.; Märk, T.D.; et al. Monitoring of lactic fermentation driven by different starter cultures via direct injection mass spectrometric analysis of flavour-related volatile compounds. *Food Res. Int.* **2015**, *76*, 682–688. [CrossRef] [PubMed]
34. Richter, T.M.; Silcock, P.; Algarra, A.; Eyres, G.T.; Capozzi, V.; Bremer, P.J.; Biasioli, F. Evaluation of PTR-ToF-MS as a tool to track the behavior of hop-derived compounds during the fermentation of beer. *Food Res. Int.* **2018**, *111*, 582–589. [CrossRef] [PubMed]
35. Khomenko, I.; Stefanini, I.; Cappellin, L.; Cappelletti, V.; Franceschi, P.; Cavalieri, D.; Märk, T.D.; Biasioli, F. Non-invasive real time monitoring of yeast volatilome by PTR-ToF-MS. *Metab. Off. J. Metab. Soc.* **2017**, *13*, 118. [CrossRef]
36. Capozzi, V.; Makhoul, S.; Aprea, E.; Romano, A.; Cappellin, L.; Sanchez Jimena, A.; Spano, G.; Gasperi, F.; Scampicchio, M.; Biasioli, F. PTR-MS Characterization of VOCs Associated with Commercial Aromatic Bakery Yeasts of Wine and Beer Origin. *Molecules* **2016**, *21*, 483. [CrossRef]
37. Makhoul, S.; Romano, A.; Capozzi, V.; Spano, G.; Aprea, E.; Cappellin, L.; Benozzi, E.; Scampicchio, M.; Märk, T.D.; Gasperi, F.; et al. Volatile Compound Production During the Bread-Making Process: Effect of Flour, Yeast and Their Interaction. *Food Bioprocess Technol.* **2015**, *8*, 1925–1937. [CrossRef]
38. Makhoul, S.; Romano, A.; Cappellin, L.; Spano, G.; Capozzi, V.; Benozzi, E.; Märk, T.D.; Aprea, E.; Gasperi, F.; El-Nakat, H.; et al. Proton-transfer-reaction mass spectrometry for the study of the production of volatile compounds by bakery yeast starters. *J. Mass Spectrom. JMS* **2014**, *49*, 850–859. [CrossRef]
39. Campbell-Sills, H.; Capozzi, V.; Romano, A.; Cappellin, L.; Spano, G.; Breniaux, M.; Lucas, P.; Biasioli, F. Advances in wine analysis by PTR-ToF-MS: Optimization of the method and discrimination of wines from different geographical origins and fermented with different malolactic starters. *Int. J. Mass Spectrom.* **2016**, *397–398*, 42–51. [CrossRef]
40. Romano, A.; Fischer, L.; Herbig, J.; Campbell-Sills, H.; Coulon, J.; Lucas, P.; Cappellin, L.; Biasioli, F. Wine analysis by FastGC proton-transfer reaction-time-of-flight-mass spectrometry. *Int. J. Mass Spectrom.* **2014**, *369*, 81–86. [CrossRef]
41. Sémon, E.; Arvisenet, G.; Guichard, E.; Quéré, J.L.L. Modified proton transfer reaction mass spectrometry (PTR-MS) operating conditions for *in vitro* and *in vivo* analysis of wine aroma. *J. Mass Spectrom.* **2018**, *53*, 65–77. [CrossRef] [PubMed]
42. Muñoz-González, C.; Canon, F.; Feron, G.; Guichard, E.; Pozo-Bayón, M.A. Assessment Wine Aroma Persistence by Using an *in Vivo* PTR-ToF-MS Approach and Its Relationship with Salivary Parameters. *Mol. Basel Switz.* **2019**, *24*, 1277. [CrossRef] [PubMed]
43. Garofalo, C.; El Khoury, M.; Lucas, P.; Bely, M.; Russo, P.; Spano, G.; Capozzi, V. Autochthonous starter cultures and indigenous grape variety for regional wine production. *J. Appl. Microbiol.* **2015**, *118*, 1395–1408. [CrossRef] [PubMed]
44. Cappellin, L.; Biasioli, F.; Fabris, A.; Schuhfried, E.; Soukoulis, C.; Märk, T.D.; Gasperi, F. Improved mass accuracy in PTR-TOF-MS: Another step towards better compound identification in PTR-MS. *Int. J. Mass Spectrom.* **2010**, *290*, 60–63. [CrossRef]
45. Cappellin, L.; Biasioli, F.; Granitto, P.M.; Schuhfried, E.; Soukoulis, C.; Costa, F.; Märk, T.D.; Gasperi, F. On data analysis in PTR-TOF-MS: From raw spectra to data mining. *Sens. Actuators B Chem.* **2011**, *155*, 183–190. [CrossRef]
46. Lindinger, W.; Hansel, A.; Jordan, A. On-line monitoring of volatile organic compounds at pptv levels by means of proton-transfer-reaction mass spectrometry (PTR-MS) medical applications, food control and environmental research. *Int. J. Mass Spectrom. Ion Process.* **1998**, *173*, 191–241. [CrossRef]
47. Bisson, L.F.; Waterhouse, A.L.; Ebeler, S.E.; Walker, M.A.; Lapsley, J.T. The present and future of the international wine industry. *Nature* **2002**, *418*, 696–699. [CrossRef]

48. Berbegal, C.; Fragasso, M.; Russo, P.; Bimbo, F.; Grieco, F.; Spano, G.; Capozzi, V. Climate Changes and Food Quality: The Potential of Microbial Activities as Mitigating Strategies in the Wine Sector. *Fermentation* **2019**, *5*, 85. [CrossRef]
49. Bisson, L.F. Stuck and Sluggish Fermentations. *Am. J. Enol. Vitic.* **1999**, *50*, 107–119.
50. Fleet, G.H. Wine yeasts for the future. *FEMS Yeast Res.* **2008**, *8*, 979–995. [CrossRef]
51. Pretorius, I.S. Tailoring wine yeast for the new millennium: Novel approaches to the ancient art of winemaking. *Yeast* **2000**, *16*, 675–729. [CrossRef]
52. Nardi, T. Microbial Resources as a Tool for Enhancing Sustainability in Winemaking. *Microorganisms* **2020**, *8*, 507. [CrossRef]
53. Sablayrolles, J.M. Control of alcoholic fermentation in winemaking: Current situation and prospect. *Food Res. Int.* **2009**, *42*, 418–424. [CrossRef]
54. Molina, A.M.; Swiegers, J.H.; Varela, C.; Pretorius, I.S.; Agosin, E. Influence of wine fermentation temperature on the synthesis of yeast-derived volatile aroma compounds. *Appl. Microbiol. Biotechnol.* **2007**, *77*, 675–687. [CrossRef] [PubMed]
55. Russo, P.; Berbegal, C.; De Ceglie, C.; Grieco, F.; Spano, G.; Capozzi, V. Pesticide Residues and Stuck Fermentation in Wine: New Evidences Indicate the Urgent Need of Tailored Regulations. *Fermentation* **2019**, *5*, 23. [CrossRef]
56. Di Egidio, V.; Sinelli, N.; Giovanelli, G.; Moles, A.; Casiraghi, E. NIR and MIR spectroscopy as rapid methods to monitor red wine fermentation. *Eur. Food Res. Technol.* **2010**, *230*, 947–955. [CrossRef]
57. Mouret, J.R.; Perez, M.; Angenieux, M.; Nicolle, P.; Farines, V.; Sablayrolles, J.M. Online-Based Kinetic Analysis of Higher Alcohol and Ester Synthesis During Winemaking Fermentations. *Food Bioprocess Technol.* **2014**, *7*, 1235–1245. [CrossRef]
58. Mallouchos, A.; Komaitis, M.; Koutinas, A.; Kanellaki, M. Investigation of Volatiles Evolution during the Alcoholic Fermentation of Grape Must Using Free and Immobilized Cells with the Help of Solid Phase Microextraction (SPME) Headspace Sampling. *J. Agric. Food Chem.* **2002**, *50*, 3840–3848. [CrossRef]
59. Morales, M.L.; Fierro-Risco, J.; Callejón, R.M.; Paneque, P. Monitoring volatile compounds production throughout fermentation by *Saccharomyces* and non-*Saccharomyces* strains using headspace sorptive extraction. *J. Food Sci. Technol.* **2017**, *54*, 538–557. [CrossRef]
60. Silva Ferreira, A.C.; Monforte, A.R.; Teixeira, C.S.; Martins, R.; Fairbairn, S.; Bauer, F.F. Monitoring Alcoholic Fermentation: An Untargeted Approach. *J. Agric. Food Chem.* **2014**, *62*, 6784–6793. [CrossRef]
61. Buratti, S.; Benedetti, S. Chapter 28-Alcoholic Fermentation Using Electronic Nose and Electronic Tongue. In *Electronic Noses and Tongues in Food Science*; Rodríguez Méndez, M.L., Ed.; Academic Press: San Diego, CA, USA, 2016; pp. 291–299, ISBN 978-0-12-800243-8.
62. Aguilera, T.; Lozano, J.; Paredes, J.A.; Álvarez, F.J.; Suárez, J.I. Electronic Nose Based on Independent Component Analysis Combined with Partial Least Squares and Artificial Neural Networks for Wine Prediction. *Sensors* **2012**, *12*, 8055–8072. [CrossRef] [PubMed]
63. Jiang, B.; Zhang, Z. Volatile Compounds of Young Wines from Cabernet Sauvignon, Cabernet Gernischet and Chardonnay Varieties Grown in the Loess Plateau Region of China. *Molecules* **2010**, *15*, 9184–9196. [CrossRef] [PubMed]
64. Jolly, N.P.; Varela, C.; Pretorius, I.S. Not your ordinary yeast: Non-*Saccharomyces* yeasts in wine production uncovered. *FEMS Yeast Res.* **2014**, *14*, 215–237. [CrossRef] [PubMed]
65. Berbegal, C.; Spano, G.; Tristezza, M.; Grieco, F.; Capozzi, V. Microbial Resources and Innovation in the Wine Production Sector. *S. Afr. J. Enol. Vitic.* **2017**, *38*, 156–166. [CrossRef]
66. Lambrechts, M.G.; Pretorius, I.S. Yeast and its Importance to Wine Aroma—A Review. *S. Afr. J. Enol. Vitic.* **2000**, *21*, 97–129. [CrossRef]
67. Escribano, R.; González-Arenzana, L.; Garijo, P.; Berlanas, C.; López-Alfaro, I.; López, R.; Gutiérrez, A.R.; Santamaría, P. Screening of enzymatic activities within different enological non-*Saccharomyces* yeasts. *J. Food Sci. Technol.* **2017**, *54*, 1555–1564. [CrossRef]
68. Escribano, R.; González-Arenzana, L.; Portu, J.; Garijo, P.; López-Alfaro, I.; López, R.; Santamaría, P.; Gutiérrez, A.R. Wine aromatic compound production and fermentative behaviour within different non-*Saccharomyces* species and clones. *J. Appl. Microbiol.* **2018**, *124*, 1521–1531. [CrossRef]

69. Padilla, B.; Gil, J.V.; Manzanares, P. Past and Future of Non-*Saccharomyces* Yeasts: From Spoilage Microorganisms to Biotechnological Tools for Improving Wine Aroma Complexity. *Front. Microbiol.* **2016**, *7*, 411. [CrossRef]
70. Ciani, M.; Beco, L.; Comitini, F. Fermentation behaviour and metabolic interactions of multistarter wine yeast fermentations. *Int. J. Food Microbiol.* **2006**, *108*, 239–245. [CrossRef]
71. Taillandier, P.; Lai, Q.P.; Julien-Ortiz, A.; Brandam, C. Interactions between *Torulaspora delbrueckii* and *Saccharomyces cerevisiae* in wine fermentation: Influence of inoculation and nitrogen content. *World J. Microbiol. Biotechnol.* **2014**, *30*, 1959–1967. [CrossRef]
72. Contreras, A.; Hidalgo, C.; Henschke, P.A.; Chambers, P.J.; Curtin, C.; Varela, C. Evaluation of Non-*Saccharomyces* Yeasts for the Reduction of Alcohol Content in Wine. *Appl. Environ. Microbiol.* **2014**, *80*, 1670–1678. [CrossRef] [PubMed]
73. Tristezza, M.; Tufariello, M.; Capozzi, V.; Spano, G.; Mita, G.; Grieco, F. The Oenological Potential of *Hanseniaspora uvarum* in Simultaneous and Sequential Co-fermentation with *Saccharomyces cerevisiae* for Industrial Wine Production. *Front. Microbiol.* **2016**, *7*. [CrossRef] [PubMed]
74. Russo, P.; Tufariello, M.; Renna, R.; Tristezza, M.; Taurino, M.; Palombi, L.; Capozzi, V.; Rizzello, C.G.; Grieco, F. New Insights into the Oenological Significance of *Candida zemplinina*: Impact of Selected Autochthonous Strains on the Volatile Profile of Apulian Wines. *Microorganisms* **2020**, *8*, 628. [CrossRef]
75. González-Royo, E.; Pascual, O.; Kontoudakis, N.; Esteruelas, M.; Esteve-Zarzoso, B.; Mas, A.; Canals, J.M.; Zamora, F. Oenological consequences of sequential inoculation with non-*Saccharomyces* yeasts (*Torulaspora delbrueckii* or *Metschnikowia pulcherrima*) and *Saccharomyces cerevisiae* in base wine for sparkling wine production. *Eur. Food Res. Technol.* **2015**, *240*, 999–1012. [CrossRef]
76. Renault, P.E.; Albertin, W.; Bely, M. An innovative tool reveals interaction mechanisms among yeast populations under oenological conditions. *Appl. Microbiol. Biotechnol.* **2013**, *97*, 4105–4119. [CrossRef]
77. Gobert, A.; Tourdot-Maréchal, R.; Morge, C.; Sparrow, C.; Liu, Y.; Quintanilla-Casas, B.; Vichi, S.; Alexandre, H. Non-*Saccharomyces* Yeasts Nitrogen Source Preferences: Impact on Sequential Fermentation and Wine Volatile Compounds Profile. *Front. Microbiol.* **2017**, *8*. [CrossRef]
78. Hansen, E.H.; Nissen, P.; Sommer, P.; Nielsen, J.C.; Arneborg, N. The effect of oxygen on the survival of non-*Saccharomyces* yeasts during mixed culture fermentations of grape juice with Saccharomyces cerevisiae. *J. Appl. Microbiol.* **2001**, *91*, 541–547. [CrossRef]
79. Vasserot, Y.; Mornet, F.; Jeandet, P. Acetic acid removal by *Saccharomyces cerevisiae* during fermentation in oenological conditions. Metabolic consequences. *Food Chem.* **2010**, *119*, 1220–1223. [CrossRef]
80. Jackowetz, J.N.; Dierschke, S.; Mira de Orduña, R. Multifactorial analysis of acetaldehyde kinetics during alcoholic fermentation by *Saccharomyces cerevisiae*. *Food Res. Int.* **2011**, *44*, 310–316. [CrossRef]
81. Legras, J.L.; Karst, F. Optimisation of interdelta analysis for *Saccharomyces cerevisiae* strain characterization. *FEMS Microbiol. Lett.* **2003**, *221*, 249–255. [CrossRef]
82. Garofalo, C.; Berbegal, C.; Grieco, F.; Tufariello, M.; Spano, G.; Capozzi, V. Selection of indigenous yeast strains for the production of sparkling wines from native Apulian grape varieties. *Int. J. Food Microbiol.* **2018**, *285*, 7–17. [CrossRef] [PubMed]
83. Jin, X.; Chen, W.; Chen, H.; Chen, W.; Zhong, Q. Combination of *Lactobacillus plantarum* and *Saccharomyces cerevisiae* DV10 as Starter Culture to Produce Mango Slurry: Microbiological, Chemical Parameters and Antioxidant Activity. *Molecules* **2019**, *24*, 4349. [CrossRef] [PubMed]
84. Seguinot, P.; Bloem, A.; Brial, P.; Meudec, E.; Ortiz-Julien, A.; Camarasa, C. Analyzing the impact of the nature of the nitrogen source on the formation of volatile compounds to unravel the aroma metabolism of two non-*Saccharomyces* strains. *Int. J. Food Microbiol.* **2020**, *316*, 108441. [CrossRef] [PubMed]
85. Chua, J.Y.; Lu, Y.; Liu, S.Q. Evaluation of five commercial non-*Saccharomyces* yeasts in fermentation of soy (tofu) whey into an alcoholic beverage. *Food Microbiol.* **2018**, *76*, 533–542. [CrossRef]
86. Prior, K.J.; Bauer, F.F.; Divol, B. The utilization of nitrogenous compounds by commercial non-*Saccharomyces* yeasts associated with wine. *Food Microbiol.* **2019**, *79*, 75–84. [CrossRef]
87. Ivit, N.N.; Kemp, B. The Impact of Non-*Saccharomyces* Yeast on Traditional Method Sparkling Wine. *Fermentation* **2018**, *4*, 73. [CrossRef]

88. Nardi, T.; Panero, L.; Petrozziello, M.; Guaita, M.; Tsolakis, C.; Cassino, C.; Vagnoli, P.; Bosso, A. Managing wine quality using *Torulaspora delbrueckii* and *Oenococcus oeni* starters in mixed fermentations of a red Barbera wine. *Eur. Food Res. Technol.* **2019**, *245*, 293–307. [CrossRef]
89. Chandra, M.; Oro, I.; Ferreira-Dias, S.; Malfeito-Ferreira, M. Effect of Ethanol, Sulfur Dioxide and Glucose on the Growth of Wine Spoilage Yeasts Using Response Surface Methodology. *PLoS ONE* **2015**, *10*, e0128702. [CrossRef]
90. Ciani, M.; Capece, A.; Comitini, F.; Canonico, L.; Siesto, G.; Romano, P. Yeast Interactions in Inoculated Wine Fermentation. *Front. Microbiol.* **2016**, *7*. [CrossRef]
91. Pinu, F.R.; Edwards, P.J.B.; Gardner, R.C.; Villas-Boas, S.G. Nitrogen and carbon assimilation by *Saccharomyces cerevisiae* during Sauvignon blanc juice fermentation. *FEMS Yeast Res.* **2014**, *14*, 1206–1222. [CrossRef]
92. Barbosa, C.; Lage, P.; Esteves, M.; Chambel, L.; Mendes-Faia, A.; Mendes-Ferreira, A. Molecular and Phenotypic Characterization of *Metschnikowia pulcherrima* Strains from Douro Wine Region. *Fermentation* **2018**, *4*, 8. [CrossRef]
93. Escribano-Viana, R.; Portu, J.; Garijo, P.; López, R.; Santamaría, P.; López-Alfaro, I.; Gutiérrez, A.R.; González-Arenzana, L. Effect of the Sequential Inoculation of Non-*Saccharomyces*/*Saccharomyces* on the Anthocyans and Stilbenes Composition of Tempranillo Wines. *Front. Microbiol.* **2019**, *10*. [CrossRef]
94. Benito, Á.; Calderón, F.; Palomero, F.; Benito, S. Combine Use of Selected *Schizosaccharomyces pombe* and *Lachancea thermotolerans* Yeast Strains as an Alternative to the Traditional Malolactic Fermentation in Red Wine Production. *Molecules* **2015**, *20*, 9510–9523. [CrossRef] [PubMed]
95. Benito, Á.; Calderón, F.; Benito, S. The Combined Use of *Schizosaccharomyces pombe* and *Lachancea thermotolerans*—Effect on the Anthocyanin Wine Composition. *Molecules* **2017**, *22*, 739. [CrossRef] [PubMed]
96. Benito, Á.; Calderón, F.; Benito, S. Combined Use of *S. pombe* and *L. thermotolerans* in Winemaking. Beneficial Effects Determined Through the Study of Wines' Analytical Characteristics. *Molecules* **2016**, *21*, 1744. [CrossRef] [PubMed]
97. King, E.S.; Kievit, R.L.; Curtin, C.; Swiegers, J.H.; Pretorius, I.S.; Bastian, S.E.P.; Leigh Francis, I. The effect of multiple yeasts co-inoculations on Sauvignon Blanc wine aroma composition, sensory properties and consumer preference. *Food Chem.* **2010**, *122*, 618–626. [CrossRef]
98. Gustafsson, F.S.; Jiranek, V.; Neuner, M.; Scholl, C.M.; Morgan, S.C.; Durall, D.M. The Interaction of Two *Saccharomyces* cerevisiae Strains Affects Fermentation-Derived Compounds in Wine. *Fermentation* **2016**, *2*, 9. [CrossRef]
99. Berbegal, C.; Garofalo, C.; Russo, P.; Pati, S.; Capozzi, V.; Spano, G. Use of Autochthonous Yeasts and Bacteria in Order to Control *Brettanomyces bruxellensis* in Wine. *Fermentation* **2017**, *3*, 65. [CrossRef]
100. Berbegal, C.; Spano, G.; Fragasso, M.; Grieco, F.; Russo, P.; Capozzi, V. Starter cultures as biocontrol strategy to prevent *Brettanomyces bruxellensis* proliferation in wine. *Appl. Microbiol. Biotechnol.* **2018**, *102*, 569–576. [CrossRef]
101. Capozzi, V.; Fragasso, M.; Russo, P. Microbiological Safety and the Management of Microbial Resources in Artisanal Foods and Beverages: The Need for a Transdisciplinary Assessment to Conciliate Actual Trends and Risks Avoidance. *Microorganisms* **2020**, *8*, 306. [CrossRef]
102. Berbegal, C.; Borruso, L.; Fragasso, M.; Tufariello, M.; Russo, P.; Brusetti, L.; Spano, G.; Capozzi, V. A Metagenomic-Based Approach for the Characterization of Bacterial Diversity Associated with Spontaneous Malolactic Fermentations in Wine. *Int. J. Mol. Sci.* **2019**, *20*, 3980. [CrossRef] [PubMed]
103. Englezos, V.; Cachón, D.C.; Rantsiou, K.; Blanco, P.; Petrozziello, M.; Pollon, M.; Giacosa, S.; Río Segade, S.; Rolle, L.; Cocolin, L. Effect of mixed species alcoholic fermentation on growth and malolactic activity of lactic acid bacteria. *Appl. Microbiol. Biotechnol.* **2019**, *103*, 7687–7702. [CrossRef] [PubMed]
104. Balmaseda, A.; Bordons, A.; Reguant, C.; Bautista-Gallego, J. Non-Saccharomyces in Wine: Effect Upon *Oenococcus oeni* and Malolactic Fermentation. *Front. Microbiol.* **2018**, *9*, 534. [CrossRef] [PubMed]
105. Capozzi, V.; Berbegal, C.; Tufariello, M.; Grieco, F.; Spano, G.; Grieco, F. Impact of co-inoculation of *Saccharomyces cerevisiae*, *Hanseniaspora uvarum* and *Oenococcus oeni* autochthonous strains in controlled multi starter grape must fermentations. *LWT* **2019**, *109*, 241–249. [CrossRef]
106. Willaert, R.; Kasas, S.; Devreese, B.; Dietler, G. Yeast Nanobiotechnology. *Fermentation* **2016**, *2*, 18. [CrossRef]

107. Botstein, D.; Fink, G.R. Yeast: An Experimental Organism for 21st Century Biology. *Genetics* **2011**, *189*, 695–704. [CrossRef]
108. Walker, R.S.K.; Pretorius, I.S. Applications of Yeast Synthetic Biology Geared towards the Production of Biopharmaceuticals. *Genes* **2018**, *9*, 340. [CrossRef]

© 2020 by the authors. Licensee MDPI, Basel, Switzerland. This article is an open access article distributed under the terms and conditions of the Creative Commons Attribution (CC BY) license (http://creativecommons.org/licenses/by/4.0/).

Article
Antioxidant Properties of Fermented Green Coffee Beans with *Wickerhamomyces anomalus* (Strain KNU18Y3)

Mesfin Haile [1] and Won Hee Kang [1,2,*]

[1] Department of Horticulture, Kangwon National University, Chuncheon 24341, Korea; mesfinhaile97@gmail.com
[2] Convergence Program of Coffee Science, Kangwon National University, Chuncheon 24341, Korea
* Correspondence: whkang@kangwon.ac.kr

Received: 20 November 2019; Accepted: 27 January 2020; Published: 28 January 2020

Abstract: A few yeast species have been tested frequently to improve the tastes, flavors, and other important quality parameters of coffee. However, continuing evaluations of different yeast species for fermenting green coffee beans will have a significant positive contribution to the coffee industry. This experiment was conducted to evaluate the antioxidant properties, total phenol content (TPC), total flavonoid content (TFC), total tannin content (TTC), and the consumer acceptability of fermented green coffee beans with *Wickerhamomyces anomalu*. The coffee beans were roasted at different roasting conditions (light, medium, and dark). There was no significant ($p > 0.05$) difference between the yeast-fermented and non-fermented coffee with regard to the oxygen radical absorbance capacity (ORAC) values in medium and dark roasted coffee. Similarly, the superoxide dismutase-like (SOD)-like activity did not significantly differ in all roasting conditions. However, the SOD-like activity was significantly different ($p < 0.05$), particularly within light roasted and medium roasted, and between light roasted and dark roasted in both the control and fermented coffee extracts. The 2, 2-diphenyl-1-picrylhydrazyl (DPPH) radical scavenging assay and ferric reducing antioxidant power (FRAP) were improved in fermented coffee beans. There was a significant ($p \leq 0.05$) difference between the yeast-fermented and non-fermented coffee with respect to the TPC and TFC in all roasting types and the TTC in the light and dark roasting conditions. The fermentation of green coffee beans with *W. anomalus* increased the TPC and TFC. However, the TTC was lower in the fermented coffee beans compared to the non-fermented coffee beans in medium and dark roasted coffee. In general, fermentation of green coffee beans with *W. anomalus* has the potential to improve the functionality of coffee beans.

Keywords: antioxidant; coffee; fermentation; *W. anomalus*

1. Introduction

Fermentation has the potential to improve the functionality of foods. Fermentation is primarily relied upon during the wet processing of coffee to remove mucilage. However, it is also improves coffee's sensory quality attributes [1]. Fermentation has been extensively applied in the food, chemical, and pharmaceutical industries to aid in the manufacturing, extraction, and modification of bioactive compounds [2,3]. Microbial fermentation is an interesting biotechnological processing system that can improve the total phenolic content of foods and herbs by liberating their insolubly bound phenolics and hence boost their nutritional value [4,5]. Numerous research papers provide a comparison of the antioxidant properties of popular drinks such as coffee, tea, and cocoa [6–8]. Phenolic compounds are found in a broad range of regularly consumed plant foods, such as vegetables, fruits, legumes, and cereals, and in beverages of plant origin, such as coffee, tea, and wine [9,10]. Flavonoids naturally exist in plants with a variable phenolic structure. Functional hydroxyl groups in flavonoids mediate their antioxidant effects by chelating metal ions and scavenging free radicals [11,12]. Flavonoids in food are commonly responsible for the color, the taste, the prevention of fat oxidation, and the protection of enzymes and vitamins [13].

Wickerhamomyces anomalus (formerly *Pichia anomala*) is an ascomycetous heterothallic yeast of the family *Wickerhamomycetaceae* that propagates sexually through the development of hat-shaped ascospores and asexually by budding [14]. *W. anomalus* strains are found in several environments and have been isolated from cereal grains, fruits, maize silage, wine, and high-sugar food products [14]. *W. anomalus* is categorized as a biosafety level-one organism; it is deemed harmless for healthy individuals and can grow under severe environmental stress conditions, such as high and low pH, high osmotic pressure, anaerobic conditions, and low water activity [15]. As mentioned by Comitini et al. [16], *W. anomalus* toxins have been examined as antimicrobial agents against some spoilage yeasts such as *Dekkera/Brettanomyces*. Some researchers have reported that the wild species of W. anomalus strains have the potential to use maltose and grow better than other commercial brewing yeasts [17]. However, other studies have referred to its clear inability to metabolize maltose [18]. Considering these findings, *W. anomalus* is preferred as a good starter culture in sequence or inoculated together with other yeasts in winemaking [19]. *p. anomala*-mixed starters have been evaluated to improve the final quality of cider [18]. Due to its positive effect on sensory quality, a mixed starter of W. anomalus and *Saccharomyces cerevisiae* has been proposed for making Chinese baijiu [20].

Recently, studies have been conducted to improve the functionality of coffee through green bean fermentation with selected microorganisms. In these studies, the *Saccharomyces* species [21,22], *Saccharomycopsis fibuligera* [22], *Rhizopus oligosporus* [23], and *Yarrowia lipolytica* [24], have been used to ferment green coffee beans. The fermentation of green coffee beans with R. oligosporus significantly enhanced the compositions of aroma precursors [23]. As reported by Kwak et al. [21] and Mesfin and Kang [22], the fermentation of green coffee beans improves antioxidant activity, total phenol content (TPC), and total flavonoid content (TFC), and it reduces the total tannin content (TTC), which is mostly responsible for coffee's astringency. W. anomalus strain KNU18Y3 has the ability to produce a pectinase enzyme, it was selected as a starter culture for coffee fermentation in the wet processing method [25]. Since this yeast has not been tested previously to ferment green coffee beans, this study was conducted to evaluate the antioxidant activities of fermented coffee beans with W. anomalus (strain KNU18Y3).

2. Materials and Methods

2.1. Chemicals

Disodium phosphate (Na_2HPO_4), aluminium chloride ($AlCl_3·6H_2O$), monosodium phosphate (NaH_2PO_4), sodium carbonate (Na_2CO_3), D-glucose, and sodium nitrite ($NaNO_2$) were supplied by Dae-Jung Chemicals & Metals Co., Ltd., (Jeongwang-dong, Shiheung-city, Gyeonggi-do, Republic of Korea). Pyrogallol, 2,2'-azobis (2-amidinopropane) dihydrochloride (AAPH), and fluorescein sodium were purchased from Santa Cruz Biotechnology, Inc. (Dallas, Texas, USA). NaCl, ethylenediaminetetraacetic acid, yeast peptone dextrose, gallic acid, Folin-Ciocalteu's phenol reagent, quercetin, Trolox, 2,2-diphenyl-1-picrylhydrazyl, potassium ferricyanide($C_6N_6FeK_3$), trichloroacetic acid ($C_2HCl_3O_2$), and ferric chloride ($FeCl_3$) were purchased from Sigma Aldrich LLC (St. Louis, MO, USA). Methanol was supplied by Merck KGaA (Darmstadt, Germany), and HCl was purchased from Junsei Chemical Co., Ltd. (4-4-16 Nihonbashi-honcho, Chuo-ku, Tokyo, Japan).

2.2. Fermentation of Green Coffee Beans

A freeze-dried *W. anomalus* (strain KNU18Y3) yeast cells were cultivated in a sterile yeast peptone dextrose broth (0.5% *w/v* yeast extract, 2 % *w/v* dextrose and 1% *w/v* casein peptone) for 48 h at 30 °C. The pH of the media was calibrated to 5.0 with 1 M HCl. Then, the yeast cells were collected via centrifugation at 8000× g. Finally, the collected yeast cells were washed twice using the deionized distilled water (ddH$_2$O) and mixed in a similar volume of 100 mM phosphate-buffered saline solution.

The green coffee beans (*Coffea arabica* L.) that imported from Kenya were used for this study. The coffee beans (600 g) saturated in ddH$_2$O at 4 °C for 24 h and then steamed at 80 °C for 40 min to kill native yeasts that exist on the coffee beans. Next, the steamed coffee beans then allowed to cool at 20–23 °C. The heat-treated green coffee beans were divided into two lots: fermented through inoculation (1.0 × 10^4 colony-forming unit (CFU)/g of coffee beans) and non-inoculated (control). Then, each lot was divided into three parts and underwent a different roasting treatment. Fermentation of green coffee beans and roasting were done in three replications. The fermentation was carried out for 24 h at 30 °C. After that, the coffee beans were washed thoroughly using a ddH$_2$O. The fermented coffee beans were dried using oven at 45 °C until the moisture content reached between 10% and 12%.

The pH of the fermented solution (the water that used to ferment the coffee beans) was recorded using a pH meter at 0 h and 24 h. A sample solution was taken after 24 h fermentation and serially diluted and plated onto yeast peptone dextrose agar (YPDA) plates. The plates were incubated for 24 h at 30 °C to estimate the growth of yeast cells. The pH measurements and microbial counts were made in five replications.

2.3. Coffee Roasting, Grinding, Extraction

The green coffee beans were roasted at 245 °C for 11.5 (light roast), 13.5 (medium roast) and 16 (dark roast) min using a coffee roaster (GeneCafe CR-100 coffee roaster, Genesis, Ansan, Korea) to ensure uniform roasting conditions. Each roasting was done in triplicate. After roasting, the beans were allowed to cool for 30 min and then ground by adjusting in a medium level on a coffee grinder (Latina 600N electric grinder). Hot brew extraction made using a coffee maker (HD7450, Philips, Nanjing, China) by mixing 36 g of coffee powder with 500 mL of water. A filter paper was used during brewing. After each brewing, the coffee machine was cleaned prior to brewing the next sample. The extraction was made in three replications. The analysis of antioxidant activities, TPC, TFC, and TTC were measured from this extract.

2.4. Color Parameters

Chroma Meter (CR-400 Chroma Meter, Tokyo, Japan) was used to estimate the color of the roasted coffee powder. A small coffee powder (5 g) was kept into a Petri dish for reading, and a* (redness), b* (yellowness) and L* (lightness) were assessed. The color reading was made in triplicate.

2.5. Antioxidant Activity

2.5.1. Oxygen Radical Absorbance Capacity

The oxygen radical absorbance capacity (ORAC) was estimated with some modifications using the protocol described by Ou et al. [26]. Fluorescein powder was dissolved using a phosphate buffer (NaH$_2$PO$_4$-Na$_2$HPO$_4$, 10 mM, pH 7.0). The coffee extract (50 µL) was mixed with 25 mM of the fluorescein solution (150 µL) and incubated in a dark environment for about 10 min. Then, 25 µL of 120 mM AAPH solution was mixed to the coffee extract and fluorescein mixture. Meanwhile, 10 mM phosphate buffer was substituted instead of coffee extract and used as a control. The absorbance was measured using a UV/visible spectrophotometer (U-2900, Hitachi High-Tech Corporation, Tokyo, Japan). The readings were recorded at one minute interval for 30 min (excitation wavelength: 485 nm; emission wavelength: 535 nm). The absorbance readings for ORAC estimation were made in triplicate and expressed as a µM Trolox equivalent/mL of coffee (µM TE/mL). The ORAC estimated using the formula stated below;

$$\text{ORAC (µM TE/g)} = (C_{Trolox} \times (AUC_{Sample} - AUC_{Blank}) \times k)/AUC_{Trolox} - AUC_{Blank} \qquad (1)$$

where C_Trolox, k, and AUC were the concentrations of Trolox (5 µM), the sample dilution factor, and the area under the curve, respectively. AUC was computed according to the formula presented below;

$$AUC = 1 + \sum_{n=1}^{30} \frac{fn}{f0} \qquad (2)$$

where f_n was the fluorescence at time n (min).

2.5.2. Superoxide Dismutase-Like Activity

The protocol suggested by Marklund and Marklund [27] used with some modifications to estimate the SOD-like activity. Coffee extract (400 µL), Tris-HCl buffer (600 µL, 50 mM tris (hydroxymethyl) aminomethane and 10 mM ethylenediaminetetraacetic acid, pH 8.0), and 7.2 mM pyrogallol (40 µL) were mixed together and kept at 25 °C for 10 min. The reaction was terminated by mixing 0.1N HCl (20 µL). Absorbance readings were done at 420 nm using a UV/visible spectrophotometer (U-2900, Hitachi High-Tech Corporation, Tokyo, Japan). The absorbance readings for the estimation of the SOD-like activity were performed in triplicate. SOD-like activity was computed with the formula presented below:

$$\text{SOD-like activity (\%)} = (1 - A/B) \times 100 \qquad (3)$$

where A indicates the sample absorbance and B indicates the absorbance of the blank (control).

2.5.3. 2.,2-diphenyl-1-picrylhydrazyl (DPPH) Radical Scavenging Assay

The DPPH activity was measured using the protocol described by Pataro et al. [28], with some modifications. DPPH was dissolved in methanol (0.1 mM) and 3.9 mL of it transferred into a cuvette, then immediately the absorbance was measured at 515 nm (used as a control). The coffee extracts (800 µL) from each sample were mixed with 3.2 mL of methanol-DPPH solution and placed in a dark condition for 30 min. Then the absorbance was recorded. Methanol was used as a blank. The DPPH assay was done in three replications. The DPPH inhibition percentage was calculated using the following formula;

$$\text{DPPH inhibition (\%)} = (A_{control} - A_{sample})/A_{control} \times 100 \qquad (4)$$

where $A_{control}$ is the absorbance of control reaction (DPPH and Methanol), and A_{sample} is the absorbance in the presence of coffee extract.

2.5.4. Ferric Reducing Antioxidant Power (FRAP)

FRAP of the coffee extract was assessed using the method described by Oyaizu [29]. A 2.5 mL of coffee extract was mixed with 2.5 mL of 200 mM sodium phosphate buffer (pH 6.6) 2.5 mL of 1% potassium ferricyanide. The mixture was incubated for 20 min at 50 °C following 2.5 mL of 10% trichloroacetic acid was added. Then, this mixture was centrifuged (2000× g for 10 min) and 5 mL of the supernatant was taken and mixed with 5 mL of water and finally added 1 mL of 0.1% ferric chloride. The absorbance was measured at 700 nm using a spectrophotometer (U-2900, Hitachi High-Tech Corporation, Tokyo, Japan) and the FRAP was subsequently determined using ascorbic acid as standard. The FRAP analysis was done in three replications.

2.6. Total Phenol Content, Flavonoid Content, and Tannin Content

The total phenol content of the coffee extract was measured using a protocol described by Singleton's method [30], with some adjustments. The coffee extract (20 µL) was diluted with 1580 µL ddH$_2$O. Diluted coffee extract (160 µL) and Ciocalteu's phenol reagent (10 µL) were mixed and kept for 8 min. Then, 30 µL of 20% Na$_2$CO$_3$ solution was mixed and incubated in a dark environment for 2 h. The coffee extract was substituted with distilled water and used as a blank. The absorbance readings were determined at 765 nm using a UV/visible spectrophotometer (U-2900, Hitachi High-Tech Corporation, Tokyo, Japan). Gallic acid solutions (0–1 mg/mL) were used to generate a standard curve ($r^2 = 0.997$). The estimation of the TPC was done in triplicate. Results were expressed as mg gallic acid equivalent/mL (mg GAE/mL) of coffee extract.

A protocol stated by Dewanto [31], applied with some modification to estimate the total flavonoid content of the coffee extract. The mixture was prepared using 250 µL coffee extract, 1 mL of distilled water and 75 µL of 5% $NaNO_2$. The 5 min later, 10% $AlCl_3$ $6H_2O$ solution (150 µL) was mixed and incubated for 6 min. Finally, 1N NaOH (500 µL) was added and incubated for 11 min. The ddH_2O used as a blank sample. The absorbance was measured at 510 nm using a UV/visible spectrophotometer (U-2900, Hitachi High-Tech Corporation, Tokyo, Japan). The standard solution was prepared using a quercetin solution (0–1 mg/mL) to make a standard curve ($r^2 = 0.999$). The estimation of the TFC was done in triplicate. The total flavonoids in the coffee were expressed as mg quercetin equivalent/mL of coffee extract.

The total tannin content (TTC) was measured using the Folin-Ciocalteu method, with some modifications [32]. About 100 µL of the coffee extract was added to a tube (10 mL) containing 7500 µL of distilled water, 500 µL of Folin-Ciocalteu phenol reagent, and 1000 µL of 35% sodium carbonate solution and was then diluted to 10,000 µL with distilled water. The mixture was well-mixed and placed at room temperature for 30 min. A set of standard solutions of tannic acid (20, 40, 60, 80, 100 µg/mL) were made as a reference. Absorbance for the test and standard solutions was measured with a UV/Visible spectrophotometer (U-2900, Hitachi High-Tech Corporation, Tokyo, Japan) against the blank (distilled water) at 700 nm. The estimation of the TTC was done in triplicate. The tannin content was expressed in mg/mL of tannic acid in the coffee extracts.

2.7. Sensory Evaluation

The consumer responses of the coffee made from fermented and non-fermented coffee beans were evaluated with 50 people (students and staff members of the horticulture department) at Kangwon National University, Republic of Korea. The medium roasted coffee beans were selected for testing the consumer acceptability since the Korean people widely used the medium roasted coffee. The coffee was brewed using a coffee maker (HD7450, Philips, Nanjing, China) using 36 g of coffee powder with 500 mL of water. The prepared coffee was poured into a thermos to maintain the temperature. Each consumer was provided with coffee sample (20 mL, 60 ± 2 °C) in a paper cup (150 mL) and a glass of water (24 ± 2 °C) for cleansing palate after each taste. The scale was 1 to 7 and labeled as 1 = dislike very much, 2 = dislike moderately, 3 = dislike slightly, 4= neither like nor dislike, 5 = like slightly, 6 = like moderately, 7 = like very much.

2.8. Statistical Analysis

The results were compiled using Microsoft Excel 2013. Analysis of variance (ANOVA) was performed using SAS 9.4 software (SAS Institute, 100 Campus Drive, Cary, Raleigh, North Carolina, USA), to classify significant variations among samples.

3. Results and Discussion

3.1. pH and Colony Count (log CFU/mL)

The pH of the solution (the water that was used to ferment the coffee beans) was measured right after the inoculation of the yeast (0 h) and after 24 h of fermentation. A pH reduction was observed in both the yeast-inoculated and non-yeast-inoculated (control) treatments. However, the pH reduction rate was higher in the yeast-inoculated sample. The pH in the control treatment was 6.02 at 0 h and 5.78 after 24 h (Figure 1A). In the fermented treatment, the pH was reduced from 6.05 to 5.46 within 24 h of fermentation (Figure 1A). The pH of the yeast-inoculated solution showed a higher reduction compared to the non-yeast-inoculated solution. The alkalinity and acidity of the solution are expressed based on the pH values. The pH continues to decrease while fermentation takes place for various reasons. Fermentation is responsible for the production of organic acids and the absorption of basic amino acids, which substantially reduces the pH [33]. However, a small pH reduction was found in the non-yeast-inoculated treatment. This reduction was expected because the water-soluble organic acids leaked from the coffee beans into the water [21,34] and/or because of bacteria that naturally exist on the coffee beans [21]. The yeast populations were measured; initially, they were 7.66 log CFU/mL, and after 24 h, they reached 9.62 log CFU/mL (Figure 1B).

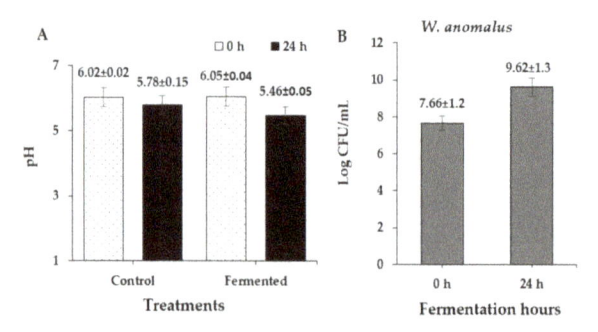

Figure 1. The pH in yeast inoculated and non-inoculated treatments (**A**) and yeast population (**B**).

3.2. Colorimeter Data

The fermented and non-fermented ground coffee after roasting is shown in Figure 2. A significant ($p < 0.05$) differences were observed between the yeast-fermented and non-fermented treatments of ground roasted (light roasting level) coffee with respect to the colorimetric parameter (L* and a*) (Figure 3A,B). The L* and a* of ground roasted coffee did not significantly differ between the fermented and control treatment in both medium and dark roasting condition. However, we observed that the L* value increased and the a* value decreased as the roasting time increased. The b* value was significantly higher in the fermented treatment at the light, medium and dark roasting level (Figure 3C). Supporting reports showed that fermentation of green coffee beans with *S. cerevisiae* strains [21,22], and *S. fibuligera* [22] resulted in high L*, a*, and b* values compared to non-fermented coffee beans. During cocoa fermentation, gradually, a color change was observed as a result of a high amount of phenolic compounds and flavonoids, which act as a substrate of polyphenol oxidase in the presence of oxygen [35].

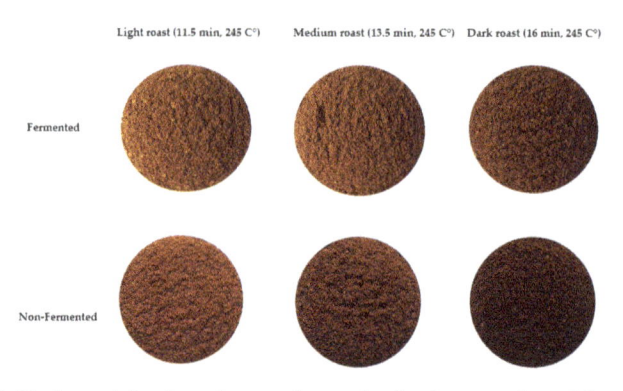

Figure 2. The fermented and non-fermented ground coffee that roasted at a different level.

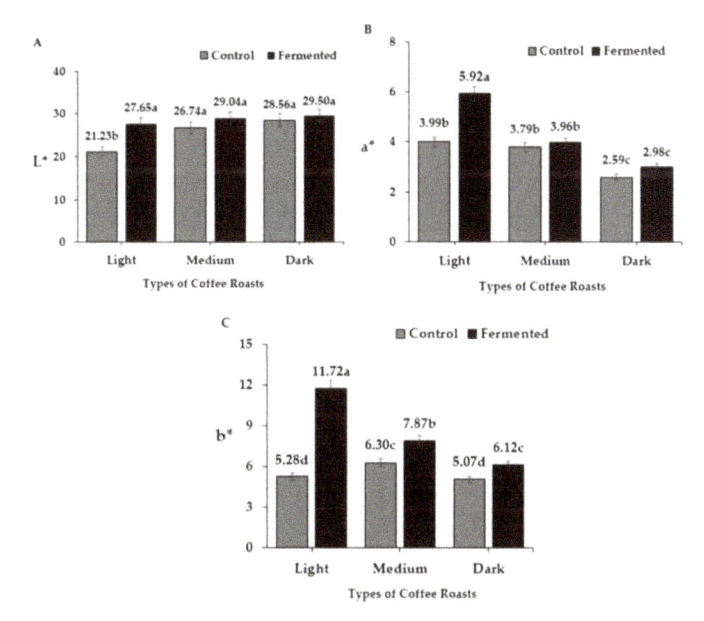

Figure 3. Colorimeter (L*, a* and b*) values of fermented roasted and ground coffee (**A–C**). Bars with different letters indicate a significant difference (*p* < 0.5) among treatments.

3.3. Antioxidant Activity

3.3.1. SOD-Like and ORAC

The antioxidants of fermented and non-fermented coffee extract were evaluated in terms of the oxygen radical absorbance capacity (ORAC) and superoxide dismutase (SOD)-like activity. The peroxyl radical scavenging potentials of aqueous soluble components present in the coffee could be estimated more accurately with a modified ORAC assay [36,37]. There was no significant ($p > 0.05$) difference between the yeast-fermented and non-fermented coffee extracts with respect to the ORAC value in medium (Figure 4A) and dark roasted coffee. The SOD-like activity was not significantly differed between fermented and non-fermented coffee in all roasting conditions (Figure 4B). However, both ORAC and SOD-like activity were significantly lower in both treatments in the light roasting condition. The means of ORAC value for the yeast-fermented coffee extracts was 28.33, 46.08, and 47.82 μ MTE/mL in the light, medium and dark roasted coffee, respectively (Figure 4A). The means of the SOD-like activity of yeast-fermented coffee extracts was 59.56, 87.31, and 89.05% at the light, medium and dark roasting levels, respectively (Figure 4B). The ORAC value was higher by 3.96 μ MTE/mL in fermented coffee compared to non-fermented coffee in the light roasting condition (Figure 4A). We found a significant improvement in the ORAC value and SOD-like activity in the medium and dark conditions compared to the light roasted coffee regardless of the fermentation. The medium and dark roasting conditions improved the ORAC by 21.93 and 23.94 μ MTE/mL in the control treatment. However, it was increased by 17.75 and 19.49 μ MTE/mL in the fermented coffee extracts compared to light roasted coffee, respectively. This is in agreement with previous reports on the influence of roasting on antioxidant activity. Nicoli et al. [38] reported higher antioxidant activity for medium and dark roast brews. Similarly, the elevated antioxidant activity for higher degrees of roasting has been confirmed in studies [39–41]. As the roasting temperature increased the antioxidant activities of the kernel, cashew nut and testa extracts increased [42]. Several publications have described variations in coffee antioxidant contents as a cause of roasting conditions [43,44]. Del Castillo et al. [43] noticed a higher antioxidant activity in the medium roasted coffee brew. However, Perrone et al. [45]

found a greater antioxidant activity for light roast brews. The results from various investigations regarding the antioxidant activity of coffee have been considerably diverse. The causes of such variable research results can be associated with the development of melanoidin compounds, which increase the antioxidant capacity despite the reductions in the polyphenol content. Nevertheless, other studies have not detected any significant effects of roasting on antioxidant activity [46], nor any indications that roasting for a longer duration at high temperature reduce antioxidant activity [44,47,48] when polyphenol degradation is not compensated by melanoidin formation [49,50]. However, amongst other coffees, this phenomenon was noticed for dark roasted Robusta coffee [40].

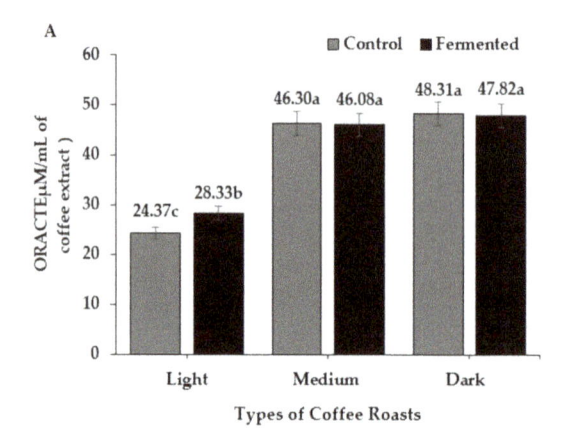

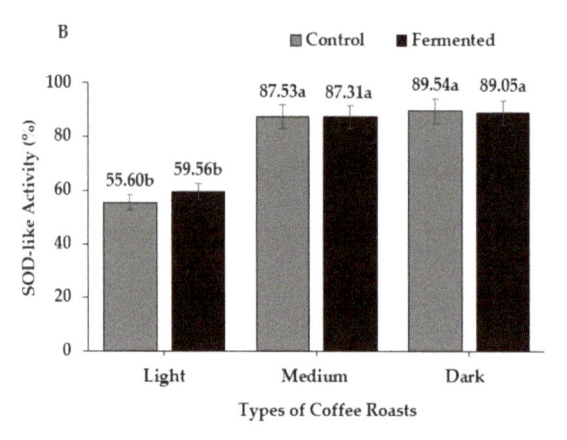

Figure 4. The oxygen radical absorbance capacity (ORAC) and superoxide dismutase-like (SOD)-like activity of fermented coffee at different degrees of roast (**A** and **B**, respectively). Bars with different letters indicate a significant difference ($p < 0.5$) among treatments.

3.3.2. DPPH and FRAP

DPPH and FRAP are among the various methods of evaluating the antioxidant properties of foods and drinks. The percentage of DPPH inhibition was higher in the fermented coffee than the non-fermented coffee in all roasting conditions (Figure 5A). Like the DPPH, the FRAP was improved in the fermented coffee in the light, medium, and dark roast types (Figure 5B). The maximum DPPH inhibition resulted from the fermented treatment in the light roasting condition (46.31%). The highest FRAP was found in the fermented coffee in the dark roasting condition (176.11 AAE µg/mL of coffee extract). In general, fermentation of green coffee beans using *W. anomalus* improved the DPPH and

FRAP. Adetuyi and Ibrahim [51] reported that fermentation enhanced the DPPH radical-scavenging ability and FRAP of okra seeds. The increase in DPPH capacity after the fermentation process shows that fermentation likely has great potential in generating some metabolites with superior radical scavenging activity [52]. The percentage of DPPH inhibition dropped as the roasting time increased (from the light to dark roasting condition). The result is in agreement with Jung et al. [53], who reported that lightly roasted coffee extract has the highest antioxidant activity with a low roasting temperature and that the DPPH decreases in the dark roasted extract. However, the FRAP improved as the roasting time increased. A supporting result published by Song et al. [54] showed that as the roasting time increased from 11 to 13 min (medium-light to medium-dark) the FRAP increased as well.

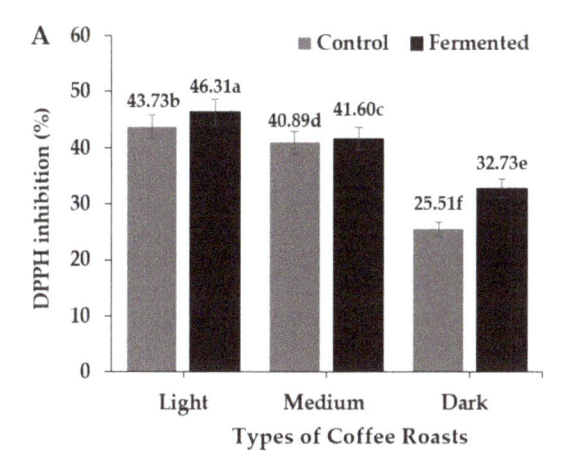

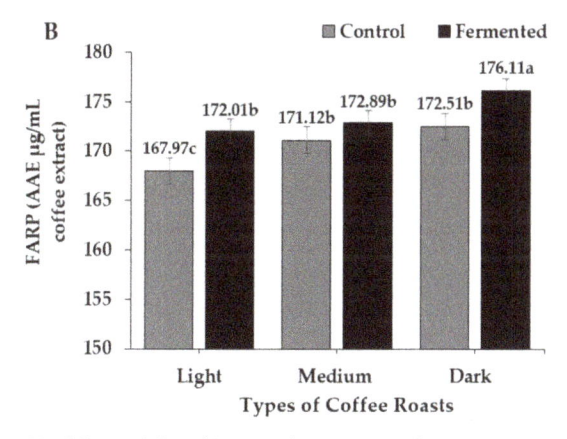

Figure 5. The DPPH inhibition (%) and Ferric reducing antioxidant power (AAE µg/mL of coffee extract) of fermented coffee. (**A** and **B**, respectively). Bars with different letters indicate a significant difference ($p < 0.5$) among treatments.

3.4. Total Phenol, Flavonoid and Tannin Content

The TPC of yeast-fermented coffee extracts had significant differences ($p < 0.05$) compared to the non-fermented coffee in all roasting types (Figure 6). During the medium roasting condition, the highest TPC difference was observed between the fermented and non-fermented coffee. The TPC in the control treatment showed a decreasing trend as the roasting time increased (light to medium to dark). The TPC in the fermented treatment was 0.98, 1.29 and 0.91 GAEmg/mL of coffee extract in the light, medium and dark roasting condition, respectively (Figure 6). The fermentation of green coffee beans with *W. anomalus* increased the TPC of the coffee extract by 0.09, 0.46, and 0.21 GAE mg/mL of coffee extract compared to non-fermented coffee at each degree of roasting (light, medium and dark roast, respectively) (Figure 6). Generally, fermentation improved the TPC compared to non-fermented coffee. These results are in agreement with several reports on fermented seeds, where fermentation induced an improvement in the phenolic content of seeds, such as legumes [55–59], okra seeds [51], soybeans [60], and coffee [21,22]. Similarly, fermentation elevated the phenolic and flavonoid contents and improved the antioxidant properties of the following plant seeds: chestnut, adlay, walnut, and lotus seed [61]. This might be associated with, proteolytic enzymes from the starter organism that hydrolyze the complexes of phenolics into simple, soluble-free phenols and biologically more active forms during fermentation, which are readily absorbed by organisms [56,62]. However, other researchers' findings indicate the proteolytic activities alone did not show a significant TPC increase; rather, when it was combined with a pectinase enzyme, the olive oil polyphenol content showed a significantly high increment [63]. These findings lead us to conclude *W. anomalus*'s ability to produce pectinase enzymes linked with the increased TPC content in this experiment. In our experiment, the TPC was decreased as the roasting time increased, except the fermented coffee in the medium roast condition which significantly increased the TPC, regardless of the fermentation. A supporting result was published by Odžaković, B., et al. [64], the TPC decreased as the roasting temperature increased. During coffee roasting the degradation of polyphenols [36,44–46], which are sensitive to heat, are affected by roasting temperature. The reduction in the polyphenol content which is linked to the extended roasting is due to the heat-sensitive nature of such compounds and the lengthened roasting duration, as well as the high processing temperature. Phenol reductions or losses during the processing steps are undesirable because of their profound effects on human health.

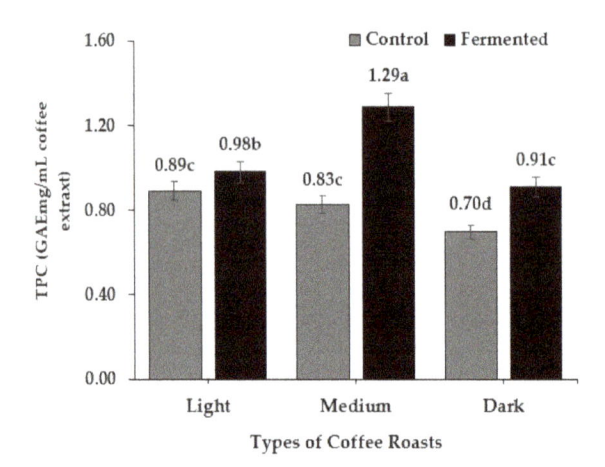

Figure 6. Total phenol contents (GAE mg/mL of coffee extract) of fermented coffee. Bars with different letters indicate a significant difference ($p < 0.5$) among treatments.

Fermentation of green coffee beans with *W. anomalous* significantly ($p < 0.05$) increased the TFC compared to non-fermented coffee in all roasting conditions (light, medium and dark) (Figure 7). TFC was approximately consistent in the control treatment in all roasting conditions. However, in the fermented coffee, the TFC was significantly higher in the light roasting compared to medium and dark roasting conditions. As shown in Figure 7, the TFC in the control treatment was 0.64, 0.67, and 0.64. The TFC in the fermented coffee at different roasting levels was 0.86 (light), 0.74 (medium), and 0.72 (dark) QE mg/mL of coffee extract. Fermentation improves the TFC content by 0.22, 0.07, and 0.08 QE mg/mL of coffee extract in the light, medium, and dark roasting conditions, respectively (Figure 7). These findings agree with several published papers on different fermented seeds, where fermentation significantly enhanced the TFC compared to the TFC in unfermented seeds, such as in soybeans [60], legumes [45,55,56], coffee beans [21,22], and okra seeds [51]. The increase in the flavonoid content might be linked to the rise in acidic values during fermentation, which releases bound flavonoid components and makes them more bioavailable [51].

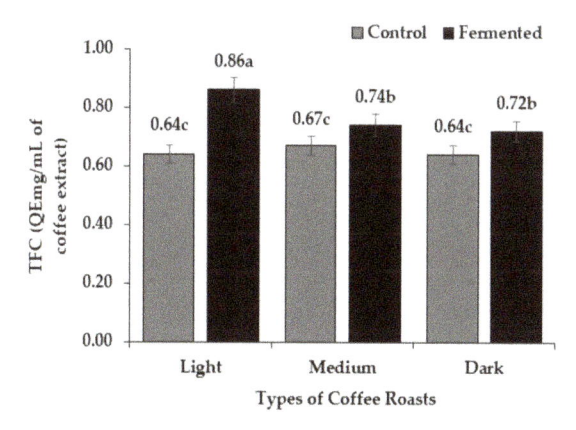

Figure 7. Total flavonoid contents (QE mg/mL of coffee extract) of fermented coffee. Bars with different letters indicate a significant difference ($p < 0.5$) among treatments.

Regardless of the fermentation, the average TFC from both treatments at each roasting condition was 0.75 ± 0.11, 0.71 ± 0.03, and 0.68 ± 0.04 QE mg/mL of coffee extract at the light, medium, and dark roasting levels, respectively. The average TFC value showed a decreasing trend as the roasting time increased, regardless of the fermentation. The result in our study coincides with Tiwari, B. K., et al. [65], who found that the TFC was reduced significantly as the roasting temperature increased from 80 °C to 120 °C and roasting duration proceeded from 20 min to 40 min. The cause for the reduction in the TFC at a higher temperature might be associated with the degradation of flavonoids. The preparation and processing of food may reduce the flavonoid contents depending on the techniques used [66].

The yeast-fermented coffee extracts had a significant difference ($p < 0.05$) compared to non-fermented coffee with regard to TTC in the light and dark roasting conditions (Figure 8). The TTC of non-fermented coffee was 0.53, 0.33, and 0.21 mg tannic acid/mL of coffee extract in light, medium and dark roasting conditions, respectively. In the fermented coffee, the TTC was 0.73, 0.28, and 0.15 mg tannic acid/mL coffee extract in the light, medium, and dark roasting conditions, respectively. In the light roasting condition, the tannin content was higher in fermented coffee than the control and it was lower than the control in the medium and dark roasting conditions (Figure 8). It has been reported that fermentation significantly decreased the tannin content of fermented sorghum cultivars [67], and in fermentation of *Xuan mugua* fruit with lactic acid bacteria, where it caused a significant decrease in the tannin content [68]. We have also found a supporting result in our previous experiment that indicates the fermentation of green coffee beans with *S. cerevisiae* and *S. fibuligera* strains reduced the TTC compared to non-fermented coffee [22]. Regardless of the fermentation, the TTC was reduced as the roasting time increased. The average TTC in each roasting condition was 0.63, 0.31 and 0.19 mg tannic acid mg/mL coffee extract in the light, medium, and dark roasting condition (Figure 8). Likewise, the tannin content was decreased in soya bean flour (0.01–0.30 g/100 g dry weight) [69], and 22% in maize [70]; because of soaking in water for 48 h and roasting. Tannin content has an astringent characteristic that contributes to the bitterness of coffee, so the reduction in tannins can be viewed as a positive aspect.

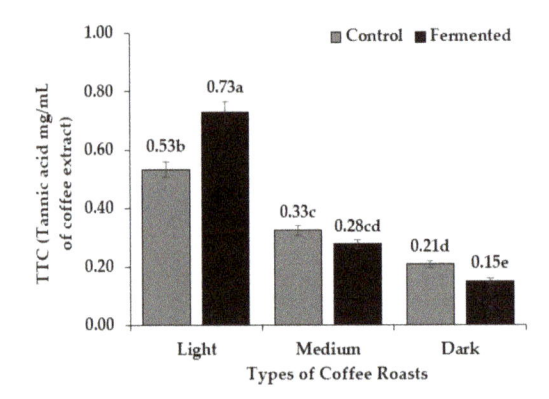

Figure 8. Total tannin contents (tannic acid mg/mL of coffee extract) of fermented coffee. Bars with different letters indicate a significant difference ($p < 0.5$) among treatments.

3.5. Sensory Evaluation

Based on the results of consumer acceptance ratings, the non-fermented coffee received a higher score than the fermented coffee in terms of its color, sourness, mouthfeel, acidity, and overall quality. However, the fermented coffee received the highest rating with respect to aroma, bitterness, and astringency. The average score for the overall quality of the coffee was 5.22 and 4.51 for non-fermented and fermented coffee, respectively (Figure 9). The average score for acidity was 3.75 for the fermented coffee and 4.15 for the non-fermented coffee. Kwak et al. [21] have evaluated the consumer acceptability of fermented green coffee beans with different yeasts, noting that one of their fermented coffee treatment acceptability ratings was lower but insignificant when compared to the control treatment. In addition, they mentioned that all the fermented coffee received lower ratings when compared to the control. Fermentation has both positive and negative influences on the flavor and aroma characteristics of coffee. Based on the overall quality rating, we have categorized the consumer responses into two groups: people who preferred the non-fermented coffee over the fermented coffee (55.48%) and people who preferred the fermented coffee over the non-fermented coffee (44.52%). A supporting result was published by Kwak et al. [21], who found that 39.4% of consumers preferred fermented coffee while 60.6% preferred non-fermented coffee by disliking the control (non-fermented) treatments based on the overall quality ratings. However, consumer acceptance evaluations are very subjective to the individual.

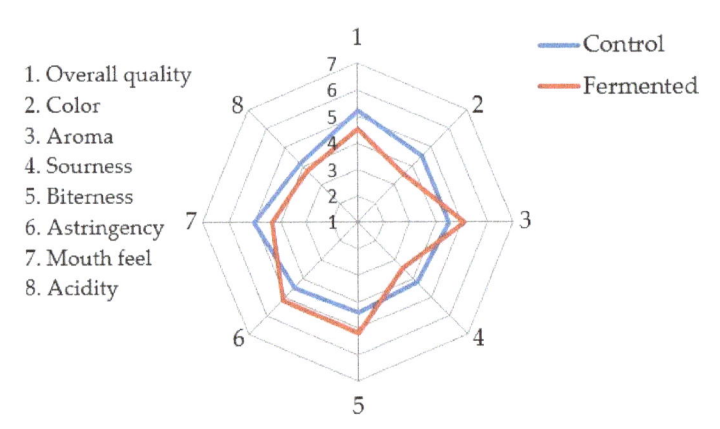

Figure 9. The consumer acceptability ratings of fermented and non-fermented coffee.

4. Conclusions

This study showed the responses of the coffee beans fermented with *W. anomalus* strain KNU18Y3. The fermentation of green coffee beans with *W. anomalus* for 24 h increased the DPPH, FRAP, TPC, and TFC when compared to non-fermented coffee, whereas the SOD-like and ORAC did not significantly differ between the fermented and non-fermented coffee beans. We have also found that the fermentation of coffee beans for 24 h is enough to modify the functionality of coffee beans. The extent of roasting had both negative and positive effects on the overall antioxidants, TPC, TFC, and TTC. The degradation of the TPC, TFC, and TTC was observed as the roasting time increased from 11.5 to 16 min, regardless of the fermentation. The fermented coffee received better ratings in terms of its aroma, bitterness, and astringency parameters. Moreover, 44.52% of participants chose the fermented coffee over the non-fermented one based on the overall quality ratings. This result shows that fermented coffee is also preferred by consumers. Since the consumer evaluation experiment was conducted blind (the participants were not informed of the coffee types), the increased antioxidants and phenolic and flavonoid compounds will probably attract more consumers when they are informed of these positive aspects of fermented coffee. While the *W. anomalus* yeast has the potential to be used for green coffee bean fermentation, further selection and evaluation of microorganisms should be continued to maximize the functionality of the coffee bean and its health benefits.

Author Contributions: Conceptualization, W.H.K., M.H.; methodology, M.H.; Data curation, M.H.; validation, W.H.K.; formal analysis, M.H.; software, M.H.; investigation, W.H.K.; resources, W.H.K.; funding acquisition, W.H.K.; writing—original draft preparation, M.H.; writing—review and editing, W.H.K., M.H.; supervision, W.H.K.; project administration, W.H.K. All authors have read and agreed to the published version of the manuscript.

Conflicts of Interest: The authors declare no conflict of interest.

References

1. Mesfin, H.; Kang, W.H. The role of microbes in coffee fermentation and their impact on coffee quality. *J. Food Qual.* **2019**. [CrossRef]
2. Martins, S.; Mussatto, S.I.; Martínez-Avila, G.; Montañez-Saenz, J.; Aguilar, C.N.; Teixeira, J.A. Bioactive Phenolic Compounds: Production and Extraction by Solid-State Fermentation. A Review. *Biotechnol. Adv.* **2011**, *29*, 365–373. [CrossRef] [PubMed]
3. Torino, M.I.; Limón, R.I.; Martínez-Villaluenga, C.; Mäkinen, S.; Pihlanto, A.; Vidal-Valverde, C.; Frias, J. Antioxidant and Antihypertensive Properties of Liquid and Solid State Fermented Lentils. *Food Chem.* **2013**, *136*, 1030–1037. [CrossRef] [PubMed]
4. Dulf, F.V.; Vodnar, D.C.; Socaciu, C. Effects of solid-state fermentation with two filamentous fungi on the total phenolic contents, flavonoids, antioxidant activities and lipid fractions of plum fruit (*Prunus domestica* L.) by-products. *Food Chem.* **2016**, *209*, 27–36. [CrossRef]
5. Zhang, X.Y.; Chen, J.; Li, X.L.; Yi, K.; Ye, Y.; Liu, G.; Wang, Z.G. Dynamic changes in antioxidant activity and biochemical composition of tartary buckwheat leaves during *Aspergillus niger* fermentation. *J. Funct. Foods* **2017**, *32*, 375–381. [CrossRef]
6. Belitz, H.D.; Grosh, W.; Schieberle, P. Coffee, Tea, Cocoa. In *Food Chemistry*; Springer-Verlag: Heidelberg/Berlin, Germany; New York, NY, USA, 2009.
7. Rawel, H.M.; Kulling, S.E. Nutritional contribution of coffee, cacao and tea phenolics to human health. *J. Verbraucherschutz Lebensmittelsicherheit* **2007**, *2*, 399–406. [CrossRef]
8. Richelle, M.; Tavazzi, I.; Offord, E. Comparison of the antioxidant activity of commonly consumed polyphenolic beverages (coffee, cocoa, and tea) prepared per cup serving. *J Agric. Food Chem.* **2001**, *49*, 3438–3442. [CrossRef]
9. Cheynier, V. Polyphenols in foods are more complex than often thought. *Am. J. Clin. Nutr.* **2005**, *81*, 223S–229S. [CrossRef]
10. Manach, C.; Scalbert, A.; Morand, C.; Rémésy, C.; Jimenez, L. Polyphenols: Food sources and bioavailability. *Am. J. Clin. Nutr.* **2004**, *79*, 727–747. [CrossRef]
11. Kumar, S.; Mishra, A.; Pandey, A.K. Antioxidant mediated protective effect of *Parthenium hysterophorus* against oxidative damage using in vitro models. *BMC Complement Altern. Med.* **2013**, *13*, 120. [CrossRef]

12. Kumar, S.; Pandey, A.K. Phenolic content, reducing power and membrane protective activities of *Solanum xanthocarpum* root extracts. *Vegetos* **2013**, *26*, 301–307.
13. Yao, L.H.; Jiang, Y.M.; Shi, J.; Tomás-Barberán, F.A.; Datta, N.; Singanusong, R.; Chen, S.S. Flavonoids in food and their health benefits. *Plant Foods Hum. Nutr.* **2004**, *59*, 113–122. [CrossRef] [PubMed]
14. Kurtzman, C.P.; Fell, J.W. *The Yeast, a Taxonomical Study*, 4th ed.; Elsevier Science: Amsterdam, The Netherlands, 1988; p. 1055.
15. De Hoog, G.S. Risk assessment of fungi reported from humans and animals. *Mycoses* **1996**, *39*, 407–417. [CrossRef]
16. Comitini, F.; De Ingenis, J.; Pepe, L.; Mannazu, I.; Ciani, M. *Pichia anomala* and *Kluyveromyces wikerhamii* killer toxins as new tool against *Dekkera/Brettanomyces* spoilage yeasts. *FEMS Microbiol. Lett.* **2004**, *238*, 235–240. [CrossRef] [PubMed]
17. Lee, Y.J.; Choi, Y.R.; Lee, S.Y.; Park, J.T.; Shim, J.H.; Park, K.H.; Kim, J.W. Screening wild yeast strains for alcohol fermentation from various fruits. *Mycobiology* **2011**, *39*, 33–39. [CrossRef] [PubMed]
18. Ye, M.; Yue, T.; Yuan, Y. Effects of sequential mixed cultures of *Wickerhamomyces anomalus* and *Saccharomyces cerevisiae* on apple cider fermentation. *FEMS Yeast Res.* **2014**, *14*, 873–882. [CrossRef] [PubMed]
19. Madrigal, T.; Maicas, S.; Tolosa, J.J.M. Glucose and ethanol tolerant enzymes produced by *Pichia* (*Wickerhamomyces*) isolates from enological ecosystems. *Am. J. Enol. Vitic.* **2013**, *64*, 126–133. [CrossRef]
20. Zha, M.; Sun, B.; Wu, Y.; Yin, S.; Wang, C. Improving flavor metabolism of Saccharomyces cerevisiae by mixed culture with *Wickerhamomyces anomalus* for Chinese Baijiu making. *J. Biosci. Bioeng.* **2018**, *126*, 189–195. [CrossRef]
21. Kwak, H.S.; Jeong, Y.; Kim, M. Effect of Yeast Fermentation of Green Coffee Beans on Antioxidant Activity and Consumer Acceptability. *J. Food Qual.* **2018**. [CrossRef]
22. Mesfin, H.; Kang, W.H. Antioxidant Activity, Total Polyphenol, Flavonoid and Tannin Contents of Fermented Green Coffee Beans with Selected Yeasts. *Fermentation* **2019**, *5*, 29.
23. Lee, L.W.; Cheong, M.W.; Curran, P.; Yu, B.; Liu, S.Q. Modulation of coffee aroma via the fermentation of green coffee beans with *Rhizopus oligosporus*: I. Green Coffee. *Food Chem.* **2016**, *211*, 916–924. [CrossRef] [PubMed]
24. Lee, L.W.; Tay, G.Y.; Cheong, M.W.; Curran, P.; Yu, B.; Liu, S.Q. Modulation of the volatile and non-volatile profiles of coffee fermented with *Yarrowia lipolytica*: I. Green coffee. *LWT- Food Sci. Technol.* **2017**, *77*, 225–232. [CrossRef]
25. Mesfin, H.; Kang, W.H. Isolation, Identification, and Characterization of Pectinolytic Yeasts for Starter Culture in Coffee Fermentation. *Microorganisms* **2019**, *7*, 401.
26. Ou, B.; Hampsch-Woodill, M.; Prior, R.L. Development and validation of an improved oxygen radical absorbance capacity assay using fluorescein as the fluorescent probe. *J. Agric. Food Chem.* **2001**, *49*, 4619–4626. [CrossRef] [PubMed]
27. Marklund, S.; Marklund, G. Involvement of the superoxide anion radical in the autoxidation of pyrogallol and a convenient assay for superoxide dismutase. *Eur. J. Biochem.* **1974**, *47*, 469–474. [CrossRef]
28. Pataro, G.; Sinik, M.; Capitoli, M.M.; Donsì, G.; Ferrari, G. The influence of post-harvest UV-C and pulsed light treatments on quality and antioxidant properties of tomato fruits during storage. *Innov. Food Sci. Emerg. Technol.* **2015**, *30*, 103–111. [CrossRef]
29. Oyaizu, M. Studies on products of browning reaction. *Jpn. J. Nutr. Diet.* **1986**, *44*, 307–315. [CrossRef]
30. Singleton, V.L.; Orthofer, R.; Lamuela-Raventós, R.M. Analysis of total phenols and other oxidation substrates and antioxidants by means of Folin-ciocalteu reagent. *Methods Enzymol.* **1999**, *299*, 152–178.
31. Dewanto, V.; Wu, X.; Liu, R.H. Processed sweet corn has higher antioxidant activity. *J. Agric. Food Chem.* **2002**, *50*, 4959–4964. [CrossRef]
32. CI, K.C.; Indira, G. Quantitative estimation of total phenolic, flavonoids, tannin and chlorophyll content of leaves of *Strobilanthes Kunthiana* (Neelakurinji). *J. Med. Plants* **2016**, *4*, 282–286.
33. Coote, N.; Kirsop, B.H. Factors responsible for the decrease in pH during beer fermentations. *J. Inst. Brew.* **1976**, *82*, 149–153. [CrossRef]
34. Rodrigues, C.I.; Marta, L.; Maia, R.; Miranda, M.; Ribeirinho, M.; Máguas, C. Application of solid-phase extraction to brewed coffee caffeine and organic acid determination by UV/HPLC. *J. Food Compos. Anal.* **2007**, *20*, 440–448. [CrossRef]

35. Afoakwa, E.O.; Paterson, A.; Fowler, M.; Ryan, A. Flavor formation and character in cocoa and chocolate: A critical review. *Food Sci. Nutr.* **2008**, *48*, 840–857. [CrossRef] [PubMed]
36. Kwon, D.Y.; Choi, K.H.; Kim, S.J.; Choi, D.W.; Kim, Y.S.; Kim, Y.C. Peroxyl radical-scavenging activity of coffee brews. *Eur. Food Res. Technol.* **2005**, *221*, 471–477.
37. Gómez-Ruiz, J.Á.; Leake, D.S.; Ames, J.M. In Vitro Antioxidant Activity of Coffee Compounds and Their Metabolites. *J. Agric. Food Chem.* **2007**, *55*, 6962–6969. [CrossRef] [PubMed]
38. Nicoli, M.C.; Anese, M.; Manzocco, L.; Lerici, C.R. Antioxidant properties of coffee brews in relation to the roasting degree. *Lebensm. Wiss. Technol.* **1997**, *30*, 292–297. [CrossRef]
39. Liu, Y.; Kitts, D.D. Confirmation that the Maillard reaction is the principle contributor to the antioxidant capacity of coffee brews. *Food Res. Int.* **2011**, *44*, 2418–2424. [CrossRef]
40. Vignoli, J.A.; Viegas, M.C.; Bassoli, D.G.; Benassi, M.T. Roasting process affects differently the bioactive compounds and the antioxidant activity of arabica and robusta coffees. *Food Res. Int.* **2014**, *61*, 279–285. [CrossRef]
41. Sánchez-González, I.; Jiménez-Escrig, A.; Saura-Calixto, F. In vitro antioxidant activity of coffees brewed using different procedures (Italian, espresso and filter). *Food Chem.* **2005**, *90*, 113–139. [CrossRef]
42. Chandrasekara, N.; Shahidi, F. Effect of roasting on phenolic content and antioxidant activities of whole cashew nuts, kernels, and testa. *J. Agric. Food Chem.* **2011**, *59*, 5006–5014. [CrossRef]
43. Del Castillo, M.D.; Ames, J.M.; Gordon, M.H. Effect of roasting on the antioxidant activity of coffee brews. *J. Agric. Food Chem.* **2002**, *50*, 3698–3703. [CrossRef] [PubMed]
44. Cämmerer, B.; Kroh, L.W. Antioxidant activity of coffee brews. *Eur. Food Res. Technol.* **2006**, *223*, 469–474. [CrossRef]
45. Perrone, D.; Farah, A.; Donangelo, C.M. Influence of coffee roasting on the incorporation of phenolic compounds into melanoidins and their relationship with antioxidant activity of the brew. *J. Agric. Food Chem.* **2012**, *60*, 4265–4275. [CrossRef] [PubMed]
46. Szymanowska, K.; Wołosiak, R. The effect of coffee roasting on selected parameters of its quality. *Res. Teach. Appar.* **2014**, *19*, 77–83.
47. Duarte, S.M.S.; Abreu, C.M.P.; Menezes, H.C.; Santos, M.H.; Gouvêa, C.M.C.P. Effect of processing and roasting on the antioxidant activity of coffee brews. *Ciênc. Tecnol. Aliment.* **2005**, *25*, 387–393. [CrossRef]
48. Hecimovic, I.; Belscak-Cvitanovic, A.; Horzic, D.; Komes, D. Comparative study of polyphenols and caffeine in different coffee varieties affected by the degree of roasting. *Food Chem.* **2011**, *129*, 991–1000. [CrossRef]
49. Sacchetti, G.; Mattia, C.D.; Pittia, P.; Mastrocola, D. Effect of roasting degree, equivalent thermal effect and coffee type on the radical scavenging activity of coffee brews and their phenolic fraction. *J. Food Eng.* **2009**, *90*, 74–80. [CrossRef]
50. Summa, C.A.; Calle, B.; Brohee, M.; Stadler, R.H.; Anklama, E. Impact of the roasting degree of coffee on the in vitro radical scavenging capacity and content of acrylamide. *LWT- Food Sci. Technol.* **2007**, *40*, 1849–1854. [CrossRef]
51. Adetuyi, F.O.; Ibrahim, T.A. Effect of fermentation time on the phenolic, flavonoid and vitamin C contents and antioxidant activities of okra (*Abelmoschus esculentus*) seeds. *Niger. Food J.* **2014**, *32*, 128–137. [CrossRef]
52. Abu, F.; Taib, M.; Norma, C.; Moklas, M.; Aris, M.; Mohd Akhir, S. Antioxidant Properties of Crude Extract, Partition Extract, and Fermented Medium of *Dendrobium sabin* Flower. *J. Evid.-Based Complement. Altern. Med.* **2017**. [CrossRef]
53. Jung, S.; Kim, M.H.; Park, J.H.; Jeong, Y.; Ko, K.S. Cellular antioxidant and anti-inflammatory effects of coffee extracts with different roasting levels. *J. Med. Food.* **2017**, *20*, 626–635. [CrossRef] [PubMed]
54. Song, J.L.; Asare, T.S.; Kang, M.Y.; Lee, S.C. Changes in Bioactive Compounds and Antioxidant Capacity of Coffee under Different Roasting Conditions. *Korean J. Plant Resour.* **2018**, *31*, 704–713.
55. Moktan, B.; Saha, J.; Sarkar, P.K. Antioxidant activities of soybean as affected by *Bacillus*-fermentation to kinema. *Food Res. Int.* **2008**, *41*, 586–593. [CrossRef]
56. Ademiluyi, A.O.; Oboh, G. Antioxidant properties of condiment produced from fermented bambara groundnut (*Vigna subterranea* L. Verdc). *J. Food Biochem.* **2011**, *35*, 1145–1160. [CrossRef]
57. Dajanta, K.; Janpum, P.; Leksing, W. Antioxidant capacities, total phenolics and flavonoids in black and yellow soybeans fermented by *Bacillus subtilis*: A comparative study of Thai fermented soybeans (thua nao). *Int. Food Res. J.* **2013**, *20*, 3125.

58. Guzmán-Uriarte, M.L.; Sánchez-Magaña, L.M.; Angulo-Meza, G.Y.; Cuevas-Rodríguez, E.O.; Gutiérrez-Dorado, R.; Mora-Rochín, S.; Reyes-Moreno, C. Solid state bioconversion for producing common bean (Phaseolus vulgaris L.) functional flour with high antioxidant activity and antihypertensive potential. *Food Nutr. Sci.* **2013**, *4*, 480.
59. Plaitho, Y.; Kangsadalampai, K.; Sukprasansap, M. The protective effect of Thai fermented pigmented rice on urethane induced somatic mutation and recombination in Drosophila melanogaster. *J. Med. Plant Resour.* **2013**, *7*, 91–98.
60. Amadou, I.; Yong-Hui, S.; Sun, J.; Guo-Wei, L. Fermented soybean products: Some methods, antioxidants compound extraction and their scavenging activity. *Asian J. Biochem.* **2009**, *4*, 68–76.
61. Wang, C.Y.; Sz-Jie, W.; Shyu, Y.T. Antioxidant Properties of Certain Cereals as Affected by Food-Grade Bacteria Fermentation. *J. Biosci. Bioeng.* **2014**, *117*, 449–456. [CrossRef]
62. Shrestha, A.K.; Dahal, N.R.; Ndungutse, V. Bacillus fermentation of soybean: A review. *J. Food Sci. Technol.* **2010**, *6*, 1–9. [CrossRef]
63. Moustakime, Y.; Hazzoumi, Z.; Joutei, K.A. Effect of proteolytic activities in combination with the pectolytic activities on extractability of the fat and phenolic compounds from olives. *Springer Plus* **2016**, *5*, 739. [CrossRef] [PubMed]
64. Odžaković, B.; Džinić, N.; Kukrić, Z.; Grujić, S. Effect of roasting degree on the antioxidant activity of different Arabica coffee quality classes. *Acta Sci. Pol. Technol. Aliment.* **2016**, *15*, 409–417. [CrossRef] [PubMed]
65. Tiwari, B.K.; Brunton, N.P.; Brennan, C. Handbook of plant food phytochemicals: Sources, stability and extraction. John Wiley and Sons: Hoboken, NJ, USA, 2013.
66. Kumar, S.; Pandey, A.K. Antioxidant, lipo-protective and antibacterial activities of phytoconstituents present in Solanum xanthocarpum root. *Int. Rev. Biophys. Chem.* **2012**, *3*, 42–47.
67. Hassan, I.A.; El Tinay, A.H. Effect of fermentation on tannin content and in-vitro protein and starch digestibilities of two sorghum cultivars. *Food Chem.* **1995**, *53*, 149–151. [CrossRef]
68. Shang, Y.F.; Cao, H.; Ma, Y.L.; Zhang, C.; Ma, F.; Wang, C.X.; Wei, Z.J. Effect of lactic acid bacteria fermentation on tannins removal in Xuan Mugua fruits. *Food Chem.* **2019**, *274*, 118–122. [CrossRef]
69. Agume, A.; Njintang, N.; Mbofung, C. Effect of soaking and roasting on the physicochemical and pasting properties of soybean flour. *Foods* **2017**, *6*, 12. [CrossRef]
70. Oboh, G.; Ademiluyi, O.A.; Akindahunsi, A.A. The effect of roasting on the nutritional and antioxidant properties of yellow and white maize varieties. *Int. J. Food Sci. Technol.* **2010**, *45*, 1236–1242. [CrossRef]

© 2020 by the authors. Licensee MDPI, Basel, Switzerland. This article is an open access article distributed under the terms and conditions of the Creative Commons Attribution (CC BY) license (http://creativecommons.org/licenses/by/4.0/).

fermentation

Article

Effect of Co-Inoculation with *Pichia fermentans* and *Pediococcus acidilactici* on Metabolite Produced During Fermentation and Volatile Composition of Coffee Beans

Alexander da Silva Vale [1], Gilberto Vinícius de Melo Pereira [1,*], Dão Pedro de Carvalho Neto [1], Cristine Rodrigues [1], Maria Giovana B. Pagnoncelli [2] and Carlos Ricardo Soccol [1,*]

[1] Department of Bioprocess Engineering and Biotechnology, Federal University of Paraná (UFPR), 19011Curitiba, Paraná 81531-980, Brazil
[2] Departament of Chemistry and Biology, Federal University of Technology—Paraná (UTFPR), Curitiba PR 80230-901, Brazil
* Correspondence: gilbertovinicius@gmail.com (G.V.d.M.P.); soccol@ufpr.br (C.R.S.); Tel.: +55-41-33-613-697 (G.V.d.M.P.); +55-41-33-613-191 (C.R.S.)

Received: 2 July 2019; Accepted: 19 July 2019; Published: 22 July 2019

Abstract: Removal of the mucilage layer of coffee fruits by a fermentation process has became an interesting strategy to improve coffee quality, which is able to assist the formation of flavored molecules. In this study, four sets of inoculation protocols were evaluated using ripe and immature coffee fruits, respectively, including (i) pure culture fermentation with *Pichia fermentans*, (ii) pure culture fermentation with *Pediococcus acidilactici*, (ii) combined fermentation with *P. fermentans* and *P. acidilactici*, and (iv) spontaneous, non-inoculated control. The initial pulp sugar concentration of ripe coffee fruits (0.57 and 1.13 g/L glucose and fructose content, respectively) was significantly higher than immature coffee pulp (0.13 and 0.26 g/L glucose and fructose content, respectively). Combined inoculation with *P. fermentans* and *P. acidilactici* of ripe coffee beans increased pulp sugar consumption and production of metabolites (lactic acid, ethanol, and ethyl acetate), evidencing a positive synergic interaction between these two microbial groups. On the other hand, when immature coffee fruits were used, only pure culture inoculation with *P. fermentans* was able to improve metabolite formation during fermentation, while combined treatment showed no significant effect. Altogether, 30 volatile compounds were identified and semi-quantified with HS- solid phase microextraction (SPME)-gas chromatography coupled to mass spectrophotometry (GC/MS) in fermented coffee beans. In comparison with pure cultures and spontaneous process, combined treatment prominently enhanced the aroma complexity of ripe coffee beans, with a sharp increase in benzeneacetaldehyde, 2-heptanol, and benzylalcohol. Consistent with the monitoring of the fermentation process, only *P. fermentans* treatment was able to impact the volatile composition of immature coffee beans. The major impacted compounds were 2-hexanol, nonanal, and D-limonene. In summary, this study demonstrated the great potential of the combined use of yeast and lactic acid bacteria to improve fermentation efficiency and to positively influence the chemical composition of coffee beans. Further studies are still required to investigate the mechanisms of synergism between these two microbial groups during the fermentation process and influence the sensory properties of coffee products.

Keywords: coffee processing; coffee fermentation; starter culture; coffee beverage; yeast

1. Introduction

Coffee plants are cultivated in more than 80 countries around the world, providing raw materials for a global industry valued at an excess of 10 billion US$ [1]. Production conditions and post-harvest operations, such as fruit harvesting, depulping, drying, and storage, have a direct impact on the quality of coffee products. Fruit harvesting is the first step in postharvest coffee processing. The heterogeneous development of coffee fruits leads to a simultaneous presence of different maturation stages in the same coffee tree, namely: (i) Green (immature) coffee fruits, presenting incomplete endosperm development and low reducing sugars content in the mucilaginous layer; (ii) cherry (ripe) fruits, presenting mucilage rich in reducing sugars, complete development of the endosperm and red or yellow exocarp color; and (iii) raisin (overripe), which are fruits showing the initial characteristics of the senescence cycle with metabolic pathway deviation for the catabolism of the nutrients accumulated in the beans [2]. In Brazil, the largest coffee producer in the world, it is estimated that 31% of the coffee fruits are harvested in the immature stage [3]. Immature coffee beans have a high content of chlorogenic acids (caffeine, trigonelline) and lower sugar content due to incomplete cycle of maturation, attributing astringency and depreciating the quality of coffee products [4,5].

After harvesting, coffee processing must begin as quickly as possible to prevent fruit spoilage by unfavorable fermentation or mold formation [6,7]. The outer layers of the coffee fruit (skin and pulp) are easily removed, while the mucilage, parchment, and silver skin are firmly attached to the beans [8]. The way that coffee growers use to remove the mucilage layer attached to the fruits classifies coffee in the international market: 'Natural coffee', where mucilage is removed by a simple method of sun-drying, known as dry processing; 'washed coffee', produced from coffee beans that undergo a relatively complex series of steps, including depulping, fermentation, and sun-drying known as wet processing; and "pulped natural coffee", which the fruits are mechanically husked and the mucilage is removed by a sun-drying process, known as semi-dry processing [9].

In the wet processing, coffee beans are submitted to underwater tank fermentation for mucilage breakdown and removal. The sugars present in the mucilage support microbial, especially yeasts and lactic acid bacteria [10]. Recent studies have been dedicated to the use of yeast and lactic acid bacteria (LAB) as pure starter cultures in post-harvest processing, in order to reduce the time required for fermentation and modulate the chemical and sensory characteristics of coffee beans [11–14]. Among the selected microorganisms, it is possible to highlight the yeast *Pichia fermentans* YC5.2 and the lactic acid bacteria *Pediococcus acidilactici* LPBC161, which are cultures with characteristics of efficient consumption of coffee pulp-sugars and adaptability to the stress factors of coffee processing [15,16]. Despite that the use of pure cultures offers advantages, recent studies in wine, meat, and dairy fermentations demonstrate that mixed starters are able to improve the sensorial and safety proprieties of the final product [17,18]. In this regard, the aim of this study was to evaluate the effects of co-inoculation with *Pichia fermentans* YC5.2 and *Pediococcus acidilactici* LPBC161 on metabolites produced during fermentation and the volatile composition of coffee beans.

2. Material and Methods

2.1. Microorganism and Inoculum Preparation

The selected yeast (*Pichia fermentans* YC5.2) and lactic acid bacteria (*Pediococcus acidilactici* LPBC161) strains used in this study were previously isolated and selected from spontaneous coffee fermentations, as detailed in Muynarsk et al. [15] and Pereira et al. [16]. The *P. fermentans* YC5.2 and *P. acidilactici* LPBC161 were reactivated in MRS (Merck Millipore, Burlington, MA) and YEPG broth (Himedia, Marg, India), respectively, at 28 °C during 24 h. Each microorganism was then grown up to a concentration of 10^9 CFU/mL. To reach this concentration, *P. acidilactici* LPBC161 was cultivated in Erlenmeyer containing 4 L of sugar cane molasses 3% (w/v) medium, enriched with yeast extract 0.5% (w/v), ammonium citrate 0.5% (w/v), ammonium phosphate 0.5% (w/v), sodium acetate 0.5% (w/v), Tween 80 0.1% (v/v), and manganese sulfate 0.005% (w/v) [19], and *P. fermentans* YC5.2 was grown in Erlenmeyer containing 4 L of sugar cane molasses 3% (w/v) medium enriched with yeast extract 0.5% (w/v). After incubation, the yeast and lactic acid bacteria (LAB) cells were separated from the medium by centrifugation at 5000 × *g* during 5 min, washed twice with sterile saline-peptone solution (0.1% [w/v] bacteriological peptone (Himedia), 0.8% (w/v; NaCl (Merck)), and resuspended in sterile saline solution (0.9% (w/v) NaCl).

2.2. Farm Experiments

The field experiments were conducted at the Fazenda Baobá (21°42′42.8″ S, 46°49′42.2″ W; 1400 m above sea level) situated in São Sebastião da Grama, São Paulo state, Brazil. Ripe and immature coffee fruits (10 kg) were, respectively, deposited in 20-L plastic buckets with 5 L of water. Four sets of inoculation protocols were performed in triplicate: (i) Pure culture fermentation with *P. fermentans*, (ii) pure culture fermentation with *P. acidilactici*, (ii) combined fermentation with *P. fermentans* and *P. acidilactici*, and (iv) spontaneous, non-inoculated control. Prior to inoculation, yeast and LAB cells were counted by a Thoma hemocytometer chamber using methylene blue dye as a marker of cell viability. Then, appropriate amounts of inoculum were used to reach an initial cell population of about 7 log CFU/mL. At the end of fermentation, coffee beans were sun dried until the value of 12% of moisture was reached.

2.3. Sampling and pH Measurement

Samples (50 mL) of the liquid fraction of the fermenting coffee pulp were collected in triplicate at intervals of 12 h to monitor sugars consumption and organic acids, ethanol, and volatile compounds production. At each sampling point, the pH was measured using a digital pH meter (Requipal, Curitiba, Brazil).

2.4. HPLC Analysis of Fermenting Coffee Pulp

The concentration of reducing sugars (glucose and fructose), organic acids (citric, succinic, lactic, acetic, and propionic acids), and ethanol in coffee pulp (liquid fraction) was determined in intervals of 12 h. Aliquots of 2 mL were centrifuged at 6000 × g for 15 min and filtered through 0.22 µm pore size filter (Millipore Corp., Billerica, MA, USA). Analysis parameters were performed according to Carvalho Neto et al. [20]. The filtered samples were injected into high-performance liquid chromatograph (HPLC) system equipped with an Aminex HPX 87 H column (300 × 7.8 mm; Bio-Rad, Richmond, CA, USA) and a refractive index (RI) detector (HPG1362A; Hewlett–Packard Company, Palo Alto, CA, USA). The column was eluted in an isocratic mode with a mobile phase of 5 mM H_2SO_4 at 60 °C and a flow rate of 0.6 mL/min.

2.5. GC Analysis of Fermenting Coffee Pulp

The formation of major volatile compounds was determined in intervals of 12 h by gas chromatography. For sample preparation, aliquots (4 mL) from the liquid fraction of the fermenting coffee pulp were placed in 20 mL hermetically sealed flasks containing NaCl 5% (w/v), followed by heating during 10 min at 60 °C. The headspace was then collected using a glass syringe (Hamilton, Bonaduz, Switzerland) and injected into a gas chromatograph (model 17A; Shimadzu, Kyoto, Japan) equipped with a flame ionization detector at 230 °C. The operation conditions were as follows: A 30 m × 0.32 mm HP-5 capillary column, column temperature of 40 to 150 °C at a rate of 20 °C/min [13]. A standard curve was constructed using authentic analytical standards purchased from Sigma and concentration of the compounds was expressed as µmol/L of headspace

2.6. GC/MS Analysis of Fermented Coffee Beans

The volatile aroma compound composition of spontaneous and inoculated coffee beans was determined by gas chromatography coupled to mass spectrophotometry (GC-MS) according to Carvalho Neto et al. [20]. The extraction of volatile compounds from the beans samples (2 g) was performed using a headspace vial coupled to a solid phase microextraction (SPME) fiber DVB/CAR/PDMS fiber (Supelco Co., Bellefonte, PA, USA). The flasks were heated at 70 °C for 10 min without agitation, followed by 15 min of exposition of the fiber in a COMBI-PAL system. The compounds were desorbed into the gas chromatograph injection system gas phase (CGMS-gun TQ Series 8040 and 2010 Plus GC-MS; Shimadzu, Tokyo, Japan) at 260 °C. The column oven temperature was maintained at 60 °C during 10 min, followed by two heating ramps of 4 and 10 °C/min until reaching the temperatures of 100 and 200 °C, respectively. The compounds were separated on a column 95% PDMS/5% PHENYL (30 m × 0.25 mm × 0.25 mm film thickness). The GC was equipped with an HP 5972 mass selective detector (Hewlett Packard, Palo Alto, CA, USA). Helium was used as carrier gas at a rate of 1.0 mL/min. Mass spectra were obtained by electron impact at 70 eV and a start and end mass-to-charge ratio (m/z) of 30 and 200, respectively. The compounds were identified by comparison to the mass spectra from library databases (Nist'98 and Wiley7N).

2.7. Statistical Analysis

The data obtained of target metabolite analysis were analyzed by post-hoc comparison of means by Duncan's test and a principal component analysis (PCA). Statistical analyses were performed using the SAS program (Statistical Analysis System Cary, NC, USA). Level of significance was established in a two-sided p-value <0.05.

3. Results and Discussion

3.1. Field Experiment

The use of mixed fermentation instead a single culture is a practice widely applied in winemaking in order to improve the aroma complexity or mouth-feel of wines [21,22]. It offers a number of advantages over conventional single-culture fermentations, including higher microbial growth rate and metabolite yield, better utilization of the substrate, and complex formation of aromatic compounds [23]. This work represents the first study on a mixed culture in the coffee beans fermentation process. We experimentally tested the impact of the combination of two selected cultures (*P. fermentans* YC5.2 and *P. acidilactici* LPBC161) in terms of the fermentation efficiency and volatile composition of coffee beans. The experiments were performed with individual and combined inoculations in ripe and immature coffee beans, compared to a spontaneous process. The changes in major non-volatiles (sugars, organic acids, and ethanol) and volatiles metabolites were quantified in the course of the fermentation time. The initial pulp sugar concentration of ripe coffee fruits (0.57 and 1.13 g/L glucose and fructose content, respectively) was significantly higher than immature coffee pulp (0.13 and 0.26 g/L glucose and fructose content, respectively; Figure 1). The low levels of sugars in the case of immature coffee are the consequence of incomplete maturation cycle fruits [24]. In all fermentation processes, the sugar concentration showed an increase during the initial 12 h. This phenomenon can be associated with the hydrolysis of pectin, cellulose, sucrose, and other coffee pulp complexes carbohydrates, into monomers of glucose and fructose [25,26]. After this increase, both glucose and fructose were partially consumed, resulting in a residual concentration of around 0.37 and 1.51 g/L (ripe coffee pulp) and 0.19 and 0.74 g/L (immature coffee pulp) of glucose and fructose, respectively. Residual pulp sugars are generally observed after the coffee fermentation process, mainly associated with the short fermentation cycle [14,27,28]. However, fructose consumption was more efficient in the treatments that the yeast was used (i.e., *P. fermentans*-pure culture and combined treatment with *P. fermentans* and *P. acidilactici*) than *P. acidilactici* pure culture and spontaneous assay (Figure 1). The yeast's ability to withstand stress tolerance factors and the production of pectinolytic enzymes confer advantages in comparison to lactic acid bacteria [16,29]. In addition, the low availability of initially willing nutrients in immature coffee pulp [24] may have restricted the development of *P. acidilactici*, showing poor sugar consumption (Figure 1B). Lactic acid bacteria are further distinguished by their limited biosynthetic abilities, being unable to synthesize multiple cofactors, vitamins, purines, pyrimidines, and other nutrients [30].

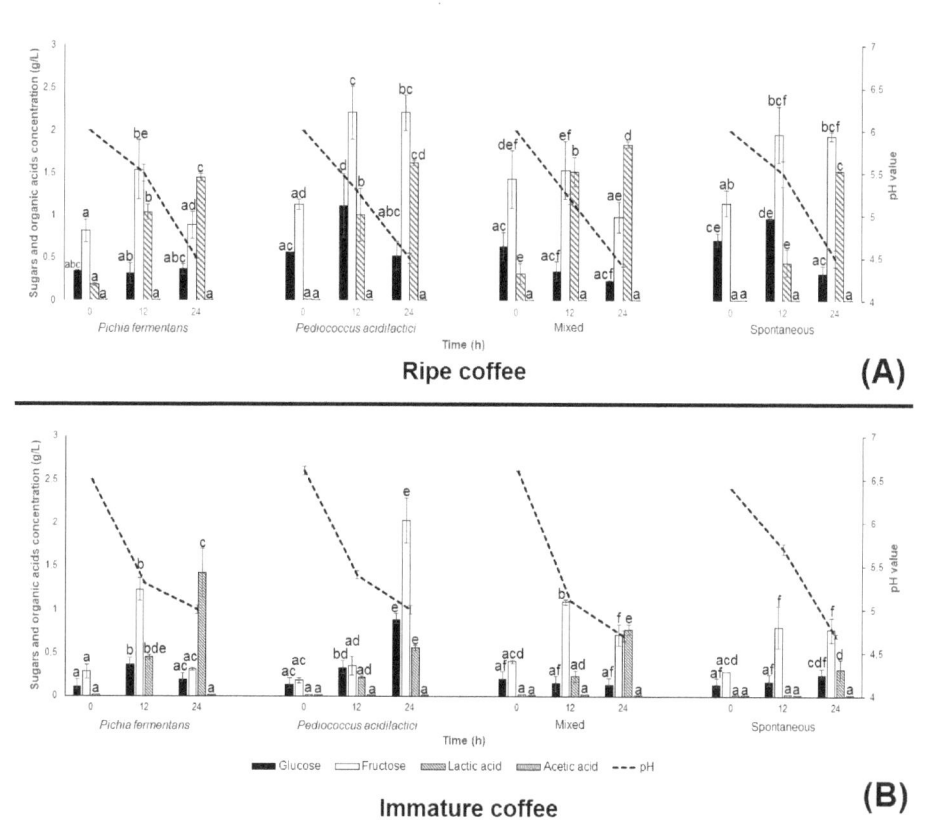

Figure 1. Pulp sugar consumption, organic acid production, and pH monitoring of inoculated (pure culture with *Pichia fermentans*, pure culture with *Pediococcus acidilactici*, and combined fermentation with *P. fermentans* and *P. acidilactici*) and spontaneous fermentation of ripe (**A**) and immature (**B**) coffee fruits. The significance of the results was assessed using an ANOVA with Duncan's post-hoc test at $p < 0.05$. Different lowercase letters (a, b, c, d, e, f) indicate significant differences within the same process (ripe and immature coffee beans) over fermentation time.

Lactic acid showed a significant increase through fermentative processes, reaching maximum concentration of ≥1.45 and ≥0.53 g/L in the ripe and immature treatments, respectively (Figure 1). Basal concentrations of acetic acid (≤0.01 g/L) were detected in all the processes, which can be associated with yeast metabolism or the heterofermentative nature of *P. acidilactici* [31,32]. Overfermentation acids, such as propionic and butyric acids, were not detected in both ripe and immature treatments. In this sense, the acidification of fermenting coffee pulp can be attributed mainly to lactic acid content. As expected, while the lactic acid concentration increased, the pH decreased progressively during all fermentation processes (Figure 1). Lactic acid is an important end-metabolite associated with coffee fermentation, which assists in the coffee-pulp acidification process without interference in the final product (Figure 1). The pH monitoring is a crucial parameter, since pH values below 4.5 are used as to indicate the end of the coffee fermentation process [33–35]. In immature coffee treatments, a pH higher than 4.5 was reported, which may be attributed to the insufficient development of *P. acidilactici*.

Ethanol and ethyl acetate were the major volatile compounds detected during fermentation processes (Figure 2). Interestingly, combined inoculations with *P. acidilactici* and *P. fermentans* resulted in a significant increase in the production of these metabolites when compared to pure cultures and a spontaneous process. The higher values of ethanol (27.04 and 14.8 μmol/L) and ethyl acetate (1.63 and 1.21 μmol/L) were reached after 24 h of ripe and immature combined fermentations, respectively.

This agrees with the findings of Sun et al. [36] and Cañas et al. [37] that demonstrated an increase of ethyl- and acetate esters as a result of the co-inoculation of LAB and yeasts in wine fermentation. The significant sugar consumption and lactic acid, ethyl acetate, and ethanol production in treatments with combined inoculations indicated an ecological interaction between these two microbial groups. The complex nature of this interaction is highlighted by the observations that (i)the autolysis of yeasts release nutrients, such as amino acids, polysaccharides and riboflavin, favorable for bacterial growth, and that (ii) the acidification of the fermentation media by LAB creates a prone environment for yeast development [9,38,39]. These positive interactions have been shown to promote desired sensory attributes in wine, sourdough, and yogurt. However, information about these mechanisms in coffee fermentation is scarce [40].

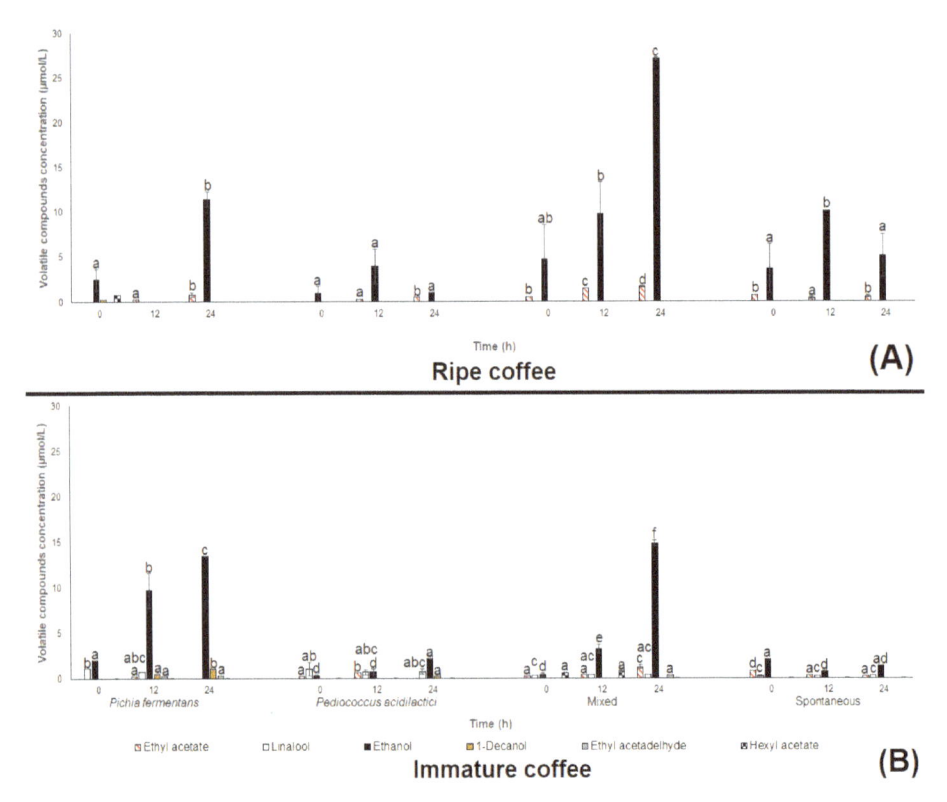

Figure 2. Concentration of volatile compounds produced in the course inoculated (pure culture with *P. fermentans*, pure culture with *P. acidilactici*, and combined fermentation with *P. fermentans* and *P. acidilactici*) and spontaneous fermentation of ripe (**A**) and immature (**B**) coffee fruits. The significance of the results was assessed using an ANOVA with Duncan's post-hoc test at $p < 0.05$. Different lowercase letters (a, b, c, d, e, f) indicate significant differences within the same process (ripe and immature coffee beans) over fermentation time.

Other minor volatile compounds that increased during the fermentation processes were 1-decanol, ethyl-acetaldehyde, and hexyl acetate (Figure 2). Yeast and lactic acid bacteria generate ethyl-acetaldehyde by a condensation reaction between fatty acids and an alcohol molecule [41], while 1-decanol can be derived from amino acid catabolism via the Ehrlich pathway [16,42]. The presence of higher concentrations of linalool using immature fruits can be associated with the inferior maturation stage of the coffee beans, since this compound has been considered a volatile marker of coffee beverages produced from immature fruits [43].

3.2. Coffee Beans Chemical Composition

Over 900 volatile compounds have already been identified in green and roasted coffee beans [27,44]. Among the major volatiles found, pyrazines, furans, ketones, aldehydes, higher alcohols, esters, and sulphur compounds can be highlighted [45–47]. Although some of these flavor-active compounds originate from the beans itself, recent studies have revealed that microbial-derived metabolites can also diffuse into the beans [11,14,27,33,48–50]. Upon characterization of the volatile composition of fermented coffee beans, it was observed that inoculation of *P. fermentans* and *P. acidilactici*, both in pure and combined treatments, resulted in the modulation of the volatile constitution of coffee beans. A total of 30 compounds were identified in fermented ripe coffee beans, including higher alcohols (seven compounds), aldehydes (six compounds), and terpenes (three compounds; Table 1). Among these, 1-hexanol, 2-heptanol, phenylethyl alcohol, and benzeneacetaldehyde were the major volatiles found. Single inoculation of *P. fermentans* and *P. acidilactici* resulted in the formation and diffusion of some volatile compounds, such as 3-octanol, 2-heptenal, benzaldehyde, dodecanal, and D-limonene, that were not detected in a spontaneous process. These compounds are strictly related to both yeast and LAB metabolism, such as aldehydes and higher alcohols formed from the catabolism of coffee pulp amino acids, and terpenes through mevalonic acid pathway or released from glycoside precursors during fermentation [10,51,52].

Table 1. Concentration of volatile compounds (Area*10⁵) in ripe coffee beans after single cultures, a combined treatment, and a spontaneous assay. Means of triplicate in each row bearing the same lowercase letters (a, b) are not significantly different ($p > 0.05$) from one another using Duncan's Test (mean ± standard variation).

Compounds	Aroma Perception	Fermented Ripe Coffee Beans			
		Spontaneous	P. fermentans	P. acidilactici	P. fermentans + P. acidilactici
Organic acids (3)					
Butanoic acid, 3-methyl	-	7.02 ± 1.00 [a]	6.88 ± 1.48 [a]	6.10 ± 1.74 [a]	6.91 ± 0.18 [a]
Butanoic acid, 2-methyl	Fruity, dirty, acidic with a dairy buttery	1.77 ± 0.35 [a]	1.98 ± 0.28 [a]	ND	1.79 ± 0.28 [a]
Hexanoic acid	Sour, fatty, sweat, cheesy	ND	1.34 ± 0.61 [a]	0.99 ± 0.23 [a]	1.67 ± 0.88 [a]
Higher alcohols (7)					
2-Heptanol	Fresh, lemon grass, herbal	10.21 ± 0.00 [a]	ND	ND	12.12 ± 0.10 [b]
5-Methyl-2-Hexanol	-	6.09 ± 2.66	ND	ND	ND
1-Hexanol	Green, fruity, apple-skin and oily	15.90± 2.36 [a]	16.95 ± 2.23 [a]	17.10 ± 2.36 [a]	18.24 ± 0.17 [a]
1-Octen-3-ol	Earthy, green, oily, umami sensation	0.82 ± 0.00 [a]	0.80 ± 0.17 [a]	ND	0.86 ± 0.03 [a]
3-Octanol	Musty, mushroom, earthy, creamy dairy	ND	0.50 ±0.14 [a]	0.58 ± 0.00 [a]	0.67 ± 0.00 [a]
Benzylalcohol	Sweet, fruity with balsamic nuances	2.58 ± 0.05 [a]	3.26 ± 0.36 [a]	3.20 ± 0.68 [a]	4.31 ± 0.17 [b]
Phenylethyl alcohol	Floral, sweet and bready	9.52 ± 0.73 [a]	10.34 ± 0.41 [a]	12.17 ± 0.25 [b]	12.60 ± 1.31 [b]
Esters (3)					
Butanoic acid, 2-methyl, ethyl ester	-	0.58 ± 0.26 [a]	0.34 ± 0.00 [a]	ND	0.54 ± 0.20 [a]
Butanoic acid, 3-methyl- ethyl ester	-	4.62 ± 0.59 [a]	3.66 ± 1.71 [a]	ND	4.70 ± 0.57 [a]
Methyl salicylate	Wintergreen, mint-like	ND	0.28 ± 0.01	ND	ND
Aldehydes (6)					
2-Heptenal	Sweet, fresh fruity apple skin nuances	ND	0.80 ± 0.00 [a]	ND	0.80 ± 0.00 [a]
Benzaldehyde	Fruity, cherry	ND	0.32 ± 0.00 [a]	2.15 ± 0.84 [a]	2.56 ± 0.69 [a]
Dodecanal	Soapy, waxy, citrus, orange mandarin	ND	ND	0.23 ± 0.14	ND
Nonanal	With a fresh green lemon peel-like nuance	0.95 ± 0.05 [a]	0.93 ± 0.35 [a]	0.81 ± 0.06 [a]	0.84 ± 0.27 [a]
Benzeneacetaldehyde	Almond, fruity, powdery, nutty	1.83 ± 0.29 [a]	2.80 ± 0.73 [a]	2.28 ± 0.04 [a]	6.94 ± 0.00 [b]
Decanal	Sweet, aldehydic, orange, waxy and citrus rind	ND	0.37 ± 0.11 [a]	0.37 ± 0.09 [a]	0.34 ± 0.07 [a]
Ketone (1)					
2-Heptanone	Fruity, spice, herbal	3.49 ± 0.69 [a]	2.92 ± 0.28 [a]	2.29 ± 0.90 [a]	4.38 ± 1.75 [a]

Table 1. *Cont.*

Compounds	Aroma Perception	Fermented Ripe Coffee Beans			
		Spontaneous	*P. fermentans*	*P. acidilactici*	*P. fermentans* + *P. acidilactici*
Pyridine (2)					
Pyridine, 2,3-dimethyl	-	ND	1.55 ± 0.94 [a]	1.44 ± 0.26 [a]	1.25 ± 0.63 [a]
Pyridine, 2,6-Lutidine	Nutty, amine, woody, bready and vegetable-like	1.85 ± 0.19	ND	ND	ND
Lactone (1)					
Butyrolactone	Creamy, oily, fatty, caramellic	5.28 ± 0.10 [a]	6.17 ± 1.04 [a]	4.86 ± 1.13 [a]	6.09 ± 0.13 [a]
Terpenes (3)					
Linalool	Citrus, orange, lemon	2.36 ± 0.29 [a]	2.27 ± 0.21 [a]	2.22 ± 0.82 [a]	2.92 ± 0.17 [b]
D-Limonene	Sweet, orange, citrus	ND	1.29 ± 0.60 [a]	1.37 ± 0.49 [a]	1.42 ± 0.35 [a]
Anethole	-	4.10 ± 1.00 [b]	1.87 ± 0.26 [a]	1.96 ± 0.66 [a]	3.13 ± 0.20 [a,b]
Hydrocarbons (2)					
Styrene	Sweet, balsamic, floral	ND	2.89 ± 0.00 [a]	ND	3.15 ± 0.00 [a]
Tetradecane	Waxy	0.91 ± 0.07 [a]	0.82 ± 0.06 [a]	0.88 ± 0.16 [a]	0.80 ± 0.06 [a]
Pyrzine (1)					
Pyrazine, 2-methoxy-3-(2-methylpropyl)	Roasted almond hazelnut peanut	0.92 ± 0.04 [a]	ND	0.93 ± 0.08 [a]	ND
Furan (1)					
Furan, 2-pentyl	Waxy, with musty, cooked caramellic nuances	1.16 ± 0.04	ND	ND	ND

Interestingly, coffee beans generated from combined treatments showed significantly increased ($p < 0.05$) of specific volatile compounds, such as benzeneacetaldehyde, 2-heptanol, and benzylalcohol. These findings are in accordance with Englezos et al. [18] and Plessas et al. [53], which evidences that mixed treatments of yeast and LAB starter cultures enable higher production of esters, aldehydes, and higher alcohols in sourdough and wine fermentations when compared to single inoculations. However, further studies are required to evidence the metabolic pathways associated with the positive interaction between these microorganisms in coffee products.

Chemical analysis of immature coffee beans revealed a composition with lower diversity and concentration of volatile compounds (Table 2). A total of 19 compounds were detected, including higher alcohols (four compounds), organic acids (three compounds), and aldehydes (three compounds). *P. fermentans*-single inoculation resulted in coffee beans with significantly higher concentrations ($p < 0.05$) of 2-hexanol, nonanal, and D-limonene when compared to the spontaneous process. These compounds are commonly attributed to *Pichia* metabolism [10,54,55]. This corroborates with results from fermentation process monitoring, which demonstrated intense microbial activity of *P. fermentans* in immature coffee pulp. On the other hand, coffee beans derived from *P. acidilactici*-pure culture and combined treatment showed no significant increase ($p > 0.05$) in the volatile constituents when compared to the control (spontaneous process). This fact can be correlated to the insufficient growth of the LAB starter culture in the nutrient-scarce environment from the pulp of immature coffee beans. The auxotrophism of several amino acids turns LAB directly dependent on a rich growth medium for its full development [31].

Table 2. Concentration of volatile compounds (Area*10^5) in immature coffee beans after single cultures, a combined treatment, and a spontaneous assay. Means of triplicate in each row bearing the same lowercase letters (a, b, c) are not significantly different ($p > 0.05$) from one another using Duncan's Test (mean ± standard variation).

Compounds	Aroma Perception	Fermented Immature Coffee Beans			
		Spontaneous	P. fermentans	P. acidilactici	P. fermentans + P. acidilactici
Organic acids (3)					
Butanoic acid, 3-methyl	-	7.77 ± 1.05 [a]	5.99 ± 2.31 [a]	8.56 ± 1.57 [a,b]	5.15 ± 0.98 [a,c]
Butanoic acid, 2-methyl	Fruity, acidic with a dairy buttery	1.32 ± 0.36 [a]	1.24 ± 0.86 [a]	1.65 ± 0.13 [a,b]	0.69 ± 0.15 [a,c]
Hexanoic acid	Sour, fatty, sweat, cheesy	ND	0.63 ± 0.15 [a]	0.50 ± 0.11 [a]	ND
Higher alcohols (4)					
1-Hexanol	Green, fruity, apple-skin and oily	10.02 ± 0.26 [a]	10.57 ± 1.33 [a]	9.14 ± 1.51 [a]	9.85 ± 1.17 [a]
2-Hexanol	-	0.36 ± 0.05 [a]	0.75 ± 0.01 [b]	ND	0.45 ± 0.10 [a]
Benzylalcohol	Sweet, fruity with balsamic nuances	0.54 ± 0.15 [a]	0.62 ± 0.08 [a]	0.51 ± 0.20 [a]	0.61 ± 0.04 [a]
Phenylethyl alcohol	Floral, sweet and bready	2.54 ± 0.38 [a]	2.78 ± 0.17 [b]	2.24 ± 0.12 [a]	2.20 ± 0.18 [a]
Aldehydes (3)					
Nonanal	Citrus, with a fresh green lemon peel-like nuance	0.30 ± 0.08 [a]	0.74 ± 0.04 [b]	0.46 ± 0.07 [c,d]	0.48 ± 0.05 [d]
Benzeneacetaldehyde	Almond, fruity, powdery, nutty	0.21 ± 0.00 [a]	0.28 ± 0.01 [a]	0.27 ± 0.13 [a]	0.38 ± 0.00 [a]
Decanal	Sweet, orange, waxy and citrus rind	0.24 ± 0.09 [a]	0.43 ± 0.06 [a]	0.30 ± 0.10 [a]	0.42 ± 0.08 [a]
Pyridines (2)					
Pyridine, 2,3-dimethyl	-	ND	1.45 ± 0.34 [a]	0.62 ± 0.14 [a]	1.10 ± 0.36 [a]
Pyridine, 2,6-Lutidine	Nutty, woody, bready and vegetable-like	0.95 ± 0.35	ND	ND	ND
Lactone (1)					
Butyrolactone	Creamy, oily, fatty, caramellic	0.92 ± 0.07 [a]	0.46 ± 0.20 [a]	0.66 ± 0.26 [a]	0.49 ± 0.17 [a]
Terpenes (1)					
D-Limonene	Sweet, orange, citrus	0.36 ± 0.06 [a]	0.72 ± 0.16 [b]	0.16 ± 0.09 [c]	0.41 ± 0.19 [a]

Table 2. Cont.

| Compounds | Aroma Perception | Fermented Immature Coffee Beans ||||
		Spontaneous	P. fermentans	P. acidilactici	P. fermentans + P. acidilactici
Furans (2)					
Furan-2-pentyl	Waxy, cooked caramellic nuances	0.57 ± 0.17 [a]	0.74 ± 0.12 [a]	0.58 ± 0.15 [a]	0.69 ± 0.13 [a]
2(3)-Furanone, dihydro-5-methyl	Creamy, waxy with a citrus fruity nuance	0.55 ± 0.14 [a]	0.38 ± 0.16 [a]	ND	0.45 ± 0.15 [a]
Hydrocarbons (2)					
Hexadecane	-	ND	0.55 ± 0.01	ND	ND
Tetradecane	Waxy	0.62 ± 0.02 [a]	0.24 ± 0.07 [b]	0.30 ± 0.06 [c]	0.76 ± 0.10 [a,c]
Pyrzine (1)					
Pyrazine, 2-methoxy-3-(2-methylpropyl)	Roasted almond hazelnut peanut	0.78 ± 0.07 [a]	0.80 ± 0.01 [a]	0.85 ± 0.12 [a]	0.92 ± 0.08 [a]

In order to explain the chemical characteristics and grouping of the samples, the parameters in Tables 1 and 2 were analyzed by a PCA (Figure 3). The first and second principal components explained, together, 73.66% of the total variability within the data. The samples were categorized into two clusters, *viz.*, ripe and immature coffee beans. This distinction was mainly related to the richer constitution of volatiles of ripe coffee beans relative to immature treatments. In addition, the presence of specific compounds (benzyl alcohol, phenylethyl alcohol, benzeneacetaldehyde, decanal, and D-limonene in ripe coffee beans, and furan-2-pentyl, 2-methyl-butanoic acid, and pyrazine, 2-methoxy-3-(2-methylpropyl) in immature coffee beans) also contributed to the separation of ripe and immature coffee beans in the PCA analysis. Interestingly, only the treatment with *P. fermentans* grouped immature coffee beans in the positive axis, which corroborates with the better yeast' adaptation and generation of volatiles in immature coffee pulp.

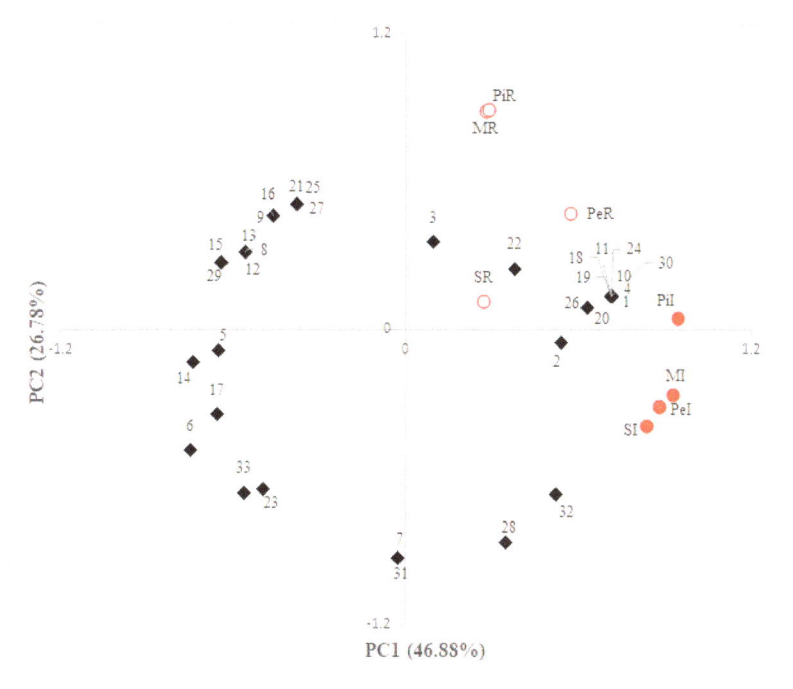

Figure 3. Principal component analysis (PCA) of volatile compounds (lozenges) identified in the different treatments of ripe (open circles) and immature (closed circles) coffee beans. Abbreviations: *SR*—spontaneous, ripe control; *PiR*—*Pichia* inoculation in ripe coffee beans; *PeR*—*Pediococcus* inoculation in ripe coffee beans; *MR*—mixed (*Pichia* plus *Pediococcus*) inoculation in ripe coffee beans; *SI*—spontaneous, immature control; *PiI*—*Pichia* inoculation in immature coffee beans; *PeI*—*Pediococcus* inoculation in immature coffee beans; *MI*—mixed (*Pichia* plus *Pediococcus*) inoculation in immature coffee beans. 1—butanoic acid, 3-methyl; 2—butanoic acid, 2-methyl; 3—hexanoic acid; 4—1-hexanol; 5—2-heptanol; 6—5-methyl-2-hexanol; 7—2-hexanol; 8—1-octen-3-ol; 9—3-octanol; 10—benzylalcohol; 11—phenylethyl alcohol; 12—butanoic acid, 2-methyl, ethyl ester; 13 - butanoic acid, 2-methyl, ethyl ester; 14—methyl salicylate; 15—2-heptenal; 16—benzaldehyde; 17—dodecanal; 18—nonanal; 19—benzeneacetaldehyde; 20—decanal; 21—2-heptanone; 22—pyridine, 2,3-dimethyl; 23—pyridine, 2,6-lutidine; 24—butyrolactone; 25—linalool; 26—D-limonene; 27—anethole; 28—furan-2-pentyl; 29—styrene; 30—tetradecane; 31—2(3)-furanone, dihydro-5-methyl; 32—pyrazine, 2-methoxy-3-(2-methylpropyl); 33—hexadecane.

4. Conclusions

This is the first study investigating the impact of co-inoculation with yeast and LAB on the fermentation of ripe and immature coffee fruits. Among the different treatments, combined inoculations with *Pichia fermentans* YC5.2 and *Pediococcus acidilactici* LPBC161 in ripe coffee fruits showed interesting features. It was possible to reach increased coffee pulp-sugar consumption and production of metabolites (lactic acid, ethanol, and ethyl acetate), evidencing a positive synergic interaction between these two microbial groups. On the other hand, when using immature coffee fruits, only *Pichia fermentans* was able to improve metabolite formation during fermentation and impact volatile composition of resulting coffee beans. This may be due to the high nutritional requirement of LAB species and poor adaptability in immature coffee pulp. Howsoever, since immature coffee beans usually have a low quality because the formation of flavored precursors is incomplete, yeast metabolism has great potential to add flavor quality to these beans.

Chemical analysis revealed a more complex volatile composition of fermented coffee beans from combined treatment in relation to pure inoculations and spontaneous process. The major compounds impacted were benzeneacetaldehyde, 1-hexanol, benzylalcohol, 2-heptanol, and phenylethyl alcohol, which are reported as important aroma-impacting compounds. Thus, this study shows the great potential of combined inoculation for the formation of desirable aroma compounds and production of specialty coffees. Further studies are still required to investigate the mechanisms of synergism between yeast and LAB and influence on sensory properties of coffee products.

Author Contributions: C.R.S. and G.V.d.M.P. designed the experiments. C.R. conducted the physicochemical characterization of fermenting coffee pulp. A.d.S.V. and D.P.d.C.N. performed the experiments and wrote the manuscript. C.R.S., G.V.d.M.P., and M.G.B.P., reviewed and edited the manuscript. All authors read and approved the final manuscript.

Funding: This work was supported by the Brazilian National Council for Scientific and Technological Development (CNPq) (project number 429560/2018-4).

Conflicts of Interest: The authors declare no conflict of interest.

References

1. Huch, M.; Franz, C.M.A.P. Coffee: Fermentation and microbiota. In *Advances in Fermented Foods and Beverages: Improving Quality, Technologies and Health Benefits*, 1st ed.; Holzapfel, W., Ed.; Woodhead Publishing: Sawston, UK, 2015; Volume 4, pp. 501–513.
2. DaMatta, F.M.; Ronchi, C.P.; Maestri, M.; Barros, R.S. Ecophysiology of coffee growth and production. *Braz. J. Plant Physiol.* **2007**, *19*, 485–510. [CrossRef]
3. Mesquita, C.M.; Rezende, J.E.; De Carvalho, J.S.; Junior, M.A.F.; Moraes, N.C.; Dias, P.T.; Carvalho, R.M.; de Araújo, W.G. *Manual do café: Colheita e preparo*; Emater: Belo Horizonte, Brazil, 2016; pp. 6–21.
4. Franca, A.S.; Oliveira, L.S.; Mendonça, J.C.F.; Silva, X.A. Physical and chemical attributes of defective crude and roasted coffee beans. *Food Chem.* **2005**, *90*, 89–94. [CrossRef]
5. Oliveira, L.S.; Franca, A.S.; Mendonça, J.C.F.; Barros-Júnior, M.C. Proximate composition and fatty acids profile of green and roasted defective coffee beans. *LWT Food Sci. Technol.* **2006**, *39*, 235–239. [CrossRef]
6. Bee, S.; Brando, C.H.J.; Brumen, G.; Carvalhaes, N.; Kölling-Speer, I.; Speer, K.; Liverani, F.S.; Teixeira, A.A.; Thomaziello, R.A.; Viani, R.; et al. The raw coffee. In *Expresso Coffee: The Science of Quality*; Illy, A., Vinai, R., Eds.; Elsevier Academic Press: London, UK, 2005; pp. 87–178.
7. Illy, E. The complexity of coffee. *Sci. Am.* **2002**, *286*, 86–91. [CrossRef] [PubMed]
8. De Bruyn, F.; Zhang, S.J.; Pothakos, V.; Torres, J.; Lambot, C.; Moroni, A.V.; Callanan, M.; Sybesma, W.; Weckx, S.; De Vuyst, L. Exploring the impacts of postharvest processing on the microbiota and metabolite profiles during green coffee bean production. *Appl. Environ. Microbiol.* **2017**, *83*, e02398-16. [CrossRef] [PubMed]
9. Pereira, G.V.M.; Soccol, V.T.; Brar, S.K.; Neto, E.; Soccol, C.R. Microbial ecology and starter culture technology in coffee processing. *Crit. Rev. Food Sci. Nutr.* **2017**, *57*, 2775–2788. [CrossRef] [PubMed]

10. Pereira, G.V.M.; Carvalho Neto, D.P.; Júnior, A.I.M.; Vásquez, Z.S.; Medeiros, A.B.P.; Vandenberghe, L.P.S.; Soccol, C.R. Exploring the impacts of postharvest processing on the aroma formation of coffee beans—A review. *Food Chem.* **2019**, *272*, 441–452. [CrossRef] [PubMed]
11. Wang, C.; Sun, J.; Lassabliere, B.; Yu, B.; Zhao, F.; Zhao, F.; Chen, Y.; Liu, S.Q. Potential of lactic acid bacteria to modulate coffee volatiles and effect of glucose supplementation: Fermentation of green coffee beans and impact of coffee roasting. *J. Sci. Food Agric.* **2019**, *99*, 409–420. [CrossRef]
12. Lee, L.W.; Cheong, M.W.; Curran, P.; Yu, B.; Liu, S.Q. Modulation of coffee aroma via the fermentation of green coffee beans with *Rhizopus oligosporus*: I. Green coffee. *Food Chem.* **2016**, *211*, 916–924. [CrossRef]
13. Pereira, G.V.M.; Carvalho Neto, D.P.; Medeiros, A.B.P.; Soccol, V.T.; Neto, E.; Woiciechowski, A.L.; Soccol, C.R. Potential of lactic acid bacteria to improve the fermentation and quality of coffee during on-farm processing. *Int. J. Food Sci. Technol.* **2016**, *51*, 1689–1695. [CrossRef]
14. Pereira, G.V.M.; Neto, E.; Soccol, V.T.; Medeiros, A.B.P.; Woiciechowski, A.L.; Soccol, C.R. Conducting starter culture-controlled fermentations of coffee beans during on-farm wet processing: Growth, metabolic analyses and sensorial effects. *Food Res. Int.* **2015**, *75*, 348–356. [CrossRef] [PubMed]
15. Muynarsk, E.S.M.; Pereira, G.V.M.; Mesa, D.; Thomaz-Soccol, V.; Carvalho, J.C.; Pagnoncelli, M.G.B.; Soccol, C.R. Draft genome sequence of *Pediococcus acidilactici* strain LPBC161, isolated from mature coffee cherries during natural fermentation. *Microbiol. Resour. Announc.* **2019**, *8*, e00332-19. [CrossRef] [PubMed]
16. Pereira, G.V.M.; Soccol, V.T.; Pandey, A.; Medeiros, A.B.P.; Lara, J.M.R.A.; Gollo, A.L.; Soccol, C.R. Isolation, selection and evaluation of yeasts for use in fermentation of coffee beans by the wet process. *Int. J. Food Microbiol.* **2014**, *188*, 60–66. [CrossRef] [PubMed]
17. Xia, X.; Luo, Y.; Zhang, Q.; Huang, Y.; Zhang, B. Mixed starter culture regulates biogenic amines formation via decarboxylation and transamination during Chinese rice wine fermentation. *J. Agric. Food Chem.* **2018**, *66*, 6348–6356. [CrossRef] [PubMed]
18. Englezos, V.; Torchio, F.; Cravero, F.; Marengo, F.; Giacosa, S.; Gerbi, V.; Rantsiou, K.; Rolle, L.; Cocolin, L. Aroma profile and composition of Barbera wines obtained by mixed fermentations of *Starmerella bacillaris* (synonym *Candida zemplinina*) and *Saccharomyces cerevisiae*. *LWT Food Sci. Technol.* **2016**, *73*, 567–575. [CrossRef]
19. Feltrin, V.P.; Sant'Anna, E.S.; Porto, A.C.S.; Torres, R.C.O. Produção de Lactobacillus plantarum em melaço de cana-de-açúcar. *Braz. Arch. Biol. Technol.* **2000**, *43*, 119–124. [CrossRef]
20. Carvalho Neto, D.P.; Pereira, G.V.M.; Tanobe, V.O.A.; Thomaz-Soccol, V.; da Silva, B.G.J.; Rodrigues, C.; Soccol, C.R. Yeast diversity and physicochemical characteristics associated with coffee bean fermentation from the Brazilian cerrado mineiro region. *Fermentation* **2017**, *3*, 11. [CrossRef]
21. Gobbi, M.; Comitini, F.; Domizio, P.; Romani, C.; Lencioni, L.; Mannazzu, I.; Ciani, M. *Lachancea thermotolerans* and *Saccharomyces cerevisiae* in simultaneous and sequential co-fermentation: A strategy to enhance acidity and improve the overall quality of wine. *Food Microbiol.* **2013**, *33*, 271–281. [CrossRef]
22. Cinai, M.; Comitini, F.; Mannazzu, I.; Domizio, P. Controlled mixed culture fermentation: A new perspective on the use of non-*Saccharomyces* yeasts in winemaking. *FEMS Yeast Res.* **2010**, *10*, 123–133. [CrossRef]
23. Gaden, E.L., Jr.; Bokanga, M.; Harlander, S.; Hesseltine, C.W.; Steinkraus, K.H. *Applications of Biotechnology to Traditional Fermented Foods*; National Academy Press: Washington, DC, USA, 1992.
24. Eira, M.T.S.; da Silva, E.A.A.; de Castro, R.D.; Dussert, S.; Walters, C.; Bewley, J.D.; Hilhorst, W.M. Coffee seed physiology. *Braz. J. Plant Physiol.* **2006**, *18*, 149–163. [CrossRef]
25. Marques, W.L.; Raghavendran, V.; Stambuk, B.U.; Gombert, A.K. Sucrose and *Saccharomyces cerevisiae*: A relationship most sweet. *FEMS Yeast Res.* **2016**, *16*, fov107. [CrossRef] [PubMed]
26. Murthy, P.S.; Naidu, M. Improvement of Robusta coffee fermentation with microbial enzymes. *Eur. J. Appl. Sci.* **2011**, *3*, 130–139.
27. Evangelista, S.R.; Miguel, M.G.C.P.; Cordeiro, C.S.; Silva, C.F.; Pinheiro, A.C.M.; Schwan, R.F. Inoculation of starter cultures in a semi-dry coffee (*Coffea arabica*) fermentation process. *Food Microbiol.* **2014**, *44*, 87–95. [CrossRef] [PubMed]
28. Evangelista, S.R.; Silva, C.F.; Miguel, M.G.P.C.; Cordeiro, C.S.; Pinheiro, A.C.M.; Duarte, W.F.; Schwan, R.F. Improvement of coffee beverage quality by using selected yeasts strains during the fermentation in dry process. *Food Res. Int.* **2014**, *61*, 183–195. [CrossRef]
29. Avallone, S.; Brillouet, J.M.; Guyot, B.; Olguin, E.; Guiraud, J.P. Involvement of pectolytic micro-organisms in coffee fermentation. *Int. J. Food Sci. Technol.* **2002**, *37*, 191–198. [CrossRef]

30. Muñoz, R.; Moreno-Arribas, M.V.; de las Rivas, B. Lactic acid bacteria. In *Molecular Wine Microbiology*, 1st ed.; Carrascosa, A.V., Muñoz, R., González, R., Eds.; Academic Press: Burlington, VT, USA, 2011; pp. 191–226.
31. Endo, A.; Dicks, L.M.T. Physiology of the LAB. In *Lactic Acid Bacteria: Biodiversity and Taxonomy*; Holzapfel, W.H., Wood, B.J.B., Eds.; Wiley Blackwell: Chichester, UK, 2014; pp. 13–30.
32. Rantsiou, K.; Dolci, P.; Giacosa, S.; Torchio, F.; Tofalo, R.; Torriani, S.; Suzzi, G.; Rolle, L.; Cocolin, L. *Candida zemplinina* can reduce acetic acid produced by *Saccharomyces cerevisiae* in sweet wine fermentations. *Appl. Environ. Microbiol.* **2012**, *78*, 1987–1994. [CrossRef] [PubMed]
33. Carvalho Neto, D.P.; Pereira, G.V.M.; Finco, A.M.O.; Letti, L.A.J.; Silva, B.J.G.; Vandenberghe, L.P.S.; Soccol, C.R. Efficient coffee beans mucilage layer removal using lactic acid fermentation in a stirred-tank bioreactor: Kinetic, metabolic and sensorial studies. *Food Biosci.* **2018**, *26*, 80–87. [CrossRef]
34. Velmourougane, K. Impact of natural fermentation on physicochemical, microbiological and cup quality characteristics of Arabica and Robusta coffee. *Proc. Natl. Acad. Sci. USA India Sect. B Biol. Sci.* **2013**, *83*, 233–239. [CrossRef]
35. Jackels, S.C.; Jackels, C.F. Characterization of the coffee mucilage fermentation process using chemical indicators: A field study in Nicaragua. *Food Chem. Toxicol.* **2005**, *70*, 321–325. [CrossRef]
36. Sun, S.Y.; Gong, H.S.; Zhao, K.; Wang, X.L.; Wang, X.; Zhao, X.H.; Yu, B.; Wang, H.X. Co-inoculation of yeast and lactic acid bacteria to improve cherry wines sensory quality. *Int. J. Food Sci. Technol.* **2013**, *48*, 1783–1790. [CrossRef]
37. Cañas, P.M.I.; Romero, E.G.; Pérez-Martín, F.; Seseña, S.; Palop, M.L. Sequential inoculation versus co-inoculation in Cabernet Franc wine fermentation. *Food Sci. Technol. Int.* **2015**, *21*, 203–212. [CrossRef] [PubMed]
38. Alexandre, H.; Guilloux-Benatier, M. Yeast autolysis in sparkling wine—A review. *Aust. J. Grape Wine Res.* **2006**, *12*, 119–127. [CrossRef]
39. Fleet, G.H. Yeast interactions and wine flavour. *Int. J. Food Microbiol.* **2003**, *86*, 11–22. [CrossRef]
40. Junqueira, A.C.O.; Pereira, G.V.M.; Medina, J.D.C.; Alvear, M.C.R.; Rosero, R.; Carvalho Neto, D.P.; Enríquez, H.G.; Soccol, C.R. First description of bacterial and fungal communities in Colombian coffee beans fermentation analysed using Illumina-based amplicon sequencing. *Sci. Rep.* **2019**, *9*, 8794. [CrossRef] [PubMed]
41. Saerens, S.M.G.; Delvaux, F.R.; Verstrepen, K.J.; Thevelein, J.M. Production and biological function of volatile esters in *Saccharomyces cerevisiae*. *Microb. Biotechnol.* **2010**, *3*, 165–177. [CrossRef] [PubMed]
42. Elías, L.G. Chemical composition of coffee-berry by-products. In *Coffee Pulp: Composition, Technology, and Utilization*; Braham, J.E., Bressani, R., Eds.; IDRC: Ottawa, ON, Canada, 1979; pp. 11–16.
43. Toci, A.T.; Farah, A. Volatile fingerprint of Brazilian defective coffee seeds: Corroboration of potential marker compounds and identification of new low quality indicators. *Food Chem.* **2014**, *153*, 298–314. [CrossRef] [PubMed]
44. Oestreich-Janzen, S. Chemistry of Coffee. In *Comprehensive Natural Products II*; Liu, H.-W., Mander, L., Eds.; Elsevier Science: Kindlington, UK, 2013; Volume 3, pp. 1085–1117.
45. Toledo, P.R.A.B.; Pezza, L.; Pezza, H.R.; Toci, A.T. Relationship between the different aspects related to coffee quality and their volatile compounds. *Compr. Rev. Food Sci. Food Saf.* **2016**, *15*, 705–719. [CrossRef]
46. Gonzalez-Rios, O.; Suarez-Quiroz, M.L.; Boulanger, R.; Barel, M.; Guyot, B.; Guiraud, J.P.; Schorr-Galindo, S. Impact of "ecological" post-harvest processing on the volatile fraction of coffee beans: I. Green coffee. *J. Food Compos. Anal.* **2007**, *20*, 289–296. [CrossRef]
47. Akiyama, M.; Murakami, K.; Hirano, Y.; Ikeda, M.; Iwatsuki, K.; Wada, A.; Tokuno, K.; Onishi, M.; Iwabuchi, H. Characterization of headspace aroma compounds of freshly brewed arabica coffees and studies on a characteristic aroma compound of Ethiopian coffee. *J. Food Sci.* **2008**, *73*, C335–C346. [CrossRef]
48. Lee, L.W.; Cheong, M.W.; Curran, P.; Yu, B.; Liu, S.Q. Modulation of coffee aroma via the fermentation of green coffee beans with *Rhizopus oligosporus*: II. Effects of different roast levels. *Food Chem.* **2016**, *211*, 925–936. [CrossRef]
49. Afriliana, A.; Pratiwi, D.; Giyarto; Belgis, M.; Harada, H.; Yushiharu, M.; Taizo, M. Volatile compounds change in unfermented Robusta coffee by re-fermentation using commercial kefir. *Nutr. Food Sci. Int. J.* **2019**, *8*, 555745. [CrossRef]
50. Peñuela-Martínez, A.E.; Zapata-Zapata, A.D.; Durango-Restrepo, D.L. Performance of different fermentation methods and the effect on coffee quality (*Coffea arabica* L.). *Coffee Sci.* **2018**, *13*, 465–476. [CrossRef]

51. Mendes-Ferreira, A.; Barbosa, C.; Falco, V.; Leão, C.; Mendes-Faia, A. The production of hydrogen sulphide and other aroma compounds by wine strains of *Saccharomyces cerevisiae* in synthetic media with different nitrogen concentrations. *J. Ind. Microbiol. Biotechnol.* **2009**, *36*, 571–583. [CrossRef] [PubMed]
52. Yvon, M.; Rijnen, L. Cheese flavour formation by amino acid catabolism. *Int. Dairy J.* **2001**, *11*, 185–201. [CrossRef]
53. Plessas, S.; Bekatorou, A.; Gallanagh, J.; Nigam, P.; Koutinas, A.A.; Psarianos, C. Evolution of aroma volatiles during storage of sourdough breads made by mixed cultures of *Kluyveromyces marxianus* and *Lactobacillus delbrueckii* ssp. *bulgaricus* or *Lactobacillus helveticus*. *Food Chem.* **2008**, *107*, 883–889. [CrossRef]
54. Patrignani, F.; Chinnici, F.; Serrazanetti, D.I.; Vernocchi, P.; Ndagijimana, M.; Riponi, C.; Lanciotti, R. Production of volatile and sulfur compounds by 10 *Saccharomyces cerevisiae* strains inoculated in trebbiano must. *Front. Microbiol.* **2016**, *7*, 1–11. [CrossRef] [PubMed]
55. Leclercq-Perlat, M.-N.; Corrieu, G.; Spinnler, H.-E. Comparison of volatile compounds produced in model cheese medium deacidified by *Debaryomyces hansenii* or *Kluyveromyces marxianus*. *J. Dairy Sci.* **2010**, *87*, 1545–1550. [CrossRef]

© 2019 by the authors. Licensee MDPI, Basel, Switzerland. This article is an open access article distributed under the terms and conditions of the Creative Commons Attribution (CC BY) license (http://creativecommons.org/licenses/by/4.0/).

fermentation

Review

The Xylose Metabolizing Yeast *Spathaspora passalidarum* is a Promising Genetic Treasure for Improving Bioethanol Production

Khaled A. Selim [1,*], Saadia M. Easa [2] and Ahmed I. El-Diwany [1]

1. Pharmaceutical and Drug Industries Research Division, National Research Centre, 33-El-Bohouth St. (former El Tahrir St.), Dokki, P.O. Box, Giza 12622, Egypt; ahmed_eldiwany@yahoo.com
2. Microbiology Department, Faculty of Science, Ain Shams University, Cairo 11566, Egypt; elkas_1111@yahoo.com
* Correspondence: Khaled.a.selim@gmail.com

Received: 15 February 2020; Accepted: 16 March 2020; Published: 18 March 2020

Abstract: Currently, the fermentation technology for recycling agriculture waste for generation of alternative renewable biofuels is getting more and more attention because of the environmental merits of biofuels for decreasing the rapid rise of greenhouse gas effects compared to petrochemical, keeping in mind the increase of petrol cost and the exhaustion of limited petroleum resources. One of widely used biofuels is bioethanol, and the use of yeasts for commercial fermentation of cellulosic and hemicellulosic agricultural biomasses is one of the growing biotechnological trends for bioethanol production. Effective fermentation and assimilation of xylose, the major pentose sugar element of plant cell walls and the second most abundant carbohydrate, is a bottleneck step towards a robust biofuel production from agricultural waste materials. Hence, several attempts were implemented to engineer the conventional *Saccharomyces cerevisiae* yeast to transport and ferment xylose because naturally it does not use xylose, using genetic materials of *Pichia stipitis*, the pioneer native xylose fermenting yeast. Recently, the nonconventional yeast *Spathaspora passalidarum* appeared as a founder member of a new small group of yeasts that, like *Pichia stipitis*, can utilize and ferment xylose. Therefore, the understanding of the molecular mechanisms regulating the xylose assimilation in such pentose fermenting yeasts will enable us to eliminate the obstacles in the biofuels pipeline, and to develop industrial strains by means of genetic engineering to increase the availability of renewable biofuel products from agricultural biomass. In this review, we will highlight the recent advances in the field of native xylose metabolizing yeasts, with special emphasis on *S. passalidarum* for improving bioethanol production.

Keywords: fermentation; xylose metabolism; genetic engineering; biofuel; *Spathaspora passalidarum*; *Pichia stipitis*

---

1. Fermentation Technology and Challenges

The modern biotechnological applications for generation of alternative and renewable sources of biofuels are receiving more attention due to global worries over the climate change, rapid global warming, and the rising of fossil fuel costs. One of such growing biotechnological trends is the fermentation technology to convert the sugar-rich agriculture waste into bioethanol by conventional or non-conventional yeasts [1–4]. In general, yeasts have advantages over bacteria for commercial fermentation due to the thickness of their cell walls, less stringent nutritional requirements, large sizes, utmost resistance to contamination, and better growth at acidic pH of bioreactor fermenters.

In nature, the second most abundant hemicellulosic sugar in fast-growing hardwoods and agricultural biomass is xylose. Xylose sugar forms up to 15–25% of all angiosperm biomass, and it could supply an alternative fuel source for its ability to be commercially fermented into ethanol. Several approaches have been employed to engineer xylose assimilation metabolism into conventional fermenting yeasts, such as *Saccharomyces cerevisiae* [4–6]. Therefore, efficient hemicellulosic sugar fermentation is crucial for the economic conversion of lignocellulose biomass to renewable biofuels [4,6–8]. The discovery of xylose-fermenting yeasts in new niches and genetic engineering of yeasts to be capable of rapid fermentation of xylose and other sugars to recoverable concentrations of bioethanol could provide alternative biofuel sources for the future (Figure 1) [4,9].

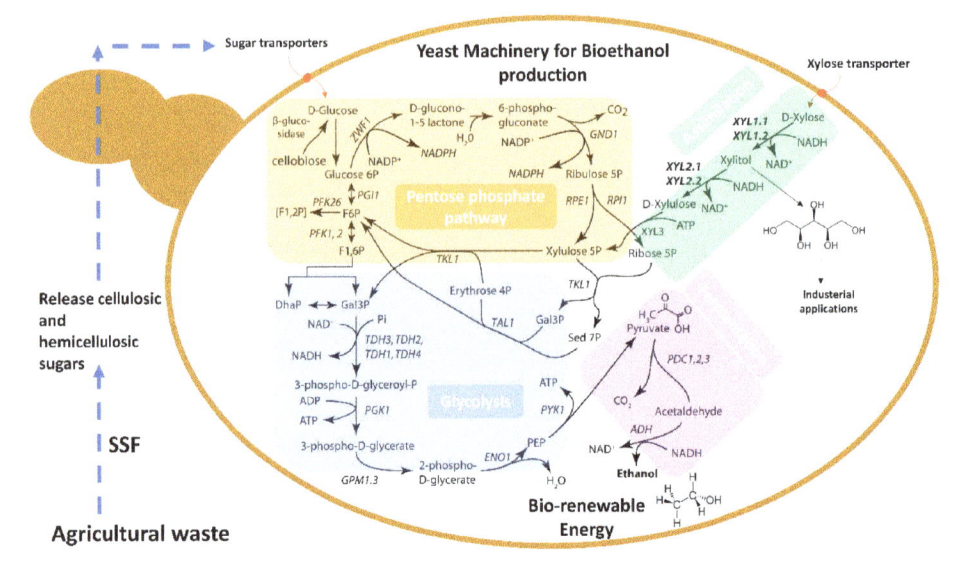

Figure 1. Model of yeast fermentation machinery for bioethanol production using the agriculture waste to feed yeasts [4]. The metabolic pathways for xylose and glucose assimilation and fermentation are indicated including the pentose phosphate pathway and glycolysis. The agriculture waste is treated through the enzymatic and chemical simultaneous saccharification and fermentation (SSF) processes to release the cellulosic and hemicellulosic sugars. The hexose and pentose sugars are transported by specific hexose and pentose sugar transporters into yeast for further metabolizing processes.

As a rule of thumb for metabolizing the xylose in most of xylose-fermenting yeasts [4], firstly the xylose is reduced by xylose reductase (XR) to xylitol. In the second step, the xylitol is oxidized by xylitol dehydrogenase (XDH) to xylulose. Afterward, the xylulose passes into the pentose phosphate pathway being metabolized into glyceraldehyde-3-P which is further reduced to pyruvate. Finally, the pyruvate is decarboxylated to acetaldehyde which is further reduced to ethanol by alcohol dehydrogenase (Figure 1). Notably, most xylose reductase enzymes have dual cofactor specificity, using both NADH and NADPH, but typically favor NADPH. However, xylitol dehydrogenase enzymes use NAD$^+$ specifically as a cofactor, which could cause imbalance between the cofactor's source for the XR-XDH pathway and xylitol accumulation under uncontrolled oxygen conditions (Figure 2) [10].

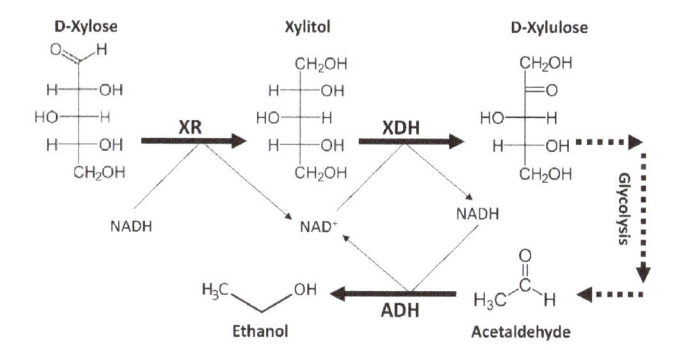

Figure 2. Schematic model of the central metabolism for bioethanol production from xylose indicating the rate limiting steps (XR: xylose reductase; XDH: xylitol dehydrogenase, and ADH: alcohol dehydrogenase) and cofactors demand/balance in most of native xylose metabolizing yeasts. Xylose assimilation reactions starts with XR to produce xylitol. The xylitol is further metabolized by XDH to produce xylulose, which further metabolized to xylulose-phosphate to enter the glycolysis (indicate by black dotted arrow and summarized in Figure 1) to produce acetaldehyde. The acetaldehyde finally converted to ethanol by ADH.

One of pioneer xylose fermenting yeasts is *Pichia stipites*. *P. stipitis* is heterothallic ascomycetous yeast, predominantly haploid and related to pentose fermenting yeasts, such as *Candida shehatae* [1,4,11–15]. *P. stipitis* was recently renamed to be *Scheffersomyces stipitis* [15] and it is natively one of the highest xylose-utilizing and fermenting yeasts. In type culture collections, the *P. stipitis* strains are among the best xylose-metabolizing microbes [16]. Under controlled low O_2 conditions, *P. stipitis* is able to consume xylose and produce up to 57 g/L of bioethanol at 30°C [4,13,14,17]. *Pichia* uses an alternative nuclear genetic code (ANGC) in which CUG encodes for Ser rather than Leu [17], which makes the genetic manipulation of *Pichia* with the commercial drug resistance markers unusually problematic because essentially all of these markers are derived from bacteria that use the universal codon system. Moreover, one of classical challenges in fermentation technology is that some of key enzymes of bioethanol production pathway are expressed relatively in low levels [1,4,13,14]. Therefore, the metabolic engineering of the bioethanol pathway in yeasts, which can ferment the sugars of the agriculture biomass with considerable and recoverable bioethanol concentrations, could enhance the productivity and sustainability of renewable biofuel sources [1,13,14]. To improve bioethanol production and xylose metabolism, a stable and manipulatable genetic system that enables overexpression or deletion of one or more of key enzymes and sugar transporters in xylose-fermenting yeast *P. stipitis*, was developed [1,3,18]. This approach comprises modelling, metabolic and flux analysis, quantitative metabolomics and transcriptomics followed by the targeted overexpression or deletion of genes of the rate-limiting steps [1,3,4,13,14]. Since there is reasonable information about the metabolic capacity of *P. stipitis* to ferment xylose on various omics levels, this makes it an attractive model system for metabolic engineering.

Recently, a new xylose-fermenting yeast *Spathaspora passalidarum* was discovered, which naturally co-ferments xylose, glucose, and cellobiose and demonstrates potentials in the effective conversion of mixed sugars from hemicellulosic hydrolysates into ethanol [9]. *S. passalidarum* was initially isolated from extremely O_2-limited and hemicellulosic sugar rich environments from the gut of a wood-boring beetle (as will be discussed below). Although the anaerobic fermentation of glucose is broadly known, the xylose fermentation typically needs a controlled oxygen condition. Uncontrolled oxygen conditions in another xylose-fermenting yeast, such as *P. stipitis*, leads to accumulation of xylitol due to insufficient amounts of NAD^+, and as consequence the xylose metabolism will be blocked (Figure 2) [9,13,14]. To solve this problem, precise controlled O_2 (very low O_2 concentrations) during the xylose fermentation is required in *P. stipitis* to generate NAD^+ from NADH.

The bioethanol production from the bioconversion of lignocellulose biomass must be achieved at high rates and yields for economically recoverable concentrations. The achieving of such targets for efficient bioethanol production are more difficult with cellulose and hemicellulose. The major barrier for cellulose utilization is enzymatic saccharification, while for hemicellulose it is the utilization of mixed sugars (hexose sugars: glucose, galactose, mannose, and rhamnose; and pentose sugars: xylose and arabinose) in the presence of ferulic and acetic acids along with other byproducts of the thermochemical pretreatment of the hydrolysates [13,14]. However, *S. passalidarum* and *P. stipitis* yeasts possess a set of unique physiological merits that make them very useful biodegradable organisms for bioconversion of lignocellulosic biomass [4,9,13,14]. *Pichia* can utilize and ferment effectively cellobiose, glucose, galactose, and mannose along with xylan high oligomeric sugars xylan and mannan, in addition to its extensively studied ability to metabolize and ferment the xylose [4,13,14]. The primary sugar released in enzymatic hydrolysis is cellobiose and, remarkably, *P. stipitis* and *S. passalidarum* have the capability to utilize the cellobiose, which make such yeasts potent organisms for simultaneous saccharification and fermentation (SSF) or hydrolysate, because most commercially available cellulase products are often deficient in β-glucosidase enzyme so the accumulation of cellobiose inhibits cellulose activities. Since *P. stipitis* and *S. passalidarum* can directly metabolize the cellobiose, they have the potential to improve SSF processes [4,9,10,13,14,19,20].

Collectively, the native ability of *P. stipitis* and *S. passalidarum* to metabolize the oligomeric sugars is of high importance as the mild acidic pretreatments of agriculture waste biomass can prevent the formation of the sugar degradation byproducts, which could inhibit significantly the fermentation process, but can release about 15–55% of soluble oligomeric sugars. Therefore, with low cost and high yield, the hemicellulosic sugars can be more readily recovered and underutilized from cellulose biomass than glucose. Although such easy recoverable sugars can be utilized for formation of a number of useful products such as xylitol, butanol, lactic acid, and other chemicals, bioethanol is still the major product with the largest potential market. Hence, bioethanol production from the lignocellulosic biomasses is receiving a lot of attention as a consequence of agriculture policies and energy demands to improve the production of alternative renewable biofuels and to reduce CO_2 emissions [2,4,13,14].

2. *Spathaspora passalidarum* a Promising Genetic Source

The Spathaspora clade contains many bioethanol producer yeasts, including *Spathaspora arborariae*, *Spathaspora brasiliensis*, *Spathaspora gorwiae*, *Spathaspora hagerdaliae*, *Spathaspora passalidarum*, *Spathaspora roraimanensis*, *Spathaspora suhii*, *Spathaspora xylofermentans*. They are usually endosymbioticly associated with wood-boring-beetles that occupy rotting wood. *Spathaspora passalidarum* (Figure 3), the first identified species of genus Spathaspora, was isolated from the gut of passalid beetle *Odontotaenius disjunctus* [9,21–24]. Notably, *S. arborariae*, *S. gorwiae*, *S. hagerdaliae*, and *S. passalidarum* ferment xylose to produce bioethanol, whereas the rest within the Spathaspora clade are thought to be xylitol producers [9].

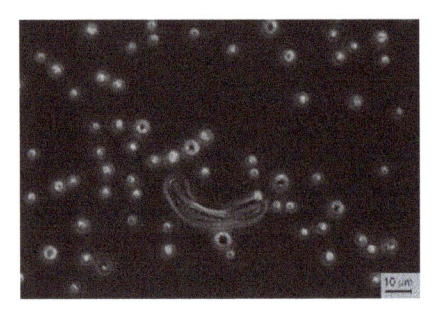

Figure 3. *Spathaspora passalidarum* budding cells with characteristic curved and elongated ascospore.

S. passalidarum was firstly described in 2006 by Nguyen et al. [21]. The authors speculated that *Spathaspora* mainly exists in the beetle's biosphere rather than the beetle's gut microbiota, and it may be only by coincidence that *O. disjunctus* beetles ingested decaying wood contaminated by yeasts. Later in 2012 and 2017, another 12 strains were described in two independent studies from wood-boring beetles and wood samples of Amazonian forest in Brazil [23,25]. Among these strains, only one isolate was obtained from the gut of *Popilus marginatus* beetle, while the rest of the strains were obtained from the woody samples inhabited by the beetles [23,25]. In 2014, two more strains were isolated from rotted wood in China [26]. Additionally, Rodrussamee and colleges in 2018 reported a new thermotolerant strain, named *S. passalidarum* CMUWF1–2, which was isolated from Thailand soil [27]. The frequency of finding *S. passalidarum* mainly among the woody samples supports the notion that those yeasts are probably associated with decaying wood niches rather than with the gut microbiota of wood-boring beetles. However, the fact of the low frequency of finding *S. passalidarum* among other yeast species keeps an open possibility that they inhabit mainly the wood-related beetles [9].

3. Fermentation Capability of *Spathaspora passalidarum*

It is believed that the beetle's gut is truly anaerobic or microaerobic, therefore it was speculated that *S. passalidarum* possess a unique adaptation capability to survive under oxygen-depleted conditions on mixtures of hemicellulosic sugars in the midgut of wood-boring beetles [10,19]. Currently, *S. passalidarum* is among the best xylose-utilizing and fermenting yeasts. Under anaerobic or microaerobic conditions, *S. passalidarum* possess rapid utilization and consumption rates for xylose and produces up to 0.48 g/g bioethanol (near to the maximum theoretical bioethanol production of 0.51 g/g), in contrast to *P. stipitis* which can hardly metabolize xylose anaerobically, accumulating xylitol and a very low yield of bioethanol [10,19,20,28,29]. Under anaerobic conditions, Hou in 2012 showed that *S. passalidarum* has a high growth rate with rapid consumption rate of sugars and can ferment xylose into a high yield of bioethanol with higher production efficiency than *P. stipitis* [10]. Similarly, Veras and colleges in 2017 showed that under anaerobic conditions, *S. passalidarum* accumulates 1.5 times more bioethanol than *S. stipitis*, while both stains accumulate around 0.44 g/g under O_2 limiting conditions [30]. The previous work by Hou (2012) defined strictly that *S. passalidarum* can metabolize and ferment xylose in tightly capped flasks [10]. In contrast to the previous report by Hou (2012) [10], under stringent O_2 limiting conditions, the *S. passalidarum* was not able virtually to utilize the sugars, indicating that native wild-type *S. passalidarum* does not ferment sugars under truly anaerobic conditions [19,20]. Therefore, it is still under debate whether *S. passalidarum* can ferment xylose truly anaerobically or whether it requires a controlled microoxygenic condition similar to *P. stipitis*.

One of the major challenges in fermentation technology is the inability of the majority of known microbes to co-ferment xylose and glucose, since glucose usually inhibits the metabolization of the other sugars in lignocellulose hydrolysate, as in the case of *P. stipitis* [13,14]. Astonishingly, in a recent study to address the metabolic profiling and fermentation capacity of *S. passalidarum*, *S. passalidarum* was found to co-ferment xylose, cellobiose, and glucose simultaneously with high bioethanol yields ranging from 0.31 to 0.42 g/g [19,20]. Moreover, an adapted *S. passalidarum* strain was found to accumulate up to 39 g/L bioethanol with a 0.37 g/g yield from a lignocellulosic hydrolysate. The specific production rate of bioethanol on xylose as a carbon source was superior with three times more than the corresponding rate on glucose, where the flux of glycolytic intermediates was meaningfully lower on glucose than on xylose and its xylose reductase enzyme had a higher affinity for NADH than NADPH [19,20]. Thus, the allosteric activation of glycolytic routes associated with the xylose utilization and the NADH-dependent xylose reductase are most likely the causes for such unique ability of *S. passalidarum* to co-ferment mixed sugars [19,23]. Later, such results were confirmed in a metabolic flux study, where *S. passalidarum* showed about 1.5–2 times high flux rate in the NADH-dependent xylose reductase reaction [31], which caused continuous recycling and reduction of xylitol levels. Such directed high flux rates to glycolytic routes and pentose phosphate pathway was the cause for high levels of bioethanol production in *S. passalidarum* [31]. In large scale fed-batch fermentation study, *S.*

passalidarum was able metabolize around 90% of xylose sugar and all of glucose of sugarcane bagasse hydrolysate, even so glucose had approximately three-fold higher xylose content; and produced a high ethanol yield of 0.46 g/g with volumetric productivity of 0.81 g/L/h in contrast to *P. stipitis* which produced 0.32 g/g ethanol with productivity of 0.36 g/L/h [32]. In follow up study, *S. passalidarum* UFMG-CM-Y473 strain was able to simultaneously utilize and co-ferment about 78% of the released sugars (xylose, glucose, and cellobiose) of pretreated sugarcane bagasse hydrolysate (delignified and enzymatically hydrolyzed) to yield up to 0.32 g/g bioethanol with productivity of 0.34 g/L/h without any nutritional supplementation [33]. Moreover, the new thermotolerant strain, *S. passalidarum* CMUWF1–2, was able to co-ferment various sugars (mannose, galactose, xylose, and arabinose) of lignocellulosic biomass, even in presence of glucose, to accumulate considerable amounts of bioethanol and low amounts of xylitol at higher temperatures. For example, it was able to accumulate 0.43, 0.40, and 0.20 g/g ethanol per xylose at 30, 37, and 40 °C, respectively [27]. Constant with absence of the glucose repression effect on the utilization of other sugars, *S. passalidarum* CMUWF1–2 exhibited a resistance to 2-deoxy glucose, the nonmetabolizable glucose analog, and tolerance to elevated levels of glucose (35.0% of w/v) and ethanol (8.0% of v/v) [27]. In contrast, the first discovered *S. passalidarum* NRRL Y-27907 strain was sensitive to 2-deoxy glucose, as 2-deoxy glucose suppressed the xylose consumption under anaerobic conditions. While under aerobic conditions, the 2-deoxy glucose inhibited, only partially, *S. passalidarum* NRRL Y-27907 [10]. Therefore, the author speculated that xylose uptake in *S. passalidarum* NRRL Y-27907 may take place by different xylose transport systems under aerobic and anaerobic conditions. Under aerobic conditions, xylose is taken up by means of ATP-dependent/high affinity xylose-proton symporter and low affinity transporter via facilitated diffusion driven only by the sugar gradient. While, under anaerobic condition, the yeasts are most likely to use only the low affinity xylose pump, as the active transport via xylose-proton symporter will deplete the ATP levels. The inhibitory effect of 2-deoxy glucose on *S. passalidarum* NRRL Y-27907 can be therefore explained by (1) blocking of the low-affinity-facilitated diffusion transporters which are occupied with transporting 2-deoxy glucose, and (2) the inhibition xylose active transport due to the depletion of the intracellular ATP levels to actively phosphorate the 2-deoxy glucose into the non-metabolizable phospho-2-deoxy glucose [10].

4. Genetic and Physiological Features of *Spathaspora passalidarum* Emphasis Special Roles for Xylose Reductase and Xylitol Dehydrogenase

These unusual unique traits of *S. passalidarum* are very attractive for studying on a molecular level. The complete genome sequence of xylose-fermenting yeast *S. passalidarum* was therefore necessary and it was accomplished and published for first time in 2011 [34]. The comparative genomic, transcriptomic, and metabolomic analysis between two of the native xylose-fermenting yeasts, the relatively newly discovered *S. passalidarum*, and the deeply studied *P. stipitis*, allowed a better understanding of the regulatory mechanisms of lignocellulose utilization, and identified the target key genes involved in xylose metabolism [9,10,19,31,34]. The comparative genomic and phylogenetic analysis clearly revealed that *S. passalidarum* is one of the CUG yeast clades, similar to *P. stipites* [34]. In addition, the transcriptome analysis indicated upregulation of the genes implicated in transporting carbohydrate and xylose- and carbohydrate-metabolisms under xylose growth. Several of genes, which are involved in regulation of redox balance and recycling of NAD(P)H/$^+$, were upregulated to probably keep the redox balance during xylose utilization. Additionally, the genes encoding for cellulases and β-glucosidases were also upregulated, which suggests a positive feedback of xylose on the upstream genes to activate its own liberation from the higher oligomeric sugars of hemicelluloses by means of the catalytic activities of cellulases and β-glucosidases [34].

The previously mentioned capabilities of *S. passalidarum* to co-ferment different sugars and accumulate high levels of bioethanol with very low concentrations of xylitol, can be explained by the presence of a set of physiological characters encoded by unique set of genes [10,19,34]. Thus, the high capacity of xylose fermentation and low levels of xylitol accumulation by *S. passalidarum* was speculated to be due to the cofactor's equilibrium between the intracellular demand and supply of the cofactors via NADH-favored xylose reductase enzyme and NAD$^+$-specific xylitol dehydrogenase enzyme, the key enzymes of the xylose utilization pathway [10,28]. Normally, the xylose metabolization occurs through the reduction of xylose to xylitol with xylose reductase, which requires NADPH or NADH as a cofactor with preference for NADPH. Only few NADH-favored xylose reductase enzymes have been described so far [10,35–37]. Then the xylitol is metabolized further by xylitol dehydrogenase which is strictly NAD$^+$ dependent (Figure 2). The unbalance between NAD$^+$ supplement and requirement can block the xylose metabolization and leads to accumulation of xylitol. Later, *S. passalidarum* was found to harbor two genes encoding for xylose reductase (*SpXYL1.1* and *SpXYL1.2*) [28]. The *SpXYL1.1* gene product is more equivalent to *XYL1* found in other yeasts. The expression levels of *SpXYL1.2* were found to be higher than *SpXYL1.1* and bioethanol production in *S. passalidarum* was attributed to higher xylose reductase activity with NADH than with NADPH [28]. The *SpXYL1.2* was found to use both NADH and NADPH with preference for NADH, while *SpXYL1.1* was stringently NADPH-dependent. Furthermore, the transformation of *S. cerevisiae* with *SpXYL1.2* of *S. passalidarum* enabled the overexpressing *S. cerevisiae*::*SpXYL1.2* strain to grow anaerobically on xylose and to ferment it to higher ethanol yield than the isogenic *S. cerevisiae* TMB 3422 strain, which overexpresses *P. stipitis XYL1*. While, the *S. cerevisiae*::*SpXYL1.1* overexpressing strain was not able to grow on xylose [28]. Similarly, in the yeast-like fungus *Aureobasidium pullulans*, the overexpression of *SpXYL1.2* xylose reductase along with *S. passalidarum* xylitol dehydrogenase encoded by *SpXYL2.2* enhanced the xylose metabolization by 17.76% and improved the fermentation capability and the pullulan production by 97.72% of the overexpressing mutants compared with the parental strain [38].

Finally, a metabolic analysis of *S. passalidarum* speculated that NADH-preferred xylose reductase and NAD$^+$-dependent xylitol dehydrogenase would tend to drive both of xylose assimilation via the oxidoreductase pathway and the acetaldehyde reduction to ethanol by the alcohol dehydrogenase enzyme [19]. Recently, a metabolic flux analysis of different xylose-fermenting yeasts confirmed a better cofactors balance within *S. passalidarum* cells during xylose catabolism to bioethanol production than within *P. stipitis* cells [31], which further supports the growth characteristics of *S. passalidarum*.

Collectively, those unique and unusual traits of *S. passalidarum* encourage using it as a source for genes to improve xylose utilization and bioethanol production from lignocellulosic biomass in the current xylose fermenting yeasts, such as *P. stipites*, or to introduce xylose metabolism genes to develop industrial strains of *Saccharomyces cerevisiae* capable of co-fermentation of pentose and hexose sugars. Or alternatively, to domesticate it given the excellent results already accomplished by wild-type representatives of that species for co-fermentation of mixed sugars. To facilitate that purpose, Li et al. (2017) developed a stable genetic expression system compatible with the CUG yeasts clade for genomic integration of Gene Of Interest (GOI) into several yeasts [39]. The developed multi-host integrative system was functional in several of the xylose-fermenting yeasts including *S. passalidarum*, *P. stipitis*, and *Candida jeffriesii* and *Candida amazonensis*, as well as in a hexose metabolizing yeast *Saccharomyces cerevisiae*, for heterologous expression of green fluorescent protein (GFP) or lactate dehydrogenase. For lactate dehydrogenase overexpressing strains, all the engineered yeast strains were able to metabolize either glucose (in case of *S. cerevisiae*) or xylose (in case of xylose-fermenting yeasts) to produce lactate [39].

5. New Adaptive Strains of *Spathaspora passalidarum* for Potential Industrial Applications

One of the unique features of those xylose metabolizing yeasts, is the ability to use not only the monomeric hexose and pentose sugars but also the high oligomeric disaccharide sugar in mixed co-fermentation [9,19], which can be an advantage for large scale industrial applications. The mild acid pretreatment of agriculture waste biomasses is relatively cheap and prevents the accumulation of harmful compounds, which inhibits the fermentation processes, but releases the sugars in higher oligomeric stats. Therefore, the ability of such native xylose fermenting strains, *P.*

stipitis, and *S. passalidarum*, to use the high oligomeric sugar can be a great advantage for various biotechnological applications.

One of the major problems that hinders the use of *S. passalidarum* for industrial bioethanol production, even with its remarkable ability for bioethanol production, is the high sensitivity of *S. passalidarum* to the chemical inhibitors, such as ferulic and acetic acids, which are released in preparation of hemicellulosic hydrolysates [9]. Several elaborative studies have focused mainly on improving the tolerability of *S. passalidarum* to the hydrolysates inhibitors with keeping in mind the bioethanol productivity of the strains [19,40–43]. Hou and Yao in (2012) reported a strong strain [40], which is able to grow on furfurals and many other inhibitors of wheat straw hydrolysate (75%) and able to accumulate up to 0.40 g/g ethanol. Such strain was generated through hybridization of a *S. cerevisiae* and a UV-mutagenized *S. passalidarum* [40]. In 2012 also, another resistant strain was developed under O_2 limiting conditions through several passage of the wild-type *S. passalidarum* NRRL Y-27907 on wood hydrolysate, followed by adaptive growth of the strain on corn stover AFEX (ammonia fiber expansion) hydrolysate [19]. Even with such efforts, the strain was not able to accumulate significant amounts of ethanol during the fermentation of the AFEX hydrolysate, despite its ability to grow in AFEX hydrolysate media. When the acetic acid was depleted from AFEX hydrolysate media, ethanol production was surprisingly observed with a yield of 0.45 g/g and most of the xylose content was consumed [19]. Later in 2017, Morales and colleagues developed an evolutionary adapted strain [41] with high tolerance toward the classical inhibitor of the fermentation processes, acetic acid, and that produces ethanol with a yield of 0.48 g/g. In a non-detoxified hydrolysate of *Eucalyptus globulus*, the authors reported also the ability of this strain to co-utilize mixed sugars of xylose, glucose, and cellobiose under microaerobic conditions [41]. This strain was generated by UV irradiation followed by successive growing of the strain under elevated acetic acid concentrations [41]. In similar way, another group also obtained a mutated *S. passalidarum* strain but via plasma mutagenesis and continuous cultivation in alkaline liquor pretreated corncob [42]. Under a simultaneous saccharification and co-fermentation, the obtained strain produced bioethanol with efficiency of 75% [42]. Finally, Su et al. in 2018 developed an adaptive *S. passalidarum* strain (named YK208-E11) [43], which is designated for resistance to AFEX hydrolysate inhibitors, from the wild-type NRRL Y-27907 through high-throughput screen via combining several approaches of batch adaptation, cell recycling, and cell mating [43]. The *S. passalidarum* YK208-E11 strain produced less biomass (about 40% compared to the wild-type), co-metabolized mixed sugars of xylose, glucose, and cellobiose, and exhibited a three-fold improvement in the ethanol production rate with a yield of 0.45 g/g. The whole genome sequence of *S. passalidarum* YK208-E11 strain revealed a deletion of about 11 kb in this strain. The ORF, which was deleted in *S. passalidarum* YK208-E11, is encoding for proteins predicted to be involved in cell division and respiration. Therefore, the authors speculated that this deletion may account for those unique adaptive/physiological features of this AFEX-acclimatized *S. passalidarum* YK208-E11 strain [43].

6. Future Perspective for Engineering New Strains for Better Bioethanol Production

The metabolic engineering approaches involve targeted overexpression and/or deletion of fermentative key genes that facilities quick and efficient conversion of sugars into bioethanol with high recoverable yields [1,3]. As we discussed above, *S. passalidarum* xylose reductase and xylitol dehydrogenase are among the promising candidates for targeted overexpression. The cumulative knowledge of the transcriptomics, metabolomics, and comparative genomics studies for *P. stipitis* and *S. passalidarum*, identified other key enzymes controlling the xylose assimilation, rather than XDH and XR (Figure 2). One of such promising key genes is *adh* that encodes for fermentative isozyme alcohol dehydrogenase (ADH), which is vital for production and/or assimilation of ethanol (Figure 2). Generally, ADH catalyzes the final (rate limiting) step in the yeast glycolytic pathway, the reduction of acetaldehyde to ethanol and NAD^+, and therefore it accepts NADH as a co-factor [44,45]. However, ADH enzymes are also able to perform the reverse reaction from ethanol to acetaldehyde, enabling the yeasts to oxidize and grow on ethanol as a carbon source. In *P. stipitis*, the ADH fermentative activities is crucial not only for ethanol production and/or consumption but also for maintenance redox balance within the yeast cell, so it is considered to be a part of the cofactor balance system in *P. stipitis* [44].

The sequencing projects of *S. passalidarum* NRRL Y-27907 and *P. stipitis* CBS6054 (JGI-MycoCosm) revealed the presence of several/different alcohol dehydrogenase (ADH) encoding genes. For example, in *P. stipitis*, sevem genes were predicted to encode for alcohol dehydrogenase (*PsADH1* to *PsADH7*) enzymes [13,17]. Among them *PsADH1* and *PsADH2* were found to be essential for xylose assimilation and ethanol production [44,46]. Each of the ADH proteins in *S. passalidarum* and *P. stipitis* are supposed to have different kinetic properties. Some of the enzymes could be mainly responsible for producing ethanol while others might be responsible for oxidizing it. In *S. passalidarum*, the gene encoding for *SpADH1* was found to be expressed at a very high level during xylose metabolization [34]. In addition, metabolic analysis and metabolic flux analysis revealed that alcohol dehydrogenase is one of key enzymes driving ethanol production in *S. passalidarum* [19,31]. Notably, owing to relative *SpADH1* abundance, the *SpADH1* promotor was used to develop a multi-host integrative system for xylose-fermenting yeast [39].

While in *P. stipitis*, the function of some ADH enzymes are better understood, in particular *PsADH1* and *PsADH2* [13,17,44,46–49]. Transcriptomic studies of the *P. stipites adh* system indicated that the *PsADH* activities are correlated with and induced under O$_2$ limited/microaerobic conditions [46,48]. Under xylose fermentation, the *PsADH1* was found to be the primary key enzyme among the *PsADH* system. The deletion of *PsADH1* caused a reduction in *P. stipites* growth rate and a notable increase in xylitol accumulation accompanied with a dramatic decrease in ethanol production, due to intracellular cofactors imbalance [44]. The *PsADH2* is not expressed under microaerobic or aerobic conditions unless *PsADH1* is deleted [44,46], which further confirms that the significant role of *PsADH1* is in sugar assimilation and ethanol production. The levels of *PsADH1* and *PsADH2* transcripts were observed, however, to be low through xylose metabolism relative to the transcript levels of other fermentative and glycolytic enzymes [13,17]. In addition, *PsADH1* and *PsADH2* were able to complement the growth of the *S. cerevisiae* Δ*adh* mutant on ethanol as a sole carbon source [47]. Moreover, *PsADH3* to *PsADH7* were speculated to keep the balance between the cofactors NADPH and NADH [17]. However, the expression patterns of the other *PsADHs* on xylose and glucose under microaerobic conditions, in particular, for *PsADH7* and *PsADH4* are not fully understood [13]. *PsADH5* was found in proximity to NADPH dehydrogenase, implying a function in maintenance the intracellular cofactors balance, however, it is not proven yet. Notably, *PsADH7* was found to be upregulated under aerobic growth on xylose [50]. *PsADH7* was described as a strictly NADP(H) dependant enzyme with broad spectrum for substrates-specificity, including variety of aromatic and linear aldehydes (e.g., acetaldehyde, butanal, propanal, and furfural) and alcohols (e.g., ethanol, butanol, pentanol, hexanol, and octanol) for forward and reverse reactions, respectively [50]. Surprisingly, *PsADH7* was able to utilize xylitol as a substrate too with moderate activity. In the same context, the overexpression of *PsADH7* into a *P. stipites* xylitol dehydrogenase mutant (Δ*PsXDH*) [18], which cannot metabolize xylitol and therefore cannot grow on xylose as a sole carbon source, was able exclusively to complement the growth of Δ*PsXDH* on xylose, in contrast to *PsADH1, 2, 4*, and 5 [50]. Hence, there is a need to understand the kinetic characteristics of each of *PsADH* and *SpADH* enzymes in order to target the correct genes for overexpression and/or deletion. Finally, we would like to state that genes encoding for *adh* isozymes are worth studying, especially of *S. passalidarum*, owning to their significant functions in bioethanol production/consumption and/or intracellular cofactor balance.

7. Conclusions

Taken together, the advances in fermentation performance by *S. passalidarum* pave the way for engineering the conventional and the nonconventional fermenting yeasts, such as *S. cerevisiae* and *P. stipites*, for economical fermentation of hexose and pentose sugars in hemicellulosic hydrolysates on industrial scales. Keeping in mind that the efficient metabolization and fermentation of xylose is essential for the bioconversion of lignocellulosic biomasses into biofuels and chemicals, but the conventional wildtype strains like *S. cerevisiae* cannot use the xylose. Therefore, researchers keep trying to engineer the xylose utilization pathway into the conventional yeast. The genomes of the natural xylose-fermenting yeasts, in particular of *P. stipitis* and *S. passalidarum,* are of huge importance, as their genomics features and regulatory patterns can serve as guides and genomic resources for further genetic engineering development in those native xylose-metabolizing yeasts or to engineer non-xylose fermenting yeasts. Therefore, *S. passalidarum* and *P. stipitis* can be considered as genomic treasure sources for various genes to engineer the xylose metabolism and to improve the bioethanol production [1,24,34].

Author Contributions: K.A.S. undertook extensive literature review and wrote the manuscript. S.M.E. and A.I.E.-D. supervised the research. All authors have read and agreed to the published version of the manuscript.

Funding: This research received no external funding.

Acknowledgments: The authors would like to thank Thomas Jeffries (University of Wisconsin Madison), Laura Willis (Forest Products Laboratory, USDA), Amal rabeá, Ali Selim, and Katerina Peros for the invaluable and unlimited support to complete this project. K.A.S. received the Parwon scholarship at Institute for Microbial and Biochemical Technology, FPL, USDA.

Conflicts of Interest: The authors declare no conflict of interest.

References

1. Jeffries, T.W. Engineering yeasts for xylose metabolism. *Curr. Opin. Biotechnol.* **2006**, *17*, 320–326. [CrossRef] [PubMed]
2. Balat, M.; Balat, H. Recent trends in global production and utilization of bio-ethanol fuel. *Appl. Energy* **2009**, *86*, 2273–2282. [CrossRef]
3. Jarboe, L.R.; Zhang, X.; Wang, X.; Moore, J.C.; Shanmugam, K.T.; Ingram, L.O. Metabolic Engineering for Production of Biorenewable Fuels and Chemicals: Contributions of Synthetic Biology. *J. Biomed. Biotechnol.* **2010**, *2010*, 761042. [CrossRef] [PubMed]
4. Selim, K.A.; El-Ghwas, D.E.; Easa, S.M.; Hassan, M.I.A. Bioethanol a Microbial Biofuel Metabolite; New Insights of Yeasts Metabolic Engineering. *Fermentation* **2018**, *4*, 16. [CrossRef]
5. Jeffries, T.W.; Shi, N.Q. Genetic engineering for improved xylose fermentation by yeasts. *Adv. Biochem. Eng. Biotechnol.* **1999**, *65*, 117–161.
6. Ruchala, J.; Kurylenko, O.O.; Dmytruk, K.V.; Sibirny, A.A. Construction of advanced producers of first- and second-generation ethanol in *Saccharomyces cerevisiae* and selected species of non-conventional yeasts (*Scheffersomyces stipitis*, *Ogataea polymorpha*). *J. Ind. Microbiol. Biotechnol.* **2020**, *47*, 109–132. [CrossRef]
7. Hinmann, N.D.; Wright, J.D.; Hoagland, W.; Wyman, C.E. Xylose fermentation—An economic analysis. *Appl. Biochem. Biotechnol.* **1989**, *20*, 391–401. [CrossRef]
8. Saha, B.C.; Dien, B.S.; Bothast, R.J. Fuel ethanol production from corn fiber—Current status and technical prospects. *Appl. Biochem. Biotechnol.* **1998**, *70*, 115–125. [CrossRef]
9. Cadete, R.M.; Rosa, C.A. The yeasts of the genus *Spathaspora*: Potential candidates for second-generation biofuel production. *Yeast* **2018**, *35*, 191–199. [CrossRef]
10. Hou, X. Anaerobic xylose fermentation by *Spathaspora passalidarum*. *Appl. Microbiol. Biotechnol.* **2012**, *94*, 205–214. [CrossRef]
11. Kurtzman, C.P. *Candida shehatae*—Genetic diversity and phylogenetic relationships with other xylose-fermenting yeasts. *Antonie Leeuwenhoek* **1990**, *57*, 215–222. [CrossRef] [PubMed]
12. Melake, T.; Passoth, V.V.; Klinner, D. Characterization of the genetic system of the xylose-fermenting yeast *Pichia stipitis*. *Curr. Microbiol.* **1996**, *33*, 237–242. [CrossRef] [PubMed]
13. Jeffries, T.W.; Van Vleet, J.R.H. *Pichia stipitis* genomics, transcriptomics, and gene clusters. *FEMS Yeast Res.* **2009**, *9*, 793–807. [CrossRef] [PubMed]
14. Van Vleet, J.H.; Jeffries, T.W. Yeast metabolic engineering for hemicellulosic ethanol production. *Curr. Opin. Biotechnol.* **2009**, *20*, 300–306. [CrossRef] [PubMed]
15. Kurtzman, C.P.; Suzuki, M. Phylogenetic analysis of the ascomycete yeasts that form coezyme Q-9 and the proposal of the new genera *Babjeviella*, *Meyerozyma*, *Millerozyma*, *Priceomyces* and *Scheffersomyces*. *Mycoscience* **2010**, *51*, 2–14. [CrossRef]

16. Van Dijken, J.P.; van den Bosch, E.; Hermans, J.J.; de Miranda, L.R.; Scheffers, W.A. Alcoholic fermentation by 'nonfermentative' yeasts. *Yeast* **1986**, *2*, 123–127. [CrossRef]
17. Jeffries, T.W.; Grigoriev, I.V.; Grimwood, J.; Laplaza, J.M.; Aerts, A.; Salamov, A.; Schmutz, J.; Lindquist, E.; Dehal, P.; Shapiro, H.; et al. Genome sequence of the lignocellulose-bioconverting and xylose fermenting yeast *Pichia stipitis*. *Nat. Biotechnol.* **2007**, *25*, 319–326. [CrossRef]
18. Laplaza, J.M.; Torres, B.R.; Jin, Y.S.; Jeffries, T.W. *Sh ble* and Cre adapted for functional genomics and metabolic engineering of *Pichia stipitis*. *Enzym. Microb. Technol.* **2006**, *38*, 741–747. [CrossRef]
19. Long, T.M.; Su, Y.K.; Headman, J.; Higbee, A.; Willis, L.B.; Jeffries, T.W. Cofermentation of glucose, xylose, and cellobiose by the beetle-associated yeast, *Spathaspora passalidarum*. *Appl. Environ. Microbiol.* **2012**, *78*, 5492–5500. [CrossRef]
20. Su, Y.K.; Willis, L.B.; Jeffries, T.W. Effects of aeration on growth, ethanol and polyol accumulation by *Spathaspora passalidarum* NRRL Y-27907 and *Scheffersomyces stipitis* NRRL Y-7124. *Biotechnol. Bioeng.* **2015**, *112*, 457–469. [CrossRef]
21. Nguyen, N.H.; Suh, S.O.; Marshall, C.J.; Blackwell, M. Morphological and ecological similarities: Wood-boring beetles associated with novel xylose-fermenting yeasts, *Spathaspora passalidarum* gen. sp. nov. and *Candida jeffriesii* sp. nov. *Mycol. Res.* **2006**, *110*, 1232–1241. [CrossRef] [PubMed]
22. Cadete, R.M.; Santos, R.O.; Melo, M.A.; Mouro, A.; Gonçalves, D.L.; Stambuk, B.U.; Gomes, F.C.O.; Lachance, M.A.; Rosa, C.A. *Spathaspora arborariae* sp. nov., a D-xylose-fermenting yeast species isolated from rotting wood in Brazil. *FEMS Yeast Res.* **2009**, *9*, 1338–1342. [PubMed]
23. Cadete, R.M.; Melo, M.A.; Dussán, K.J.; Rodrigues, R.C.; Silva, S.S.; Zilli, J.E.; Vital, M.J.; Gomes, F.C.; Lachance, M.A.; Rosa, C.A. Diversity and physiological characterization of D-xylose-fermenting yeasts isolated from the Brazilian Amazonian Forest. *PLoS ONE* **2012**, *7*, e43135. [CrossRef] [PubMed]
24. Cadete, R.M.; Melo, M.A.; Zilli, J.E.; Vital, M.J.; Mouro, A.; Prompt, A.H.; Gomes, F.C.; Stambuk, B.U.; Lachance, M.A.; Rosa, C.A. *Spathaspora brasiliensis* sp. nov., *Spathaspora suhii* sp. nov., *Spathaspora roraimanensis* sp. nov. and *Spathaspora xylofermentans* sp. nov., four novel (D)-xylose-fermenting yeast species from Brazilian Amazonian forest. *Antonie Leeuwenhoek* **2013**, *103*, 421–431. [CrossRef] [PubMed]
25. Souza, G.F.L.; Valentim, L.T.C.N.; Nogueira, S.R.P.; Abegg, M.A. Efficient production of second generation ethanol and xylitol by yeasts from Amazonian beetles (Coleoptera) and their galleries. *Afr. J. Microbiol. Res.* **2017**, *11*, 814–824.
26. Ren, Y.; Chen, L.; Niu, Q.; Hui, F. Description of *Scheffersomyces henanensis* sp. nov., a new D-xylose-fermenting yeast species isolated from rotten wood. *PLoS ONE* **2014**, *9*, e92315. [CrossRef]
27. Rodrussamee, N.; Sattayawat, P.; Yamada, M. Highly efficient conversion of xylose to ethanol without glucose repression by newly isolated thermotolerant *Spathaspora passalidarum* CMUWF1-2. *BMC Microbiol.* **2018**, *18*, 73. [CrossRef]
28. Cadete, R.M.; de las Heras, A.M.; Sandström, A.G.; Ferreira, C.; Gírio, F.; Gorwa-Grauslund, M.-F.; Rosa, C.A.; Fonseca, C. Exploring xylose metabolism in *Spathaspora* species: *XYL1.2 from Spathaspora passalidarum* as the key for efficient anaerobic xylose fermentation in metabolic engineered Saccharomyces cerevisiae. *Biotechnol. Biofuels* **2016**, *9*, 167.
29. Kwak, S.; Jin, Y.S. Production of fuels and chemicals from xylose by engineered *Saccharomyces cerevisiae*: A review and perspective. *Microb. Cell Fact.* **2017**, *16*, 82. [CrossRef]
30. Veras, H.C.T.; Parachin, N.S.; Almeida, J.R.M. Comparative assessment of fermentative capacity of different xylose-consuming yeasts. *Microb. Cell Fact.* **2017**, *16*, 153. [CrossRef]
31. Veras, H.C.T.; Campos, C.G.; Nascimento, I.F.; Abdelnur, P.V.; Almeida, J.R.M.; Parachin, N.S. Metabolic flux analysis for metabolome data validation of naturally xylose-fermenting yeasts. *BMC Biotechnol.* **2019**, *19*, 58. [CrossRef] [PubMed]
32. Nakanishi, S.C.; Soares, L.B.; Biazi, L.E.; Nascimento, V.M.; Costa, A.C.; Rocha, G.J.M.; Ienczak, J.L. Fermentation strategy for second generation ethanol production from sugarcane bagasse hydrolyzate by *Spathaspora passalidarum* and *Scheffersomyces stipitis*. *Biotechnol. Bioeng.* **2017**, *114*, 2211–2221. [CrossRef] [PubMed]
33. De Souza, R.D.F.R.; Dutra, E.D.; Leite, F.C.B.; Cadete, R.M.; Rosa, C.A.; Stambuk, B.U.; Stamford, T.L.M.; de Morais, M.A., Jr. Production of ethanol fuel from enzyme-treated sugarcane bagasse hydrolysate using d-xylose-fermenting wild yeast isolated from Brazilian biomes. *3 Biotech*. **2018**, *8*, 312. [CrossRef] [PubMed]

34. Wohlbach, D.J.; Kuo, A.; Sato, T.K.; Potts, K.M.; Salamov, A.A.; Labutti, K.M.; Sun, H.; Clum, A.; Pangilinan, J.L.; Lindquist, E.A.; et al. Comparative genomics of xylose-fermenting fungi for enhanced biofuel production. *Proc. Natl. Acad. Sci. USA* **2011**, *108*, 13212–13217. [CrossRef] [PubMed]
35. Jeppsson, M.; Bengtsson, O.; Franke, K.; Lee, H.; Hahn-Hagerdal, B.; Gorwa-Grauslund, M.F. The expression of a *Pichia stipitis* xylose reductase mutant with higher K(m) for NADPH increases ethanol production from xylose in recombinant *Saccharomyces cerevisiae*. *Biotechnol. Bioeng.* **2006**, *93*, 665–673. [CrossRef]
36. Petschacher, B.; Nidetzky, B. Altering the coenzyme preference of xylose reductase to favor utilization of NADH enhances ethanol yield from xylose in a metabolically engineered strain of Saccharomyces cerevisiae. *Microb. Cell* **2008**, *7*, 9. [CrossRef]
37. Bengtsson, O.; Hahn-Hagerdal, B.; Gorwa-Grauslund, M.F. Xylose reductase from *Pichia stipitis* with altered coenzyme preference improves ethanolic xylose fermentation by recombinant Saccharomyces cerevisiae. *Biotechnol. Biofuels* **2009**, *2*, 9. [CrossRef]
38. Guo, J.; Huang, S.; Chen, Y.; Guo, X.; Xiao, D. Heterologous expression of *Spathaspora passalidarum* xylose reductase and xylitol dehydrogenase genes improved xylose fermentation ability of *Aureobasidium pullulans*. *Microb. Cell Fact.* **2018**, *17*, 64. [CrossRef]
39. Li, H.; Fan, H.; Li, Y.; Shi, G.Y.; Ding, Z.Y.; Gu, Z.H.; Zhang, L. Construction and application of multi-host integrative vector system for xylose-fermenting yeast. *FEMS Yeast Res.* **2017**, *17*, 6. [CrossRef]
40. Hou, X.; Yao, S. Improved inhibitor tolerance in xylosefermenting yeast *Spathaspora passalidarum* by mutagenesis and protoplast fusion. *Appl. Microbiol. Biotechnol.* **2012**, *93*, 2591–2601. [CrossRef]
41. Morales, P.; Gentina, J.C.; Aroca, G.; Mussatto, S.I. Development of an acetic acid tolerant *Spathaspora passalidarum* strain through evolutionary engineering with resistance to inhibitors compounds of autohydrolysate of Eucalyptus globulus. *Ind. Crop Prod.* **2017**, *106*, 5–11. [CrossRef]
42. Yu, H.; Guo, J.; Chen, Y.; Fu, G.; Li, B.; Guo, X.; Xiao, D. Efficient utilization of hemicellulose and cellulose in alkali liquor-pretreated corncob for bioethanol production at high solid loading by *Spathaspora passalidarum* U1–58. *Bioresour. Technol.* **2017**, *232*, 168–175. [CrossRef] [PubMed]
43. Su, Y.K.; Willis, L.B.; Rehmann, L.; Smith, D.R.; Jeffries, T.W. *Spathaspora passalidarum* selected for resistance to AFEX hydrolysate shows decreased cell yield. *FEMS Yeast Res.* **2018**, *18*. [CrossRef] [PubMed]
44. Cho, J.Y.; Jeffries, T.W. *Pichia stipitis* genes for alcohol dehydrogenase with fermentative and respiratory functions. *Appl. Environ. Microbiol.* **1998**, *64*, 1350–1358. [CrossRef]
45. Lin, Y.; He, P.; Wang, Q.; Lu, D.; Li, Z.; Wu, C.; Jiang, N. The alcohol dehydrogenase system in the xylose-fermenting yeast *Candida maltosa*. *PLoS ONE* **2010**, *5*, e11752. [CrossRef]
46. Cho, J.Y.; Jeffries, T.W. Transcriptional control of ADH genes in the xylose fermenting yeast *Pichia stipitis*. *Appl. Environ. Microbiol.* **1999**, *65*, 2363–2368. [CrossRef]
47. Passoth, V.; Schafer, B.; Liebel, B.; Weierstall, T.; Klinner, U. Molecular cloning of alcohol dehydrogenase genes of the yeast *Pichia stipitis* and identification of the fermentative ADH. *Yeast* **1998**, *14*, 1311–1325. [CrossRef]
48. Passoth, V.; Cohn, M.; Schafer, B.; Hahn-Hagerdal, B.; Klinner, U. Molecular analysis of the hypoxia induced ADH2-promoter in the respiratory yeast *Pichia stipitis*. *Yeast* **2003**, *20*, S199. [CrossRef]
49. Passoth, V.; Cohn, M.; Schafer, B.; Hahn-Hagerdal, B.; Klinner, U. Analysis of the hypoxia-induced ADH2 promoter of the respiratory yeast *Pichia stipitis* reveals a new mechanism for sensing of oxygen limitation in yeast. *Yeast* **2003**, *20*, 39–51. [CrossRef]
50. Lu, C.; Chapter, V. In Elucidating Physiological Roles of *Pichia stipitis* Alcohol Dehydrogenases in Xylose Fermentation and Shuffling Promoters for Multiple Genes in *Saccharomyces cerevisiae* to Improve Xylose Fermentation. Ph.D. Thesis, University of Wisconsin, Madison, WI, USA, 2007.

© 2020 by the authors. Licensee MDPI, Basel, Switzerland. This article is an open access article distributed under the terms and conditions of the Creative Commons Attribution (CC BY) license (http://creativecommons.org/licenses/by/4.0/).

fermentation

Article

Ustilago Rabenhorstiana—An Alternative Natural Itaconic Acid Producer

Susan Krull, Malin Lünsmann, Ulf Prüße and Anja Kuenz *

Thünen-Institute of Agricultural Technology, Bundesallee 47, 38116 Braunschweig, Germany; susan.krull@thuenen.de (S.K.); malinluensmann@ymail.com (M.L.); ulf.pruesse@thuenen.de (U.P.)
* Correspondence: anja.kuenz@thuenen.de; Tel.: +49-531-596-4265

Received: 22 November 2019; Accepted: 24 December 2019; Published: 2 January 2020

Abstract: Itaconic acid is an industrial produced chemical by the sensitive filamentous fungus *Aspergillus terreus* and can replace petrochemical-based monomers for polymer industry. To produce itaconic acid with alternative renewable substrates, such as lignocellulosic based hydrolysates, a robust microorganism is needed due to varying compositions and impurities. Itaconic acid producing basidiomycetous yeasts of the family *Ustilaginaceae* provide this required characteristic and the species *Ustilago rabenhorstiana* was examined in this study. By an optimization of media components, process parameters, and a fed-batch mode with glucose the final titer increased from maximum 33.3 g·L^{-1} in shake flasks to 50.3 g·L^{-1} in a bioreactor. Moreover, itaconic acid was produced from different sugar monomers based on renewable feedstocks by *U. rabenhorstiana* and the robustness against weak acids as sugar degradation products was confirmed. Based on these findings, *U. rabenhorstiana* has a high potential as alternative natural itaconic acid producer besides the well-known *U. maydis* and *A. terreus*.

Keywords: *Ustilago*; itaconic acid; process improvement; lignocellulosic feedstock

---

1. Introduction

Itaconic acid is an interesting chemical for the polymer industry, which is produced in a biotechnological process based on renewable substrates [1]. Petrochemical-based substances, like methacrylic or acrylic acid, can be replaced by this single unsaturated dicarbonic acid and its derivatives. Therefore, the field of products and applications is widespread, e.g., synthetic latex, styrene-butadiene rubber, superabsorbent polymers, or unsaturated polyester resins [2–6].

Since the 1960s, the filamentous fungus *Aspergillus terreus* is industrially used with a titer of 85–100 g·L^{-1}, whereas in laboratory scale, final titers of 160 g·L^{-1} itaconic acid are described [4,7–9]. *A. terreus* achieves a productivity up to 1.15 g (L·h)$^{-1}$ and a yield of 0.64 (w/w), whereby the theoretical yield with glucose is 0.72 (w/w) [7,8]. Besides pure glucose, itaconic acid was successfully produced by *A. terreus* with glycerol, starch hydrolysates, molasses, and different monosaccharides, like xylose, arabinose, galactose, and rhamnose [10]. A great cultivation challenge is caused by sugar degradation products or other impurities in lignocellulosic hydrolysates, which influence the morphology and itaconic acid production of the fungus. Due to the sensitivity of the fungus, complex purification processes are used for such hydrolysates or more resistant strains are generated by mutagenesis [11–13]. Another alternative is itaconic acid producing yeasts of the species *Candida*, *Pseudozyma*, or *Ustilago*, which are more robust and not as sensitive to metal ions as *A. terreus* [14–18]. For wildtype strains of the species *Ustilago*, low final titers of 44.5 g·L^{-1} itaconic acid, low yields up to 0.24 (w/w), and a low productivity of maximum 0.31 g (L·h)$^{-1}$ are disadvantageous [19]. This is due to a variety of by-products like other organic acids, glycolipids, and intracellular triacylglycerols, which are produced in parallel to itaconic acid [16,18,20,21]. Nevertheless, in addition to the robustness of the yeasts, the formation of haploid yeast-like cells is an advantage compared to the filamentous growth or formation of pellets of *A. terreus* with a decreased oxygen supply or increased viscosity [5,22].

In recent years, the research interest in itaconic acid production with the phytopathogenic basidiomycete *Ustilago maydis* increased. It was found, that an ammonium limitation triggers the itaconic acid overproduction in *U. maydis* [5,19] and itaconic acid is synthesized in the cytosol via the intermediate cis-aconitate and trans-aconitate and can be further converted to 2-hydroxyparaconic acid [23–25]. The itaconic acid gene cluster was also characterized and relevant enzymes, transporters, and promoters were found [23,25,26], whereby a summary of metabolic aspects is given by Wierckx et al. [27]. Based on these findings, metabolic engineering strategies and process optimization of *U. maydis* resulted in a reduction of by-product concentrations of malic acid and 2-hydroxyparaconic acid with a significant increased itaconic acid titer of 63.2 g·L^{-1} and a yield of 0.48 (w/w) [23]. All in all, detailed examinations are available for itaconic acid production of *U. maydis*, but also other wildtype strains of the family *Ustilaginaceae* could offer advantages of less sensibility or a yeast-like morphology for using second-generation feedstocks. This family is well-known for organic acid production [16,18,28,29], but the level of knowledge about alternative itaconic acid producer, like *U. cynodontis* or *U. rabenhorstiana*, are low.

This study considers the cultivation of *Ustilago rabenhorstiana* for itaconic acid production and its potential as alternative natural producer. Although the used strain is known as natural itaconic acid producer [18,30], the microorganism was not examined in literature more precisely. Only the growth of the organism with glycerol as substrate was described, whereby non-formation of organic acids was detected [28]. Concerning itaconic acid production based on renewable resources, usability of different sugar monomers and robustness towards influence of sugar degradation products were examined in this study. Moreover, the influences of media and fermentation parameters on the production of itaconic acid and by-products as well as the morphology of the yeast were investigated.

2. Materials and Methods

2.1. Microorganism

The basidiomycete *Ustilago rabenhorstiana* NBRC 8995 was purchased from the National Institute of Technology and Evaluation (Tokyo, Japan) and was stored at −80 °C as 50% (v/v) glycerol stock culture.

2.2. Media Compositions

YEPS-medium was used for the preparation of agarplates and preculture (20 g·L^{-1} sucrose, 10 g·L^{-1} yeast extract, 20 g·L^{-1} peptone, optional 20 g·L^{-1} agar-agar).

If not mentioned otherwise, the production media was a Tabuchi-medium [18] containing 120 g·L^{-1} glucose, 0.5 g·L^{-1} KH$_2$PO$_4$, 1.6 g·L^{-1} NH$_4$Cl, 0.2 g·L^{-1} MgSO$_4$·7 H$_2$O, 10 mg·L^{-1} FeSO$_4$·7 H$_2$O, 1 g·L^{-1} yeast extract, and 30 g·L^{-1} CaCO$_3$. All components were prepared separately in stock solutions; the pH-value was adjusted to pH 6.0 for all solutions with 0.5 M H$_2$SO$_4$ or 1 M NaOH and autoclaved. The pH-value of the iron-solution was not corrected, and the solution was sterile filtered. CaCO$_3$ was weighed in the glassware and autoclaved.

All media components were p.a. quality and purchased from Merck (Darmstadt, Germany), Sigma-Aldrich (St. Louis, MO, USA) or Roth (Karlsruhe, Germany). In fed-batch cultivation the glucose concentration was monitored during the cultivation. If necessary, glucose was added in solid form without previous sterilization to prevent a glucose limitation.

2.3. Cultivation

Precultures were conducted in a 250 mL shake flask with three baffles, a filling volume of 50 mL, and inoculated with a single colony from a YEPS-agar plate (30 °C, 3 days). The preculture was cultivated at 30 °C and 120 rpm (50 mm shaking diameter) for 24 h until an optical density of 10 at 605 nm was achieved. All experiments were inoculated with 1% (v/v) of the preculture.

The main cultures in shake flasks were carried out at 30 °C and 120 rpm in 250 mL shake flasks with three baffles and a filling volume of 100 mL Tabuchi-medium.

To identify potential impurities and the utilization of monosaccharides based on lignocellulosic feedstock, test tubes with Kapsenberg caps were used and a working volume of 2 mL (ø 16 mm × 100 mm). The test tubes were incubated for 4 days at 30 °C and 120 rpm in an inclined test tube holder with an inclination angle of 30°. Shake flasks and test tubes were continuously rotated by hand while sampling avoiding inhomogeneity.

The cultivation in bioreactors were conducted in four parallel 1 L-bioreactors, equipped with a Rushton impeller and an L-sparger (model SR0700ODLS, DASGIP GmbH, Jülich, Germany). DASGIP Control software (DASGIP GmbH, Jülich, Germany) was used for the regulation of gassing, temperature, pH-value, and stirring rate, as well as recording the data of dissolved oxygen (DO) and pH. The pH regulation to pH 6.0 was carried out with 4 M NaOH, if not otherwise mentioned. At the beginning of the cultivation, 0.5 mL antifoam solution (Ucolup N-115, Brenntag, Mühlheim/Ruhr, Germany) was added to the broth. The experiments were carried out at 30 °C, 500 rpm, a filling volume of 500 mL, and an aeration of 0.1 vvm, unless otherwise mentioned. All cultivations were carried out in minimum duplicates, whereby the deviation from the mean value was <5%. All results are presented as mean values without error bars on account of readability.

2.4. Analytical Methods

The samples were centrifuged at 21,000 g for 20 min at 20 °C and the supernatant was used for further analysis. A Shimadzu HPLC (Shimadzu Corp., Kyoto, Japan) with a HPX-87H column (BioRad, Munich, Germany) with a refractive index detector (RI) and UV detector at 210 nm was used to analyze the concentrations of sugars and organic acids. The column was tempered at 40 °C and as mobile phase a 5 M H_2SO_4 solution at a flow rate of 0.6 mL·min^{-1} was used. The concentration of an unknown product was estimated by the peak area of the RI-signal compared to a calibration of succinic acid. For samples of bioreactor experiments, the pellet was washed twice with deionized water and dried to a constant weight at 105 °C for at least 48 h to determine the cell dry weight (CDW).

The composition of fatty acids was analyzed by transesterification of the fatty acids as described by Lewis et al. [31]. The biomass of reactor cultivation was separated from the broth by centrifugation (21,000 g for 20 min at 20 °C). The supernatant was discarded, and the pellet was washed twice with 0.9% (v/v) NaCl-solution and suspended in 0.9% (v/v) NaCl-solution. The cells were disrupted by an ultrasonic-homogenisator on ice (4 cycles: 15 s at 65%, break 30 s; Sonopuls HD2200 with sonotrodetype UW2200, Bandelin electronic, Berlin, Germany). The suspension was stored at −80 °C and freeze-dried (Alpha 1-2 LD, Christ, Osterode, Germany). The fatty acids were derivatized to fatty acid methyl esters (FAME) [31] and analyzed by GC-MS on a GC-17A (Shimadzu Corp., Kyoto, Japan) with a Zebron^{TM} ZB-WAX plus column (60 m × 0.25 mm × 0.25 µm), using 1.4 mL·min^{-1} helium as carrier gas. The temperature gradient of 60 °C was increased to 150 °C at a rate of 30 °C·min^{-1}, and afterwards increased up to 240 °C at a rate of 13 °C·min^{-1}. The temperature of 240 °C was kept for 30 min and raised to 255 °C for 5 min. The FAMEs were identified with the software LabSolutions (Shimadzu Corp., Kyoto, Japan) and the mass spectral data were compared with the database of the National Institute of Standards and Technology (Gaithersburg, MD, USA).

2.5. Microscopy

The cells were examined using a phase-contrast microscope (Axioplan, Carl Zeiss AG, Jena, Germany) with the software analysis pro (Analysis 5.1, Olympus Soft Imaging Solutions GmbH, Münster, Germany). Intracellular lipids were visualized after coloring with nil-red by fluorescence microscopy [32].

3. Results

3.1. Standard Cultivation in Shake Flasks

A standard cultivation of *U. rabenhorstiana* with pure glucose as substrate was performed in shake flasks (Figure 1). After one day, the itaconic acid production started, and additionally, succinic acid and malic acid were produced. Furthermore, an unknown product accumulated after 48 h approximately in a concentration range of <1 g·L^{-1}. Malic acid was consumed in further course of cultivation, and after four days, α-ketoglutaric acid was formed and increased parallel with the itaconic acid concentration. Glucose was completely consumed after 9.7 days, resulting in 31.3 g·L^{-1} itaconic acid, 13.6 g·L^{-1} α-ketoglutaric acid, 2.3 g·L^{-1} malic acid, and traces of an unknown metabolite, followed by a further production of α-ketoglutaric acid. The overall productivity was 0.13 g (L·h)$^{-1}$ with a yield of 0.26 (w/w) after 9.8 days. Despite the use of CaCO$_3$ as buffer, the pH-value constantly decreased from pH 6.8 to 4.9 throughout the cultivation. The morphology of *U. rabenhorstiana* changed from yeast-like single cells (0–2 days; Figure 1A) via a development of pseudomycel (2–7 days, Figure 1B) to filamentous growth like long branched mycel (7–11 days, Figure 1C). Moreover, intracellular lipids deposits were visible under the microscope, which became smaller in size after the glucose limitation at day 9.7.

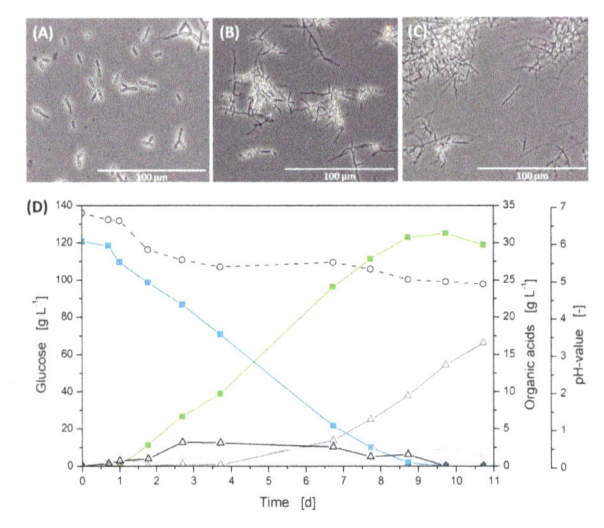

Figure 1. Cultivation of *U. rabenhorstiana* in 250 mL shake flasks in standard Tabuchi-medium (**D**) and its corresponding morphology (**A**): 0–2 days; (**B**): 2–7 days; (**C**): 7–11 days. Glucose (blue square), itaconic acid (green square), α-ketoglutaric acid (grey triangle), succinic acid (black triangle), malic acid (light grey triangle), pH (black circle), 120 rpm, 30 °C, 1% (v/v) inoculum.

3.2. Influence of Media Components

To determine the influence of media components, titer, yield, and productivity of a cultivation with 120 g·L^{-1} initial glucose, performed in shake flasks after 7.8 days, are shown in Figure 2. In the case of the variating initial glucose concentrations (Figure 2A), the point in time of glucose limitation was analyzed. 50 g·L^{-1} glucose were consumed in 3.7 days, 100 g·L^{-1} in 6.7 days, 120 g·L^{-1} in 8.7 days, and 150 g·L^{-1} in 10.7 days. Further, 200 g·L^{-1} glucose was not completely consumed by *U. rabenhorstiana* and 35 g·L^{-1}, and remained while the concentration of itaconic acid was constant after 15 days. With increasing initial glucose concentration, the titer of itaconic acid increased, but the productivity and yield decreased slightly from 0.16 g (L·h)$^{-1}$ to 0.09 g (L·h)$^{-1}$ and from 0.27 (w/w) to 0.20 (w/w) for glucose concentrations larger than 100 g·L^{-1}. The amount of α-ketoglutaric, succinic, and malic acid of the total organic acid concentration was raised from 15.5% (50 g·L^{-1} glucose) over 33.7% (120 g·L^{-1} glucose) to 50.1% (200 g·L^{-1} glucose). In case of different ammonia chloride concentrations, the highest titer of 31.8 g·L^{-1} itaconic acid, a productivity of 0.17 g (L·h)$^{-1}$, and a yield of 0.26 (w/w) was achieved using 1.6 g·L^{-1} NH$_4$Cl, which corresponded to the used concentration in standard Tabuchi-medium (Figure 2B). With a lower concentration of 1 g·L^{-1} NH$_4$Cl and higher concentrations between 3 and 7 g·L^{-1} NH$_4$Cl the titer, productivity and yield decreased up to 35%. Also, the chosen concentration of 0.2 g·L^{-1} MgSO$_4$·7 H$_2$O in Tabuchi-medium was optimal for itaconic acid production with *U. rabenhorstiana*, and a titer of 28.9 g·L^{-1} with a productivity of 0.16 g (L·h)$^{-1}$ was reached (Figure 2C). Lower or higher levels of magnesium resulted in a decrease of all target values. In the concentration range of 0.1–1 g·L^{-1} KH$_2$PO$_4$, there were no significant differences between the titer, yield, and productivity. All cultivations yielded in titers of 29.3 g·L^{-1} ± 1.2 g·L^{-1} with a productivity between 0.15–0.16 g (L·h)$^{-1}$ and a yield of 0.24–0.25 (w/w) (Figure 2D). In the range of 0.5–25 mg·L^{-1} FeSO$_4$·7 H$_2$O, the itaconic acid decreased from 32.4 g·L^{-1} to 24.2 g·L^{-1} (Figure 2E). The productivity of 0.17 g (L·h)$^{-1}$ was reduced by 17% and the yield of 0.3 (w/w) itaconic acid by 25%. With increasing yeast extract concentration (0.25–1.5 g·L^{-1}) the titer increased to 27.2 g·L^{-1} with a productivity of 0.15 g (L·h)$^{-1}$ at a concentration of 1.5 g·L^{-1} yeast extract (Figure 2F). A further increase in the yeast extract concentration up to 2 g·L^{-1} resulted in a decreased titer of 24 g·L^{-1} and a lowered productivity of 0.13 g (L·h)$^{-1}$. None of the media components had an influence on the filamentous growth.

3.3. Monosaccharide Utilization

It is intended to produce itaconic acid based on renewable feedstocks, e.g., lignocellulosic biomass, biomass with a high starch content, or molasses. The usability of the monosaccharides based on those feedstocks (arabinose, fructose, galactose, glucose, mannose, rhamnose, and xylose) were investigated with 100 g·L^{-1} of each sugar in test tubes (Table 1). For the precultivation, sucrose was used. Filamentous growth occurred on glucose, fructose, mannose, and xylose. An accumulation of long hyphae, a buildup of pellets with a diameter of 50 µm grew with arabinose as substrate. The yield and productivity were very different depending on the substrate. For the reference cultivation with glucose, the productivity was 0.16 g (L·h)$^{-1}$ with a yield of 0.24 (w/w). The productivity of 0.09 g (L·h)$^{-1}$ of itaconic acid with mannose was 44% lower, while the yield was in the same range with 0.22 (w/w). Using fructose, the same productivity compared to mannose was achieved, but with a lower yield of 0.17 (w/w). *U. rabenhorstiana* was able to use both pentoses for itaconic acid production, whereby the productivity with arabinose with 0.04 g (L·h)$^{-1}$ was twice as high as with xylose. In the cultivation with glactose, only traces of itaconic acid were detected. The yeast was not able to produce itaconic acid or even grow with rhamnose as single substrate.

Table 1. Cultivation of *U. rabenhorstiana* in test tubes with different monosaccharides as substrate (four days, 30 °C, 120 rpm, inclination angle of 30°, and 1% (v/v) inoculum).

Monosaccharide	Productivity [g (L·h)$^{-1}$]	Y$_{P/S}$ [w/w]
Glucose	0.16	0.24
Mannose	0.09	0.22
Fructose	0.09	0.17
Arabinose	0.04	0.06
Xylose	0.02	0.04
Galactose	<0.01	<0.01
Rhamnose	-	-

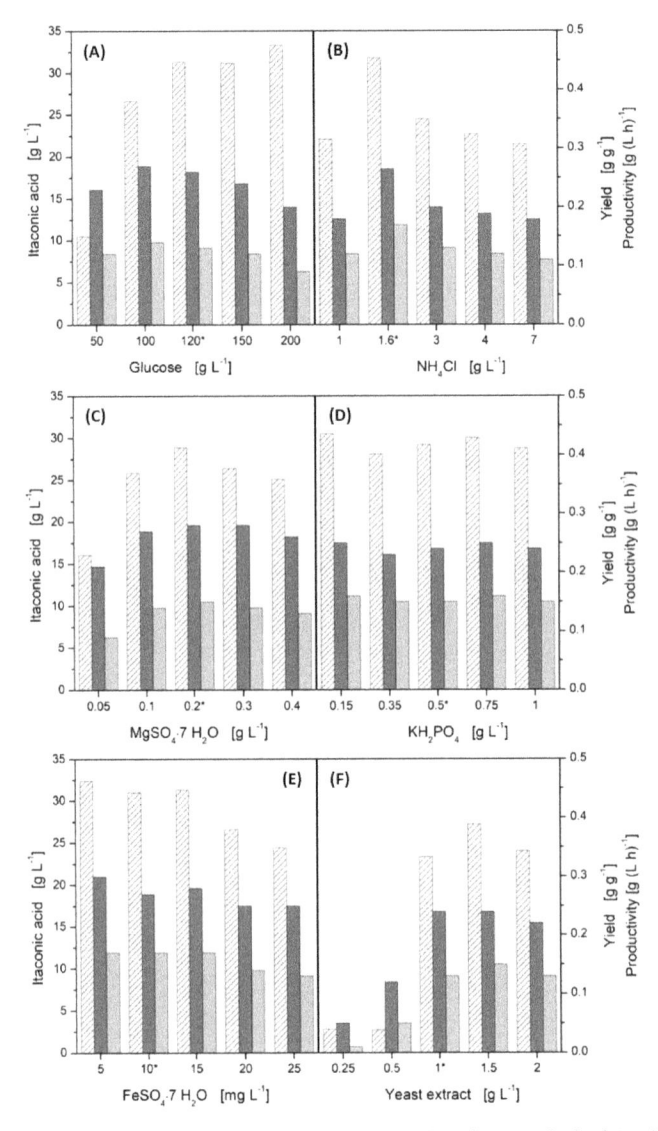

Figure 2. Influence of Tabuchi-medium components in 250 mL shake flasks on the final titer (dashed bar), yield (black bar), and productivity (grey bar) of *U. rabenhorstiana* at 120 rpm and 30 °C after 7.8 days (**B–F**). Cultivation time for different initial glucose concentrations depended on the point of glucose limitation (**A**). Asterisks highlight the standard media composition.

3.4. Influence of Sugar Degradation Products

In case of lignocellulosic feedstocks, different sugar degradation products are formed due to the harsh conditions in the pretreatment. To test the inhibitory effect of sugar degradation products like weak acids or furan derivates, the components were added by the lowest expected concentration levels to the media. The effect of 0–2 g·L^{-1} acetic acid, formic acid, furfural, or hydroxymethylfurfural (HMF) was carried out in test tubes (Figure 3). A productivity of 0.15 g (L·h)$^{-1}$ with standard Tabuchi-medium without the addition of inhibitory components was reached. Up to a concentration of 0.5 g·L^{-1} formic acid, the productivity did not differ. The addition of 1 g·L^{-1} resulted in an increased productivity of approximately 1.4 times and decreased to 0.13 g (L·h)$^{-1}$ with 2 g·L^{-1} formic acid. The result was very similar with the addition of acetic acid. The standard productivity increased up to 0.19 g (L·h)$^{-1}$ by adding 0.5 g·L^{-1} acetic acid and was reduced to 0.13 g (L·h)$^{-1}$ by increasing the acetic acid concentration. Both furan derivates influenced the microorganism very strongly, amounts of 0.1 g·L^{-1} of HMF or 0.5 g·L^{-1} furfural already resulted in a growth inhibition. If the growth was not inhibited, the yeast grew filamentous and stored intracellular lipid droplets comparable with the cultivation without addition of inhibitors.

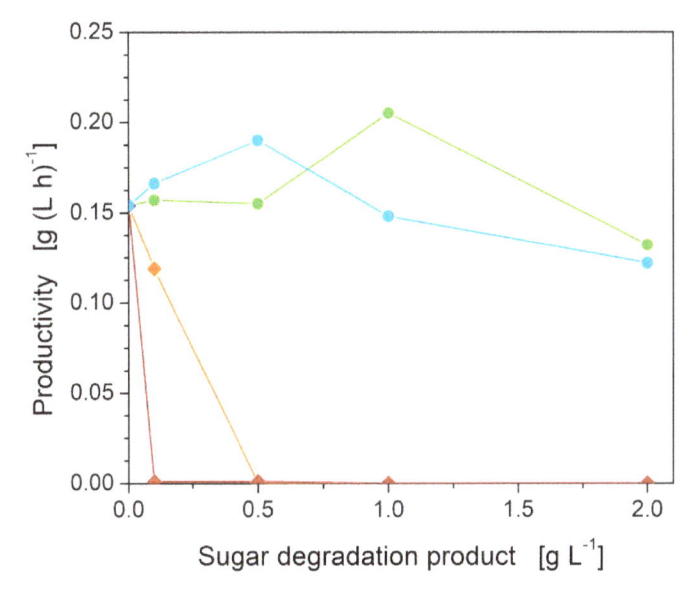

Figure 3. Inhibition effects of sugar degradation products on the itaconic acid productivity with *U. rabenhorstiana* in test tubes with Tabuchi-medium after four days, 30 °C, 120 rpm, inclination angle of 30°, pH > 5.5, and 1% (v/v) inoculum. Acetic acid (blue circle), formic acid (green circle), HMF (red diamond), furfural (orange diamond).

3.5. Influence of the pH-Value in 1 L-Bioreactor

The pH-value dropped in shake flask cultivations from pH 6.7 to pH 4.9 with CaCO$_3$ as buffer. To estimate the influence of the pH-value, the cultivation was transferred in 1 L-bioreactors with pH-control using 4 M NaOH (Table 2). Moreover, the modified Tabuchi-medium with 100 g·L^{-1} glucose and 1.5 g·L^{-1} yeast extract was used, based on the findings regarding the tested media components. The highest titer of 31.7 g·L^{-1} itaconic acid with a productivity of 0.23 g (L·h)$^{-1}$ and a yield of 0.34 (w/w) was reached with a controlled pH of 6.0. Beside itaconic acid, 0.4 g·L^{-1} α-ketoglutaric acid, 2 g·L^{-1} malic acid, 2.9 g·L^{-1} succinic acid, and the unknown product (<1 g·L^{-1}) were produced. The rate of the byproducts did not differ among the tested pH-values; also, the pH-value did not have any influence on the filamentous growth of the yeast and formation of intracellular lipids.

Table 2. Cultivation results of *U. rabenhorstiana* with pH-control in modified Tabuchi-medium with 100 g·L^{-1} glucose and 1.5 g·L^{-1} yeast extract in 1 L-bioreactor (30 °C, 500 rpm, 0.1 vvm, 4 M NaOH).

pH-Value [-]	Itaconic Acid [g·L^{-1}]	Productivity [g (L·h)$^{-1}$]	Y$_{P/S}$ [w/w]
5.5	23.7	0.15	0.25
6.0	31.7	0.23	0.34
6.5	18.6	0.12	0.19
7.0	15.9	0.10	0.16

3.6. Influence of Aeration in 1 L-Bioreactor

The effect of aeration was tested in 1 L-bioreactors by different aeration rates between 0.1–1 vvm, the stirring rate was kept constant, and the pH-value was regulated to pH 6.0 (Table 3). An increasing aeration rate from 0.1 to 1 vvm resulted in a decreased yield and titer of 20%, as well as in a 45% lower productivity. In contrast, the formed biomass increased from 15.7 g·L^{-1} at 0.1 vvm to 21.3 g·L^{-1} at 1 vvm. There were no significant differences between the by-product concentrations depending on the aeration rate, which corresponds to the concentrations described in Section 3.5.

Table 3. Influence of aeration on the cultivation of *U. rabenhorstiana* in modified Tabuchi-medium with 100 g·L^{-1} glucose and 1.5 g·L^{-1} yeast extract in 1 L-bioreactor (30 °C, 500 rpm, pH 6.0).

Aeration [vvm]	Itaconic Acid [g·L^{-1}]	Productivity [g·(L·h)$^{-1}$]	Y$_{P/S}$ [w/w]	CDW [g·L^{-1}]
0.1	29.8	0.22	0.30	15.7
0.5	26.1	0.16	0.26	17.5
1.0	23.6	0.12	0.24	21.3

3.7. Fed-Batch Mode in 1 L-Bioreactor with Glucose

Glucose concentrations larger than 150 g·L^{-1} resulted in a decreased yield and productivity in shake flasks (Figure 2A). For this reason, a fed batch with glucose was realized at a constant pH of pH 6.0 in a 1 L-bioreactor (Figure 4). An initial glucose concentration of 100 g·L^{-1} was chosen. After five days, 73 g·L^{-1} and after 10 days, 25 g·L^{-1} glucose were added into the cultivation broth, in which the average glucose consumption rate was 0.73 g (L·h)$^{-1}$ for the first batch (0–5 days), 0.53 g (L·h)$^{-1}$ for the second batch (5–10 days), and 0.28 for the third batch (10–15 days). The yield amounted to 0.31 (w/w) in the first batch and was constant with 0.26 (w/w) in the second and third batches. The DO decreased to 2% within the first day and varied between 2%–20% during the further cultivation. After one day, the itaconic acid production started and rose to a final titer of 50.3 g·L^{-1} within 15 days. Beside itaconic acid, 3.6 g·L^{-1} malic acid, 13.6 g·L^{-1} succinic acid, 2.5 g·L^{-1} α-ketoglutaric acid, and the unknown product (<10 g·L^{-1}) were formed by 17.2 g·L^{-1} filamentous biomass (Appendix A, Figure A2). This cultivation resulted in a productivity of 0.14 g (L·h)$^{-1}$ with an overall yield of 0.27 (w/w) after 15 days. Further, 175 mL of a 4 M NaOH was used to keep the pH-value constant at pH 6.0. After 15 days, the cell dry weight was analyzed regarding the fatty acids (Appendix A, Table A1); C16:0, C18:0, and C18:2 were the main elements.

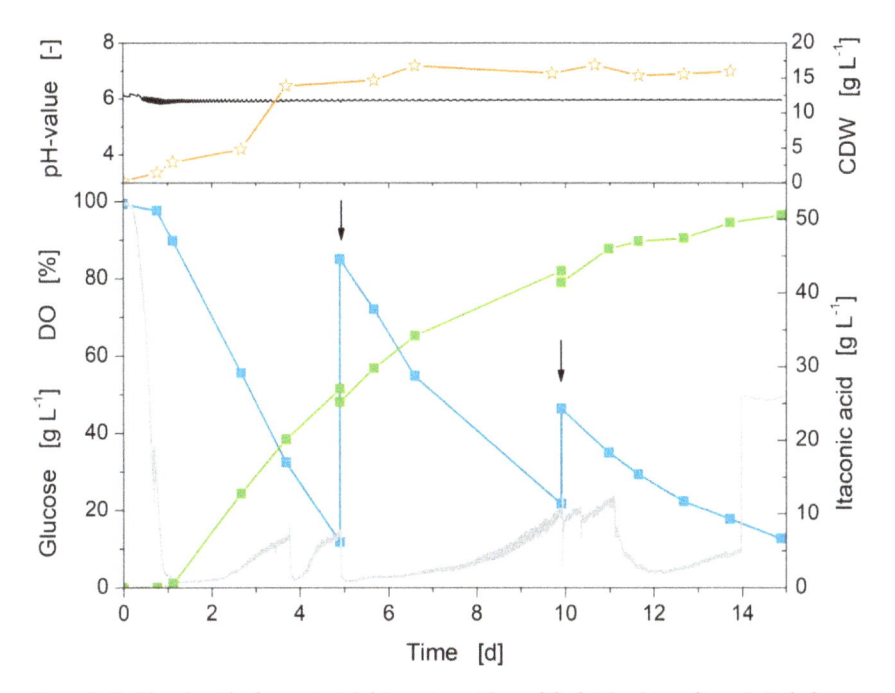

Figure 4. Fed batch with glucose in 1 L-bioreactor with modified Tabuchi-medium (initial glucose concentration 100 g·L^{-1} and 1.5 g·L^{-1} yeast extract) at 30 °C, 500 rpm, 0.1 vvm, pH 6.0 with 4 M NaOH as base and 1% (v/v) inoculum. Glucose (blue square), itaconic acid (green square), pO$_2$ (grey line), pH (black line), cell dry weight (CDW) (orange asterisk), arrows symbolize the addition of glucose.

4. Discussion

In a standard cultivation in shake flasks, a final titer of 31.3 g·L^{-1} itaconic acid was achieved after 9.7 days without media and process optimization of *U. rabenhorstiana*. Additionally, succinic acid, malic acid, α-ketoglutaric acid, an unknown product, and intracellular lipids were formed. Guevarra and Tabuchi reached a titer of about 16 g·L^{-1} itaconic acid in the same media after seven days and verified 2-hydroxyparaconic acid, itatartaric acid, and erythritol as byproducts with a total concentration of 30 g·L^{-1} [18]. Furthermore, they reduced the byproduct concentration to 19 g·L^{-1} with a constant itaconic acid titer by using an unbuffered media, whereby the final pH-value was pH 2.8 [30]. Erythritol as the unknown byproduct could be excluded in this work. In comparison with the literature, the unknown product (<1 g·L^{-1}) could either be 2-hydroxyparaconic acid or itatartaric acid (Appendix A, Figure A1) [18,23,30] and must be characterized precisely in a further study. Moreover, the formation of different organic acids and intracellular lipid bodies by *Ustilaginaceae* besides itaconic acid production are very well known [5,19], as well as the production of cellobiose lipids and mannosylerythritol lipids [20,29], which were not proved in this study.

A point-by-point analysis of each media component indicated that the chosen concentrations of each component in the Tabuchi-medium were nearly optimal for itaconic acid production with *U. rabenhorstiana*. Only the glucose and yeast extract concentrations were adjusted from 120 g·L^{-1} to 100 g·L^{-1} for glucose and from 1 g·L^{-1} to 1.5 g·L^{-1} for yeast extract in the modified Tabuchi-medium. Increased glucose concentration resulted in an increased final titer, but the yield and productivity decreased with glucose concentrations >100 g·L^{-1}. The decreased yield and productivity can be explained by a higher osmotic stress and a higher number of byproducts. A very similar result was obtained by the production of itaconic acid with *U. maydis* [19]. This microorganism also showed lower yields at higher initial glucose concentrations, explained by the formation of other organic acids, polyols, and glycolipids. The increase of yeast extract by 0.5 g·L^{-1} raised the titer, productivity, and yield. A further increase of yeast extract did not result in an improved titer, suggesting that a limitation of vitamins, amino acids, salts, trace elements or nucleic acids [33] is prevented with a concentration >1.5 g·L^{-1} yeast extract. The itaconic acid overproduction by *U. maydis* or *P. antarctica* are mainly triggered by an ammonium limitation, higher concentrations of NH$_4$Cl > 4 g·L^{-1} resulted in lower itaconic acid yields and an increase in biomass for *U. maydis* [5,14,19]. What is more, in this study, the ammonium concentration had the strongest influence on the fermentation performance of *U. rabenhorstiana*. A concentration of 1.6 g·L^{-1} NH$_4$Cl was optimal for itaconic acid production, and higher concentrations resulted in reduced yields and productivities as described already for *U. maydis*. For this reason, it can be assumed that an ammonium limitation caused the itaconic acid overproduction of *U. rabenhorstiana*. Further, the nitrogen limitation caused the accumulation of intracellular lipid droplets, which are mostly triacylglycerols [21,34]. A secretion of cellobiose lipids and mannosylerythritol lipids in the form of needle-like crystals or oily droplets of *U. maydis* [20], could not be verified for *U. rabenhorstiana*.

To use alternative, low-cost, or lignocellulosic feedstocks for itaconic acid production, it is important to use a microorganism, which is able to consume different monosaccharides and is robust towards varying impurities. In this study, it was worked out that *U. rabenhorstiana* is able to grow and produce itaconic acid from different monosaccharides like glucose, fructose, mannose, xylose, arabinose, and galactose. The highest productivity was reached with glucose, followed by fructose and mannose. Because the precultivation was based on sucrose, composed of glucose and fructose, it can be assumed that the cultivation with fructose was therefore such successful. An adaption to the used monosaccharide of the microorganisms in the preculture or a mixture of several monosaccharides with glucose would probably lead to higher productivity and yield using that single monosaccharide in the main culture. Furthermore, the plant pathogen *U. rabenhorstiana* is supposed to degrade a range of biomass-based polymers [35–37]. For industrial itaconic acid production, the filamentous fungus *A. terreus* is used, which is very sensitive to weak acids, furan derivates, metal ions, and other impurities, which are contained in such substrates [11,12,38]. *U. maydis* is described as a very robust microorganism [5,22,39]. *U. rabenhorstiana* was also not influenced by the addition of weak acid concentrations up to 2 g·L^{-1} in the main culture. A major advantage of the cultivation of *Ustilaginaceae* is the pH range of 5.0–6.5, whereby the dissociated weak acids cannot cross the plasma membrane into the cytosol and affect the intracellular pH-value [40,41]. A positive effect on the itaconic acid productivity was even achieved with the addition of 0.5 g·L^{-1} acetic acid or 1 g·L^{-1} formic acid. This positive influence of low weak acid concentrations in cultivation media is already known from itaconic acid production with *A. terreus*, ethanol production with *S. cerevisiae*, or enzyme production with *T. reesei* [12,42–44]. In contrast, low concentrations of 0.1 g·L^{-1} HMF or 0.5 g·L^{-1} furfural are growth limiting factors; both furan derivates reduce the activity of a number of important intracellular enzymes of the maintenance metabolism, e.g., pyruvate dehydrogenase [45]. In particular, the activity of the pyruvate dehydrogenase is essential for cells, because this enzyme links glycolysis and citric acid cycle, which supplies the cell with energy intermediates. When using lignocellulosic biomass as feedstock for *U. rabenhorstiana*, it should be taken into account, that some robustness in relation to weak acids exists, but furan derivatives influence the microorganisms mostly up to growth inhibition.

For further characterization of *U. rabenhorstiana*, the fermentation was transferred in 1 L-bioreactors to investigate the influence of pH-value and aeration. A constant pH of 6.0 and modified Tabuchi-medium yielded in the highest itaconic acid titer of 31.7 g·L^{-1}, which is comparable with the titer in standard shake flask cultivation with Tabuchi-medium. However, the overall productivity was 1.7 times higher at a constant pH-value than in shake flask experiments with CaCO$_3$, whereby the pH-value continuously decreased to pH 4.9. Moreover, the total concentration of by-products decreased by 66%, which resulted in an increased itaconic acid yield. These suggested that the buffer capacity in shake flasks is insufficient. Buffer systems like CaCO$_3$ or MES and its buffer capacity have a significant impact on the organic acid production of *Ustilaginaceae* not only on titer, but also on the ratio of products [16,24,30]. The higher the buffer capacity, the better is the itaconic acid titer in small-scale experiments, but *U. maydis* achieved the highest itaconic acid titer of 45.5–63.2 g·L^{-1} in cultivations with a constant pH of 6.0–6.5 in bioreactors [19,23]. Consequently, *U. maydis* and *U. rabenhorstiana* have nearly the same requirements in pH and productivity. Titer and yield are positively influenced by a constant pH-value.

In the literature, no detailed studies regarding influence of oxygen levels on the organic acid production with *Ustilaginaceae* exist. Only cultivation parameters like high shaking frequencies or stirrer speeds led to the conclusion that a high input of oxygen is necessary [10]. Contrary results were achieved for itaconic acid production with *U. rabenhorstiana*; the lowest aeration rate of 0.1 vvm and a constant stirring rate of 500 rpm yielded the best result regarding titer, productivity, and yield. Presumably, the increase in these values is related to the formation of 36% more biomass at higher aeration rates, because of a better supply of oxygen. In batch experiments, a maximum itaconic acid concentration of 33.3 g·L^{-1} with an initial glucose concentration of 200 g·L^{-1} was achieved, proving that initial glucose concentration ≥100 g·L^{-1} and a constant pH-value of pH 6.0 have a significant impact on the itaconic acid production with *U. rabenhorstiana*. Therefore, a cultivation in fed-batch mode with glucose was realized at pH 6.0 and resulted in 50.3 g·L^{-1} itaconic acid. Comparing the batch cultivation with 200 g·L^{-1} glucose with the fed-batch cultivation in a bioreactor, the productivity and yield were 1.4 times and the final titer 1.5 times higher. Also, the formation of organic acid as byproducts was reduced. The number of byproducts of the overall organic acid concentration decreased from 50% in batch cultivation to 28% in fed-batch mode. Thereby, the amount of organic acids shifted from α-ketoglutaric acid as the main byproduct in batch mode to succinic acid in fed-batch mode, due to the pH-value of each cultivation [16]. The final titer and yield of this study are slightly increased compared to a wildtype strain of *U. maydis*, which reached 44.5 g·L^{-1} [19]. A higher final titer up to 63.2 g·L^{-1} or yield of 0.48 (w/w) was only obtained by a genetical modification of *U. maydis* [23]. All in all, the fermentation broth was diluted by addition of 175 mL NaOH as base. Considering the dilution, the wildtype of *U. rabenhorstiana* demonstrates the potential to produce up to 68 g·L^{-1} itaconic acid. However, in all main cultures, the unicellular growth of *U. rabenhorstiana* shifted to filamentous cells with depots of intracellular lipids. Neither the variation of media components nor the investigation of process parameters influenced the filamentous growth. The typical unicellular yeast-like growth of *U. maydis* for itaconic acid could not be achieved for *U. rabenhorstiana* in this study. This morphology would be a great advantage compared to the filamentous *A. terreus* regarding oxygen supply or viscosity, especially in large-scale fermentations [5,22], but was not focused in this study. In general, it is possible to generate a stable unicellular growth by deleting several genes [24]. The filamentous growth involved the accumulation of intracellular lipid droplets, which is initiated by nitrogen limitation, which in turn is needed for itaconic acid overproduction [5,19,34]. Furthermore, 89% of all fatty acids in *U. rabenhorstiana* were long-chain fatty acids C16:0, C18:0, and C18:2 and suggested the accumulation of triacylglycerols in the cells. The lipid bodies in *U. maydis* mainly contain triacylglycerols consisting of palmitic, linoleic, and oleic acids [21].

5. Conclusions

This study describes a known, but so far unspecified, itaconic acid producer—*U. rabenhorstiana*. The cultivation in shake flasks with a maximal final titer of 33.3 g·L^{-1} itaconic acid was transferred in a bioreactor. With a controlled pH-value, a low initial glucose concentration, and fed-batch mode, a final titer of 50.3 g·L^{-1} was achieved, which is comparable with titer of other wildtype strains of *Ustilago* described in literature. However, the productivity and yield are rather low compared to *U. maydis*, which was studied very precisely in the last years regarding cultivation and process parameters as well as metabolic engineering strategies for further improvements in itaconic acid production. Transferring this knowledge from *U. maydis* to *U. rabenhorstiana* could result in a further increased final titer and improved yield and productivity. Particular attention should be paid to the morphology of the yeast and minimization of byproducts, mainly the formation of intracellular lipid droplets. Moreover, the wildtype strain *U. rabenhorstiana* turned out to be a robust and promising alternative itaconic acid producer based on renewable resources. All in all, this study serves a basis for further promising research regarding lignocellulosic hydrolysates.

Author Contributions: S.K. and M.L. designed and performed the experiments. U.P. and A.K. supervised the experiments. S.K. wrote the manuscript. All authors have read and agreed to the published version of the manuscript.

Funding: This work was carried out in the framework of the European Research Area Network for Industrial Biotechnology (ERA-IB project "Production of Organic Acids for Polyester Synthesis (POAP)") and was funded by the German Federal Ministry of Food and Agriculture, following a decision of the German Bundestag, via the Agency of Renewable Resources (Grant No. 22029312).

Conflicts of Interest: The authors declare no conflict of interest. The funder had no role in the design of the study; in the collection, analyses, or interpretation of data; in the writing of the manuscript, or in the decision to publish the results.

Appendix A

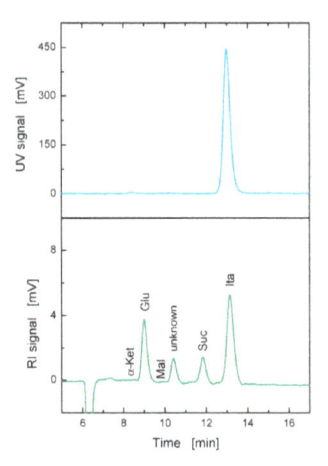

Figure A1. Representative HPLC analysis of a final sample (15 days) of a fed-batch cultivation in 1 L-bioreactor (Section 3.7). α-ketoglutaric acid (α-Ket), glucose (Glu), malic acid (Mal), unknown product, succinic acid (Suc), itaconic acid (Ita).

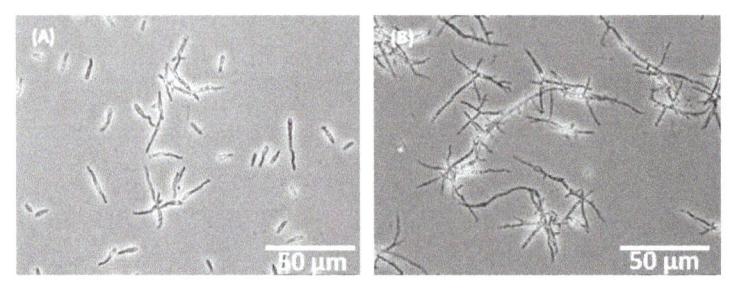

Figure A2. Corresponding morphology of *U. rabenhorstiana* in a fed-batch cultivation in 1 L-bioreactor (Section 3.7) after 1.1 days (**A**) and 6.6 days (**B**).

Table A1. Analysis of fatty acids in intracellular lipid droplets in *U. rabenhorstiana* of a fed-batch cultivation in 1 L-bioreactor (Section 3.7).

Fatty Acid	Concentration [mg·g^{-1}$_{CDW}$]
C14:0	1.46
C16:0	30.04
C16:1	1.55
C18:0	16.94
C18:1	1.83
C18:2	48.38
C22:0	4.42
C24:0	2.45

References

1. Werpy, T.; Petersen, G. *Top Value Added Chemicals from Biomass: Volume I—Results of Screening for Potential Candidates from Sugars and Synthesis Gas*; National Renewable Energy Lab.: Golden, CO, USA, 2004.
2. Magalhaes, A.I.; de Carvalho, J.C.; Medina, J.D.C.; Soccol, C.R. Downstream process development in biotechnological itaconic acid manufacturing. *Appl. Microbiol. Biotechnol.* **2017**, *101*, 1–12. [CrossRef]
3. Robert, T.; Friebel, S. Itaconic acid—A versatile building block for renewable polyesters with enhanced functionality. *Green Chem.* **2016**, *18*, 2922–2934. [CrossRef]
4. Okabe, M.; Lies, D.; Kanamasa, S.; Park, E.Y. Biotechnological production of itaconic acid and its biosynthesis in *Aspergillus terreus*. *Appl. Microbiol. Biotechnol.* **2009**, *84*, 597–606. [CrossRef]
5. Klement, T.; Milker, S.; Jäger, G.; Grande, P.M.; de Maria, P.D.; Büchs, J. Biomass pretreatment affects *Ustilago maydis* in producing itaconic acid. *Microb. Cell Factories* **2012**, *11*, 43. [CrossRef]
6. Willke, T.; Vorlop, K.D. Biotechnological production of itaconic acid. *Appl. Microbiol. Biotechnol.* **2001**, *56*, 289–295. [CrossRef]
7. Hevekerl, A.; Kuenz, A.; Vorlop, K.D. Influence of the pH on the itaconic acid production with *Aspergillus terreus*. *Appl. Microbiol. Biotechnol.* **2014**, *98*, 10005–10012. [CrossRef]
8. Karaffa, L.; Diaz, R.; Papp, B.; Fekete, E.; Sandor, E.; Kubicek, C.P. A deficiency of manganese ions in the presence of high sugar concentrations is the critical parameter for achieving high yields of itaconic acid by *Aspergillus terreus*. *Appl. Microbiol. Biotechnol.* **2015**, *99*, 7937–7944. [CrossRef]
9. Krull, S.; Hevekerl, A.; Kuenz, A.; Prüße, U. Process development of itaconic acid production by a natural wild type strain of *Aspergillus terreus* to reach industrially relevant final titers. *Appl. Microbiol. Biotechnol.* **2017**, *101*, 4063–4072. [CrossRef]
10. Kuenz, A.; Krull, S. Biotechnological production of itaconic acid-things you have to know. *Appl. Microbiol. Biotechnol.* **2018**, *102*, 3901–3914. [CrossRef]
11. Kobayashi, T. Production of itaconic acid from wood waste. *Process Biochem.* **1978**, *13*, 15–22.
12. Krull, S.; Eidt, L.; Hevekerl, A.; Kuenz, A.; Prüße, U. Itaconic acid production from wheat chaff by *Aspergillus terreus*. *Process Biochem.* **2017**, *63*, 169–176. [CrossRef]
13. Wu, X.F.; Liu, Q.; Deng, Y.D.; Li, J.H.; Chen, X.J.; Gu, Y.Z.; Lv, X.J.; Zheng, Z.; Jiang, S.T.; Li, X.J. Production of itaconic acid by biotransformation of wheat bran hydrolysate with *Aspergillus terreus* CICC40205 mutant. *Bioresour. Technol.* **2017**, *241*, 25–34. [CrossRef]
14. Levinson, W.E.; Kurtzman, C.P.; Kuo, T.M. Production of itaconic acid by *Pseudozyma antarctica* NRRL Y-7808 under nitrogen-limited growth conditions. *Enzym. Microb. Technol.* **2006**, *39*, 824–827. [CrossRef]
15. Specht, R.; Andreas, A.; Kreyß, E.; Barth, G.; Bodinus, C. Verfahren zur Biotechnologischen Herstellung von Itaconsäure. DE Patent 102008011854 B4, 20 February 2014.
16. Geiser, E.; Wiebach, V.; Wierckx, N.; Blank, L.M. Prospecting the biodiversity of the fungal family *Ustilaginaceae* for the production of value-added chemicals. *Fungal Biol. Biotechnol.* **2014**, *1*, 2. [CrossRef]
17. Tabuchi, T.; Sugisawa, T.; Ishidori, T.; Nakahara, T.; Sugiyama, J. Itaconic Acid Fermentation by a Yeast Belonging to the Genus Candida. *Agric. Biol. Chem.* **1981**, *45*, 475–479. [CrossRef]
18. Guevarra, E.D.; Tabuchi, T. Accumulation of itaconic, 2-hydroxyparaconic, itatartaric, and malic-acids by strains of the genus *Ustilago*. *Agric. Biol. Chem.* **1990**, *54*, 2353–2358. [CrossRef]

19. Maassen, N.; Panakova, M.; Wierckx, N.; Geiser, E.; Zimmermann, M.; Bolker, M.; Klinner, U.; Blank, L.M. Influence of carbon and nitrogen concentration on itaconic acid production by the smut fungus *Ustilago maydis*. *Eng. Life Sci.* **2014**, *14*, 129–134. [CrossRef]
20. Spoeckner, S.; Wray, V.; Nimtz, M.; Lang, S. Glycolipids of the smut fungus *Ustilago maydis* from cultivation on renewable resources. *Appl. Microbiol. Biotechnol.* **1999**, *51*, 33–39. [CrossRef]
21. Aguilar, L.R.; Pardo, J.P.; Lomeli, M.M.; Bocardo, O.I.L.; Oropeza, M.A.J.; Sanchez, G.G. Lipid droplets accumulation and other biochemical changes induced in the fungal pathogen *Ustilago maydis* under nitrogen-starvation. *Arch. Microbiol.* **2017**, *199*, 1195–1209. [CrossRef]
22. Regestein, L.; Klement, T.; Grande, P.; Kreyenschulte, D.; Heyman, B.; Maßmann, T.; Eggert, A.; Sengpiel, R.; Wang, Y.; Wierckx, N. From beech wood to itaconic acid: Case study on biorefinery process integration. *Biotechnol. Biofuels* **2018**, *11*, 1–11. [CrossRef]
23. Geiser, E.; Przybilla, S.K.; Engel, M.; Kleineberg, W.; Büttner, L.; Sarikaya, E.; Den Hartog, T.; Klankermayer, J.; Leitner, W.; Bölker, M. Genetic and biochemical insights into the itaconate pathway of *Ustilago maydis* enable enhanced production. *Metab. Eng.* **2016**, *38*, 427–435. [CrossRef]
24. Tehrani, H.H.; Tharmasothirajan, A.; Track, E.; Blank, L.M.; Wierckx, N. Engineering the morphology and metabolism of pH tolerant *Ustilago cynodontis* for efficient itaconic acid production. *Metab. Eng.* **2019**, *54*, 293–300. [CrossRef]
25. Geiser, E.; Przybilla, S.K.; Friedrich, A.; Buckel, W.; Wierckx, N.; Blank, L.M.; Bolker, M. *Ustilago maydis* produces itaconic acid via the unusual intermediate trans-aconitate. *Microb. Biotechnol.* **2016**, *9*, 116–126. [CrossRef]
26. Zambanini, T.; Hartmann, S.K.; Schmitz, L.M.; Buttner, L.; Hosseinpour Tehrani, H.; Geiser, E.; Beudels, M.; Venc, D.; Wandrey, G.; Buchs, J.; et al. Promoters from the itaconate cluster of *Ustilago maydis* are induced by nitrogen depletion. *Fungal Biol. Biotechnol.* **2017**, *4*, 11. [CrossRef]
27. Wierckx, N.; Agrimi, G.; Lubeck, P.S.; Steiger, M.G.; Mira, N.P.; Punt, P.J. Metabolic specialization in itaconic acid production: A tale of two fungi. *Curr. Opin. Biotechnol.* **2019**, *62*, 153–159. [CrossRef]
28. Zambanini, T.; Hosseinpour Tehrani, H.; Geiser, E.; Merker, D.; Schleese, S.; Krabbe, J.; Buescher, J.M.; Meurer, G.; Wierckx, N.; Blank, L.M. Efficient itaconic acid production from glycerol with *Ustilago vetiveriae* TZ1. *Biotechnol. Biofuels* **2017**, *10*, 131. [CrossRef]
29. Haskins, R.; Thorn, J.; Boothroyd, B. Biochemistry of the Ustilaginales: XI. Metabolic products of *Ustilago zeae* in submerged culture. *Can. J. Microbiol.* **1955**, *1*, 749–756. [CrossRef]
30. Guevarra, E.D.; Tabuchi, T. Production of 2-hydroxyparaconic and itatartaric acids by *Ustilago cynodontis* and simple recovery process of the acids. *Agric. Biol. Chem.* **1990**, *54*, 2359–2365. [CrossRef]
31. Lewis, T.; Nichols, P.D.; McMeekin, T.A. Evaluation of extraction methods for recovery of fatty acids from lipid-producing microheterotrophs. *J. Microbiol. Methods* **2000**, *43*, 107–116. [CrossRef]
32. Klose, J.; Kronstad, J.W. The multifunctional beta-oxidation enzyme is required for full symptom development by the biotrophic maize pathogen *Ustilago maydis*. *Eukaryot. Cell* **2006**, *5*, 2047–2061. [CrossRef]
33. Sommer, R. Yeast extracts: Production, properties and components. *Food Aust.* **1998**, *50*, 181–183.
34. Zavala-Moreno, A.; Arreguin-Espinosa, R.; Pardo, J.P.; Romero-Aguilar, L.; Guerra-Sánchez, G. Nitrogen source affects glycolipid production and lipid accumulation in the phytopathogen fungus *Ustilago maydis*. *Adv. Microbiol.* **2014**, *4*, 934. [CrossRef]
35. Gibson, D.M.; King, B.C.; Hayes, M.L.; Bergstrom, G.C. Plant pathogens as a source of diverse enzymes for lignocellulose digestion. *Curr. Opin. Microbiol.* **2011**, *14*, 264–270. [CrossRef]
36. Geiser, E.; Reindl, M.; Blank, L.M.; Feldbrügge, M.; Wierckx, N.; Schipper, K. Activating intrinsic carbohydrate-active enzymes of the smut fungus *Ustilago maydis* for the degradation of plant cell wall components. *Appl. Environ. Microbiol.* **2016**, *82*, 5174–5185. [CrossRef]
37. Couturier, M.; Navarro, D.; Olivé, C.; Chevret, D.; Haon, M.; Favel, A.; Lesage-Meessen, L.; Henrissat, B.; Coutinho, P.M.; Berrin, J.-G. Post-genomic analyses of fungal lignocellulosic biomass degradation reveal the unexpected potential of the plant pathogen *Ustilago maydis*. *BMC Genom.* **2012**, *13*, 57. [CrossRef]
38. Gyamerah, M. Factors affecting the growth form of *Aspergillus terreus* NRRL 1960 in relation to itaconic acid fermentation. *Appl. Microbiol. Biotechnol.* **1995**, *44*, 356–361. [CrossRef]
39. Benito, B.; Garciadeblás, B.; Pérez-Martín, J.; Rodríguez-Navarro, A. Growth at high pH and sodium and potassium tolerance in media above the cytoplasmic pH depend on ENA ATPases in *Ustilago maydis*. *Eukaryot. Cell* **2009**, *8*, 821–829. [CrossRef]

40. Stratford, M.; Nebe-von-Caron, G.; Steels, H.; Novodvorska, M.; Ueckert, J.; Archer, D.B. Weak-acid preservatives: pH and proton movements in the yeast *Saccharomyces cerevisiae*. *Int. J. Food Microbiol.* **2013**, *161*, 164–171. [CrossRef]
41. Lambert, R.; Stratford, M. Weak-acid preservatives: Modelling microbial inhibition and response. *J. Appl. Microbiol.* **1999**, *86*, 157–164. [CrossRef]
42. Larsson, S.; Palmqvist, E.; Hahn-Hägerdal, B.; Tengborg, C.; Stenberg, K.; Zacchi, G.; Nilvebrant, N.-O. The generation of fermentation inhibitors during dilute acid hydrolysis of softwood. *Enzym. Microb. Technol.* **1999**, *24*, 151–159. [CrossRef]
43. Maiorella, B.; Blanch, H.W.; Wilke, C.R. By-product inhibition effects on ethanolic fermentation by *Saccharomyces cerevisiae*. *Biotechnol. Bioeng.* **1983**, *25*, 103–121. [CrossRef]
44. Szengyel, Z.; Zacchi, G. Effect of acetic acid and furfural on cellulase production of *Trichoderma reesei* RUT C30. *Appl. Biochem. Biotechnol.* **2000**, *89*, 31–42. [CrossRef]
45. Modig, T.; Liden, G.; Taherzadeh, M.J. Inhibition effects of furfural on alcohol dehydrogenase, aldehyde dehydrogenase and pyruvate dehydrogenase. *Biochem. J.* **2002**, *363*, 769–776. [CrossRef]

© 2020 by the authors. Licensee MDPI, Basel, Switzerland. This article is an open access article distributed under the terms and conditions of the Creative Commons Attribution (CC BY) license (http://creativecommons.org/licenses/by/4.0/).

MDPI
St. Alban-Anlage 66
4052 Basel
Switzerland
Tel. +41 61 683 77 34
Fax +41 61 302 89 18
www.mdpi.com

Fermentation Editorial Office
E-mail: fermentation@mdpi.com
www.mdpi.com/journal/fermentation

Lightning Source UK Ltd.
Milton Keynes UK
UKHW021541280822
407834UK00002B/94